IDENTIFICATION OF MICROORGANISMS BY MASS SPECTROMETRY

CHEMICAL ANALYSIS

A SERIES OF MONOGRAPHS ON ANALYTICAL CHEMISTRY AND ITS APPLICATIONS

Edited by
J. D. WINEFORDNER

VOLUME 169

IDENTIFICATION OF MICROORGANISMS BY MASS SPECTROMETRY

Edited by

CHARLES L. WILKINS
JACKSON O. LAY, JR.
University of Arkansas, Fayetteville, AR

A JOHN WILEY & SONS, INC., PUBLICATION

Published by John Wiley & Sons, Inc., Hoboken, New Jersey.
Published simultaneously in Canada.

Library of Congress Cataloging-in-Publication Data:

Identification of microorganisms by mass spectrometry / edited by Charles L. Wilkins, Jackson O. Lay, Jr.
p. cm.
Includes bibliographical references and index.
ISBN-13 978-0-471-65442-1 (cloth)
ISBN-10 0-471-65442-6 (cloth)
1. Microorganisms—Identification. 2. Mass spectrometry. I. Wilkins, Charles L. (Charles Lee), 1938– II. Lay, Jackson O.
QR67.I344 2006
579—dc22 2005008550

Printed in the United States of America

10 9 8 7 6 5 4 3 2 1

IDENTIFICATION OF MICROORGANISMS BY MASS SPECTROMETRY

Edited by

CHARLES L. WILKINS
JACKSON O. LAY, JR.
University of Arkansas, Fayetteville, AR

A JOHN WILEY & SONS, INC., PUBLICATION

Published by John Wiley & Sons, Inc., Hoboken, New Jersey.
Published simultaneously in Canada.

Library of Congress Cataloging-in-Publication Data:

Identification of microorganisms by mass spectrometry / edited by Charles L. Wilkins, Jackson O. Lay, Jr.
p. cm.
Includes bibliographical references and index.
ISBN-13 978-0-471-65442-1 (cloth)
ISBN-10 0-471-65442-6 (cloth)
1. Microorganisms—Identification. 2. Mass spectrometry. I. Wilkins, Charles L. (Charles Lee), 1938– II. Lay, Jackson O.
QR67.I344 2006
579—dc22 2005008550

Printed in the United States of America

10 9 8 7 6 5 4 3 2 1

CONTENTS

PREFACE

The analysis of microorganisms by mass spectrometry is relatively new by mass spectrometry standards, having begun some 60 years after the first reports of mass spectrometry studies in the last years of the nineteenth century by J. J. Thomson and others. Nevertheless, from a modern perspective, it could be argued that mass spectrometrists have been trying to sell microbiologists (and others) on the idea of classifying bacteria using an instrumental approach since at least the 1960s. Although there were some promising early studies, many in the microbiology world believed that the capabilities of mass spectrometry were overstated then, and perhaps now. The early optimism was not really unwarranted in that mass spectrometry studies came very close to making real headway in the development of rapid and reliable methods for the characterizing of many disease-causing organisms. Two prominent lines of mass spectrometry research illustrate this point. One technique that showed much promise was pyrolysis mass spectrometry (PyMS). This approach was rapid, rugged, and seemed to provide taxonomically specific spectral profiles, especially in well-controlled laboratory settings. On this promise major research efforts were put into the development of PyMS systems, especially for battlefield detection of microbes used in germ warfare. Although these systems showed some promise for the detection of chemical agents, they were less practical for the analysis of microorganisms in real-world settings.

It is easy to look back from today's vantage point and assert that the spectral patterns produced by pyrolysis were not sufficiently dissimilar for different organisms, or that pattern recognition (or computational) approaches were not advanced enough to provide the specificity needed for routine application of PyMS to bacterial characterization. Such criticism misses the important point that these methods did work within certain limitations. Indeed, research in PyMS and related areas continues today as investigators attempt to increase the diversity of spectral patterns for bacteria using modifications of traditional Currie-point pyrolysis electron impact mass spectrometry, such as metastable atom bombardment and advanced methods for assembling libraries of pyrolysis data. Another approach that very nearly succeeded was the application of laser desorption ionization mass spectrometry to bacteria. Early pioneers in the field were attempting "whole-cell" studies using electron impact and laser desorption ionization, prior to the advent of modern desorption ionization techniques. These studies are the direct linear predecessors of techniques such as "whole-cell MALDI-TOF MS." Had these investigators been working after, rather that before, the introduction of the now-common techniques developed to study proteins and other biopolymers, their work would have been as successful as today's. Unfortunately, their laser desorption studies of bacteria were attempted without the benefit of a matrix to facilitate the interaction between the sample and the laser, or to promote the desorption/ionization process.

Much of the very important early work in this field is documented in a landmark *ACS Symposium Series* book edited by Catherine Fenselau. The editors and authors of the present book hope that they can likewise reflect much of the important work conducted in the last decade. We anticipate that many in the microbiology community will be skeptical about current claims that mass spectrometry can play an important role in the analysis of bacteria, especially rapid characterization. Several chapters in this book were prepared by microbiologists. Even though they have had a hand in the development of mass spectrometry approaches to bacterial analysis, they are nevertheless microbiology-centered investigators, and their contributions should provide both microbiology background to the mass spectrometry specialist and a perspective on mass spectrometry approaches to the microbiologist. This book includes an introduction to basic concepts in mass spectrometry, and a brief history of the development of mass spectrometry methods for analysis of bacteria. The most commonly used mass spectral method for the rapid analysis of bacteria is whole-cell MALDI time-of-flight (TOF) mass spectrometry. In addition to coverage of the widely used whole-cell MALDI-TOF MS approach, a newer approach involving very high resolution Fourier transformation mass spectrometry (FTMS) is also reported. Although many analyses are based on spectral comparisons, this book reflects the parallel development of proteome-based approaches to bacterial identification. By either approach it is now relatively routine to classify target bacteria down to the strain level using mass spectrometry. A number of taxonomically important proteins have now been identified, many of them being ribosomal proteins. These are detected because they are both abundant and easily detected analytes in MALDI mass spectrometry studies.

The most common criticism of all instrumental (chemical) methods for characterizing bacteria is the relatively large number of bacteria required. Typically 100,000 bacteria or more or needed for analysis. Not withstanding that pre-analysis amplification of bacteria is a common procedure for most methods (including standard methods used by regulatory agencies and hospitals), this remains a valid criticism, especially for those developing rapid analysis methods. Several investigators are working on approaches to concentrate bacteria or bacterial biomarkers prior to MS analysis without resort to lengthy re-growth of cells in culture media. Another criticism of mass spectrometry methods is the cost of instrumentation. For mass spectrometry to receive widespread adoption by microbiologists, a smaller footprint and reduced cost would be very desirable. Considerable progress has been made in both of these areas and is reported some of the following chapters.

Electrospray (ESI) ionization mass spectrometry also plays in important role in bacterial characterization. Because it typically includes a chromatographic separation step, the approach is not considered as rapid as MALDI approaches, which do not incorporate a separation. However, compared to the times needed to grow bacteria in culture prior to analysis, the time frame is not lengthy, and the addition of chromatographic separation provides many opportunities to increase specificity. ESI/MS has been used to characterize cellular biomarkers for metabolic, genomic, and proteomics fingerprinting of bacteria, and these approaches are reported in two chapters.

Finally, methods developed for rapid detection of microorganisms by mass spectrometry may also produce biomarker data that can be used in the design of better molecular probes. As noted above, many of the strain-differentiating proteins have been identified, and this information can be used to infer information about differences in the corresponding genes at the strain level. One chapter reports the use of strain-specific protein biomarkers, detected in mass spectrometry studies of *Vibrio* strains from a human outbreak in the Pacific Northwest that eventually led to an understanding of the genomic changes and have rendered existing gene-based probes less effective over time. This is but one example of the many mechanistic and biochemical advances that will likely come from the development of methods for bacterial speciation. It simply reflects a link between the recognition of the molecular species that differentate cells, and the elucidation of biochemical pathways that follows the determination of molecular structures.

Jackson O. Lay, Jr.
Charles L. Wilkins

Fayetteville, Arkansas

CONTRIBUTORS

Randy J. Arnold Department of Chemistry, Indiana University, Bloomington, IN 47405

Michel J. Bertrand National Center for Toxicological Research, Food and Drug Administration, Jefferson, AR 72079

Dan A. Buzatu National Center for Toxicological Research, Food and Drug Administration, Jefferson, AR 72079

Plamen A. Demirev Applied Physics Laboratory, Johns Hopkins University, Laurel, MD 20723

Todd Eschler National Center for Toxicological Research, Food and Drug Administration, Jefferson, AR 72079

Catherine Fenselau Department of Chemistry and Biochemistry, University of Manchester, College Park, MD 20742

Alvin Fox University of South Carolina, Columbia, SC 26208

Royston Goodacre School of Chemistry, University of Manchester, Manchester M60 1QD

Wayne H. Griest Oak Ridge National Laboratory, Oak Ridge, TN 37831

Les Hosey National Center for Toxicological Research, Food and Drug Administration, Jefferson, AR 72079

Kristin H. Jarman Pacific Northwest National Laboratory, Richland, WA 99352

Jeffrey J. Jones Department of Chemistry, University of Arkansas, Fayetteville, AR 72701

Jonathan A. Karty Department of Chemistry, Indiana University, Bloomington, IN 47405

Stephen A. Lammert Oak Ridge National Laboratory, Oak Ridge, TN 37831-6120

Jackson O. Lay, Jr. University of Arkansas, Fayetteville, AR 72701

Rohana Liyanage University of Arkansas, Fayetteville, AR 72701

Gary Miertschin National Center for Toxicological Research, Food and Drug Administration, Jefferson, AR 72079

Steven R. Monday Center for Food Safety and Nutrition, Food and Drug Administration, College Park, MD 20740

Steven M. Musser Center for Food Safety and Nutrition, Food and Drug Administration, College Park, MD 20740

Patrick Pribil Department of Chemistry and Biochemistry, University of Maryland, College Park, MD 20742

Fatemeh Rafii Division of Microbiology, National Center for Toxicological Research, Food and Drug Administration, Jefferson, AR 72079

Jon C. Rees Department of Chemistry and Geochemistry, Colorado School of Mines, Golden CO 80401

James P. Reilly Department of Chemistry, Indiana University, Bloomington, IN 47405

Larry Rushing National Center for Toxicological Research, Food and Drug Administration, Jefferson, AR 72079

Michael J. Stump Department of Chemistry, University of Arkansas, Fayetteville, AR 72701

John B. Sutherland Division of Microbiology, National Center for Toxicological Research, Food and Drug Administration, Jefferson, AR 72079

Éadaoin Timmins National Center for Biomedical Engineering Science, National University of Ireland-Galway, Galway, Ireland

Seetharaman Vaidyanathan School of Chemistry, University of Manchester, Manchester M60 1QD

Kent J. Voorhees Department of Chemistry and Geochemistry, Colorado School of Mines, Golden CO 80401

Karen L. Wahl Pacific Northwest National Laboratory, Richland, WA 99352

Jon G. Wilkes National Center for Toxicological Research, Food and Drug Administration, Jefferson, AR 72079

Charles L. Wilkins Department of Chemistry, University of Arkansas, Fayetteville, AR 72701

Tracie L. Williams Center for Food Safety and Nutrition, Food and Drug Administration, College Park, MD 20740

1

CULTURAL, SEROLOGICAL, AND GENETIC METHODS FOR IDENTIFICATION OF BACTERIA

John B. Sutherland and Fatemeh Rafii

Division of Microbiology, National Center for Toxicological Research, Food and Drug Administration, Jefferson, AR 72079

1.1 INTRODUCTION

The successful identification of unknown bacteria requires the intelligent use of several different techniques. Although bacteria have been classified into a large number of groups and subgroups, the essential terms used for reporting the identification of an individual bacterium are the *genus* (plural: genera) and *species*. Some clinically important species of bacteria have been further divided into serotypes distinguished by specific antibodies, genotypes distinguished by DNA sequence analysis, or other subgroups that are relevant in epidemiological studies.

A *strain* is a pure culture containing only the descendants of a single parent cell; since there is no standardized system for giving numbers to strains, each laboratory uses its own system. Many strains of bacteria are maintained for study in international culture collections but some bacteria have not yet been grown in the absence of other organisms.

When a species of bacterium is identified in a research report, its microscopic appearance and its cultural and biochemical characteristics should be compared if possible with those of an official type strain obtained from a culture collection. Determining the nucleotide sequence of a single gene,

Identification of Microorganisms by Mass Spectrometry, Edited by Charles L. Wilkins and Jackson O. Lay, Jr.

unless it codes for 16S ribosomal RNA, is rarely sufficient for identification of a bacterial species. Horizontal gene transfer may have introduced genes derived from distantly related or totally unrelated organisms.[1] Indeed, even ribosomal RNA sequence analysis may yield a wrong identification of a microorganism when there is genetic heterogeneity or the reference database is incomplete.[2,3] Several rapid techniques derived from immunology and molecular biology methods have been developed to shorten the time required for identification of a bacterium.

In *clinical practice*, a physician generally begins treatment of an infectious disease in a new patient by prescribing a broad-spectrum antibiotic considered likely to be effective against most common pathogenic bacteria.[4,5] If the pathogen and its antibiotic susceptibility could be identified before the antibiotic is prescribed, the prevalence of antibiotic resistance among pathogenic bacteria could be decreased greatly, although unfortunately not to zero.[6,7] Bacteria that cause blood infections and pneumonia require especially rapid identification for effective treatment.[5] Rapid identification of toxin-producing serotypes of bacteria, including the notorious *Escherichia coli* O157:H7,[8–10] is also desirable because most tests do not distinguish toxin-producers from harmless serotypes of *E. coli*.[4] Detection of potential bioterror agents also requires the utmost in speed; these agents may include infectious bacteria, such as cells of the anthrax-causing *Bacillus anthracis*, toxin-producing bacterium *Clostridium botulinum*, or hoax materials. They can be identified by using antibodies or DNA probes specific for the possible targets.[11–13] Caution is always required with any rapid method to prevent reporting either false positives or false negatives.[2]

The prompt identification of bacteria that are *animal pathogens* is important to veterinarians maintaining the health of pets, livestock, and poultry.[14] It is also important to physicians if the animal pathogens are transmissible to humans. In animal husbandry, information on the species found in specialized microbial habitats, such as the bovine rumen, can even be used to improve the efficiency of feed conversion.[15]

Bacterial *plant pathogens* are usually identified by microscopic observation and comparison of the symptoms on the plant with published descriptions and photographs, but various rapid serological and molecular methods are now being developed for plant disease diagnosis.[16,17]

To ensure the safety of food products, representative samples must be inspected so that *foodborne bacteria* can be identified.[15,18,19] Bacteria producing heat-stable enterotoxins, such as *Staphylococcus aureus*, may be identified by biochemical and serological techniques.[20,21] Molecular methods are now widely used for the identification of many pathogenic foodborne bacteria,[15,22,23] In addition bacteria used as starter cultures for cheese, yogurt, other fermented foods and beverages, and probiotic dietary supplements may be identified for quality assurance.[22,24,25]

Studies in *microbial ecology* require identification of the constituent microorganisms of an ecosystem as well as assessment of their activities.[26–28]

For instance, soil microbiological studies may involve not only plate counts of bacterial colony-forming units but also the extraction of messenger RNA produced by nonculturable bacteria.[29,30] Water-quality management requires the identification of pathogenic bacteria[31–33] and determination of their likely sources.[34,35] Even the identification of noninfectious airborne bacteria found in agricultural and industrial workplaces may be useful in determining the causes of occupational allergies suffered by workers.[36]

In this chapter a variety of conventional and rapid methods that are currently used for the identification of bacteria (Table 1.1) will be discussed.

1.2 IDENTIFICATION OF BACTERIA BY CULTURAL METHODS

The identification of bacteria has traditionally required the establishment of a *pure culture* before any other steps are taken. Pure cultures of bacteria may sometimes be obtained from blood and spinal fluid, which are normally sterile, or from extreme environments like hot springs. However, because there are few such situations in nature, individual bacteria must generally be isolated from other cells and grown for one to five days to obtain pure cultures before identification. Some pathogenic bacteria are obligate intracellular parasites that are difficult or impossible to grow outside their mammalian host cells;[37] for these, pure cultures are not feasible.

As observed by Bossis et al., "Over decades, bacteria have been characterised, clustered and identified according to a few phenotypic characters such as morphology, pigmentation, reaction to dyes, the presence or absence of spores, nutritional requirements, ability to produce acids from sugars, and sensitivity to inhibitors."[38] Most of these *phenotypic characters* are observed best in pure cultures.[38–40] The appearance of bacterial colonies on agar media is often helpful. Among the other cultural characteristics of bacteria that are useful for identification are the ability to grow in air or without it, to use inorganic compounds as energy sources, to grow at low or at high pH, to decompose H_2O_2, to reduce sulfate, to grow in high concentrations of NaCl, and to produce simple or branched filaments.

The traditional bacteriological techniques of streaking agar plates and serial dilution of liquids are still extremely useful for obtaining pure cultures. When relatively low numbers of cells are present, as in drinking water or air, membrane filters may be used to concentrate bacteria for use in starting pure cultures. Various types of impactors, filters, and liquid impingers are used for concentrating cells from air samples.[36,41] When higher numbers of cells are present, dilution will be necessary. Some newer methods separate cells by capillary electrophoresis or flow cytometry.[11,42]

The *morphology* of some bacteria, especially those that form spores, is distinctive enough under the light microscope to have value for identification. This means that differential staining techniques, such as the Gram stain or acid-fast stain, and fluorescence microscopy may help to determine the iden-

TABLE 1.1 Methods for Identification of Bacteria and Some Recent Examples Using Them

Method	References
Cultural methods (mostly requiring several days)	
Morphology	43, 44
Selective and differential media	19, 24, 25
Chemotaxonomic methods	38–40, 45, 47
Biochemical tests	28, 34, 40
Serological methods (mostly requiring one day)	
Latex agglutination	39, 53, 54
Enzyme-linked immunosorbent assays (ELISA)	52, 55–59
Immunofluorescence assays (IFAs)	10, 51, 60, 61
Immunomagnetic separation	10, 57, 62
Genetic methods (mostly requiring one day)	
Nucleic acid hybridization	38, 45, 60, 63, 65, 68, 69, 71
Fluorescent in situ hybridization (FISH)	76, 78–82
Polymerase chain reaction (PCR)	32, 83, 84, 86, 88
Reverse transcriptase PCR (RT-PCR)	30, 92
Multiplex PCR	60, 94, 95
Nested PCR	16, 96, 97
Arbitrarily primed (AP) PCR	17, 98
Enterobacterial repetitive intergenic consensus (ERIC) PCR	98, 99
Randomly amplified polymeric DNA (RAPD) PCR	79, 99, 100
Repetitive extragenic palindromic (REP) PCR	14, 99, 101
BOX PCR	101, 102
Restriction fragment length polymorphism (RFLP) PCR	17, 45, 103
Real-time PCR	9, 12, 104–106
DNA microarrays	15, 110–112, 115
Other methods (requiring one to several days)	
Phage typing	35, 48
Capillary electrophoresis	42
Flow cytometry	11, 42
Infrared spectroscopy, including FTIR	5, 25, 113
Polyacrylamide gel electrophoresis (PAGE)	25, 38, 39
Amplified ribosomal DNA restriction analysis (ARDRA)	17, 79
Ribotyping	14, 34, 79
Pulsed-field gel electrophoresis (PFGE)	25, 34, 79, 100, 114

tity of a bacterium by showing the structure even if the organism cannot be grown in common culture media.[43,44] Use of the Gram stain allows bacterial cultures to be placed into major groups (Gram-positive and Gram-negative). However, since the cells of most bacteria are ordinary cocci or rods that resemble many others, additional biochemical tests are also needed for identification.

Selective media containing chemical inhibitors are used in agar plates or culture tubes to favor the growth of certain groups of bacteria while inhibiting others.[19,24,25] These media, however, generally also allow the growth of some nontarget bacteria.[25] *Differential media* allow other species to grow but contain indicator dyes that cause target organisms to appear as differently colored colonies.[24,25] Genes for pathogenicity, toxin production, and other physiological characteristics may be expressed in some media but not in others.[45]

Biochemical tests are usually performed after pure cultures have been obtained. The standard indole, methyl red, Voges-Proskauer, citrate, and litmus milk tests may be used to show important physiological characteristics. To study the functional diversity of bacteria, the utilization of carbohydrates, amines, amides, carboxylic acids, amino acids, polymers, and other carbon and nitrogen sources can be tested.[28] Dilution-based most-probable number (MPN) techniques with phospholipid fatty acids as biomarkers have been employed for studying different bacterial species in lakes.[40] The patterns of antibiotic resistance in bacteria isolated from natural waters have been useful for identifying sources of water pollution.[34]

A variety of *commercial kits* and automated systems are available to test the abilities of bacteria to assimilate, ferment, decarboxylate, or cleave selected organic compounds.[46] Their reliability for species identification is usually greater with cultures from clinical samples, where a limited number of bacteria are commonly encountered, and less with environmental soil and water samples, where a great many uncommon or previously unidentified species not in the database are likely to be present.[29,45] Additional tests beyond those found in the commercial kits may be necessary; for example, the hydrolysis of various nitriles and amides is useful for identifying *Rhodococcus* spp.[47] Some commercial kits for clinical use feature antimicrobial susceptibility testing.[21]

Chemotaxonomic methods, based on differences in the chemical composition of bacterial cells, are widely used for identification. Some of them have been automated.[45] The lipid composition of the cell membrane, including phospholipids, lipopolysaccharides, and lipoteichoic acids, differs greatly among bacteria.[40] Because of this, when the saturated, unsaturated, branched, cyclopropane, and hydroxy fatty acids in the lipids are saponified and then esterified with methanol, the fatty acid methyl esters produced can be extracted and analyzed by gas chromatography for use in identification.[38,39,45] This method is widely used, although some groups, including the lactic acid bacteria, cannot be reliably identified with it.[25] Phospholipids and other polar lipids can be analyzed by two-dimensional thin-layer chromatography and stained with specific reagents.[45] Sugars from whole-cell extracts can be separated by thin-layer chromatography and then stained.[45] The content of mycolic acids, isoprenoid quinones, polyamines, and diamino acids is useful in identifying species of *Mycobacterium*, *Corynebacterium*, *Rhodococcus*, and a few other genera.[45,47,48]

Because the time required for growth of pure cultures delays identification, serological and molecular techniques that bypass this step are often employed. These methods often are able to show the presence of target organisms in mixed samples.

1.3 IDENTIFICATION OF BACTERIA BY SEROLOGICAL METHODS

Serological methods for bacterial identification are based on techniques derived from immunology. Animals that have been injected with bacterial antigens, which usually are cell proteins or lipopolysaccharides but may be other cell components, produce antibodies in their blood serum that recognize these specific antigens and bind to them with high affinity.[49–51] Either polyclonal or monoclonal antibodies may be used for bacterial identification. *Polyclonal antibodies*, which may bind to several antigens, are produced in the blood sera of animals injected with bacteria or purified antigens. *Monoclonal antibodies*, which only bind to a specific epitope (the part of the antigen molecule that is recognized by the immune system as foreign and generates a response) of the target bacterium, are produced in hybridoma cells grown in the laboratory.[52] Depending on the antigen and the specificity of the antibody, serological methods can be used for the detection of genera, species, or serotypes of bacteria.[49,50] Antibodies are used in numerous methods for bacterial identification. For instance, kits utilizing *latex agglutination*, with antibodies attached to latex particles, are used for the rapid presumptive identification of bacteria including campylobacteria, staphylococci, and streptococci.[39,53,54] The two most commonly used types of serological methods for bacterial identification, however, are enzyme-linked immunosorbent assays and immunofluorescence assays.

Enzyme-linked immunosorbent assays (ELISA) are rapid, simple, and sensitive tests that may allow the detection of less than 10 ng of antigenic protein from a bacterial culture.[55,56] Many different samples, with replicates and controls, can be tested in a 96-well microtiter plate. The bacteria to be identified are added to different wells, where they adhere to the plate. A primary antibody for detecting the bacterium of interest then is added and binds to the bacterial antigen in the well. The excess antibody is removed with several washes and then a second antibody is applied. This second antibody, which is selected to bind specifically to the first antibody, is coupled to an enzyme. After several more washes, the second antibody remains in the well only if the first antibody against the target bacterium has remained bound to the cells. A dye substrate that changes color when cleaved by the enzyme attached to the second antibody is then added. The change of color due to cleavage of the substrate by the enzyme shows the binding of the first antibody to the antigen and thus the presence of the target bacterium.

The limitations of ELISA methods include the specificity of antibodies, the concentrations of primary antibody and antigen, and the type of reaction solution. Nonspecific binding of either of the antibodies to related antigens, unrelated proteins of other bacteria, or even the microtiter plate may lead to false positive reactions.[49,52,57] Use of a monoclonal antibody may decrease cross-reactivity with other antigens. For detection of low numbers of bacteria, as in drinking water, the sample may be filtered to concentrate the cells or cultured in a selective broth until it reaches the minimum detection limit for ELISA.[49,58] Commercial test kits using dipsticks, immunoblots, and sandwich ELISA methods have been developed for the identification of pathogenic bacteria.[58,59]

Immunofluorescence assays (IFAs) may be sensitive enough to identify even a single bacterial cell in a contaminated sample.[51,60] There are two types of IFAs, direct and indirect. In a direct assay, an antibody for a specific bacterium is attached to a fluorochrome and added to an unknown sample. After the unbound antibody has been removed by washing, the sample is examined for fluorescence. In an indirect assay, a primary antibody is added to the bacterial antigen. After interfering particles have been removed, samples are filtered and mixed with a fluorochrome-labeled secondary antibody that binds to the primary antibody. Indirect IFAs are usually preferred, since secondary antibodies that have been conjugated to different fluorochromes are commercially available.[60]

Visualization of bacteria detected by IFAs may be by epifluorescence microscopy, in which fluorescent particles are detected against a dark background, or by flow cytometry, in which fluorescent particles bound to specific bacteria in the mixture are detected and enumerated.[10,51] IFAs also can be used to differentiate dead and living cells by using fluorochrome-labeled monoclonal antibodies against proteins that are expressed at different levels in the different stages of bacterial growth.[60,61]

When the bacteria to be detected are less than 1% of the total population in a sample, IFAs cannot be used because of interference from unrelated particles that are concentrated when large volumes of sample are filtered. To overcome this problem, the organism of interest may be concentrated by *immunomagnetic separation*.[10,57,62] For this procedure magnetic beads coated with monoclonal or polyclonal antibodies are mixed with the sample. The beads are collected with a magnet, and the cells attached to the beads then are removed, enumerated, and identified by IFAs.

Serological techniques can detect target bacteria rapidly in mixtures, but their accuracy depends on the specificity of the antibody used. The use of monoclonal instead of polyclonal antibodies may increase specificity.[49,52,58] However, because the same epitope can be present in more than one species, a monoclonal antibody against one species may cross-react with other bacteria.[50] For this reason serological methods are not always successful for detection of bacteria in environmental samples and nucleic acid-based methods are now commonly used.

1.4 IDENTIFICATION OF BACTERIA BY GENETIC METHODS

Two principal types of nucleic acid-based methods, nucleic acid hybridization and polymerase chain reaction (PCR), are commonly used for the rapid identification of bacteria. A few other nucleic acid-based methods will also be mentioned.

Nucleic acid hybridization methods use oligonucleotide DNA probes with sequences complementary to a portion of the nucleic acid of the target bacterium[38,60,63] and designed to hybridize with immobilized DNA or RNA on a membrane. After any unbound probe has been washed off, the hybridized probe can be detected.[64–66]

Some of the most widely used oligonucleotide probes for bacteria are based on the sequences of 16S and 23S ribosomal RNA.[66–68] Commercially available probes are useful for determining the degree of genetic similarity among different strains of bacteria.[69,70] Sequence comparison of ribosomal RNA[45,47,71] shows that some regions of the genome are always identical within serotypes, species, or even genera. Data from the Ribosomal Database Project[72] and from commercial sources can be used to design probes to detect a specific bacterium. Probes can be labeled by methods using color,[73] radioactivity,[74] fluorescence,[75] or chemiluminescence. In this last method visible light is produced by a chemical reaction.[76]

For nucleic acid hybridization the oligonucleotide probe should be designed to bind only to the target of interest and not to produce false positive results by nonspecific binding. The accuracy of bacterial identification depends on the similarity of the nucleotide sequences of probe and target, since a probe may bind not only to the intended bacterium but also to others with partial sequence similarity.[65] The length of the probe, the hybridization temperature, and the salt concentration all affect hybridization. It is possible to manipulate the conditions for hybridization so that the probe can bind only to DNA that is identical or nearly so. Depending on the probe used, nucleic acid hybridization may provide information about the genus and species or even the strain of bacterium.[77] If the number of cells is sufficient, the target bacterium can be detected within hours instead of the days that are required for culture-based methods.[68]

A variation on this method, called *fluorescent in situ hybridization* (FISH), uses fluorescent-labeled DNA and RNA probes for detection and visualization of single cells by microscopy or flow cytometry.[78–80] The FISH technique is popular because of its sensitivity and speed of visualization; fluorescent dyes can be used to produce probes with different colors for simultaneous detection of several organisms.[76,81,82]

Polymerase chain reaction (PCR) is one of the most important techniques for rapid bacterial identification. It consists of repeated cycles of enzymatic reactions in a thermal cycler (PCR machine) that copies DNA strands many times. The DNA amplified in one PCR cycle is used as a template for the next cycle. This results in an exponential increase of the desired target

DNA,[83] which produces one million or more copies of amplified DNA in a short time. For identification of bacteria, PCR can be used to amplify DNA either after extraction from a sample or after lysis of the cells.[83,84] Methods using washing, filtration, or magnetic beads with specific antibodies can be used to collect bacterial cells for PCR.[85,86] PCR can be modified for the detection of bacteria from various sources[32] and can even amplify DNA from dead cells.[87]

In PCR a particular fragment from the target DNA is amplified by several cycles of replication.[83] The reaction mixture contains bacterial DNA used as a template, the four deoxyribonucleotide triphosphates (the DNA building blocks), *Taq* polymerase (an enzyme from the heat-resistant bacterium *Thermus aquaticus*, which copies DNA by using a single strand from the template as a guide and adding the complementary nucleotides to make double-stranded DNA), and oligonucleotide primers (two short sequences with 15–30 base pairs each that are complementary and can bind to the beginning and the end of the DNA fragment to be amplified). The concentrations of the components of the reaction mixture are selected to allow optimum amplification. The reaction mixture is put into a thermal cycler that is programmed to increase and decrease the temperature many times in a set pattern to amplify a particular DNA sequence.

Amplification of DNA by PCR proceeds as follows: At the denaturation temperature (approximately 95°C), the DNA of the template is denatured from double-stranded to single-stranded and the DNA strands separate from each other. The temperature then is reduced to the annealing temperature (empirically determined, but approximately 60°C), so the primers will hybridize with appropriate segments of DNA. Since the primers consist of two short single-stranded oligonucleotides, they realign themselves with the complementary sequences on the template, forming double strands. When the temperature is raised to the extension temperature (approximately 72°C), the primers are extended by *Taq* polymerase. This enzyme extends the primers by adding the correct nucleotides from the reaction mixture to the ends of the primers, using the sequence of the template to form a DNA strand that is complementary to the adjacent template DNA. The temperature then is raised further to the denaturation temperature, so that the templates and newly synthesized strands are separated from each other and both can be used as templates for the next cycle. This process is repeated for 20 to 40 cycles,[83] each lasting for a few minutes. In each PCR cycle the segment of DNA flanked by the primers is amplified; since the amplification is repeated in each cycle, the amount of the PCR product, known as an amplicon, increases exponentially. The amplicons consist of DNA fragments identical to the sequence between the two primers.[83]

The specificity of bacterial detection by PCR depends on both the degree of homology between the primers and the target DNA, which dictates how well the primers bind to the target, and the annealing temperature at which the target hybridizes to the primers.[83] The primers are designed from the nucleotide sequences of the flanking regions (two small regions located at the beginning and the end) of a segment of DNA with a sequence unique to the bacterium of interest, and the amplicon should be a DNA segment spanning those two primers.[83] The primers should be specific enough not to anneal to the DNA sequences of closely related bacteria. The amplicons are detected either by electrophoresis on an agarose gel and staining by a fluorochrome dye or by hybridization with a labeled DNA probe that can be detected by its chemical, radioactive, or fluorescence properties. Exponential amplification of the target DNA fragment should allow the detection of even a small number of target bacteria.[84,88]

Although PCR is considered highly specific for identification of bacteria, there are limitations to the use of this method for detecting bacteria directly in samples. PCR detects viable cells, injured cells, viable but nonculturable cells that do not grow under normal conditions used for growth of bacteria,[89] and dead cells.[87] It does not provide information on either their physiological state or their similarity to other bacteria. Also most samples contain substances that interfere with *Taq* polymerase. The inhibitors may inhibit PCR amplification completely, change the conditions required for optimum amplification, or decrease the amplicon yield.[60,90,91] They also may keep the primers from binding to the target sequence, resulting in a false negative determination, or allow them to bind to regions that are similar but not identical to the target sequence, interfering with data interpretation.[91] In some instances, the amplicons are sequenced after PCR to verify the identification of the bacterium.

To detect only the growing cells that are metabolically active, in a modified method called *reverse transcriptase PCR* (RT-PCR), an enzyme uses messenger RNA as a template to make DNA with the complementary sequence (cDNA).[30,92] Messenger RNA is produced only by living cells that are synthesizing proteins for growth. The complementary DNA produced by reverse transcriptase is then amplified, using oligonucleotide primers. RT-PCR also can be used to distinguish toxin-producing bacterial strains from similar strains that do not produce toxins.[92]

Multiplex PCR is a variation of PCR used to detect several bacteria simultaneously, using multiple primers that recognize the sequences of different target molecules.[93–95] This method also can be employed with a single bacterial strain to increase the specificity of detection by simultaneously amplifying different segments of the same DNA molecule. The length and base composition of the primers and the sizes of the DNA fragments to be amplified influence amplification in multiplex PCR.[60,95]

Another PCR variation used for detecting bacteria is *nested PCR*, which uses two successive rounds of amplification with different sets of primers.[16,96,97] After amplification of one DNA fragment in the first round, the amplicon is used as a template for a second round of amplification, together with another primer set that amplifies an internal segment of the first amplicon.[96,97] Nested PCR is especially useful when the concentration of the target bacterium is too low to generate a detectable amount of amplicon in the first round of amplification. The use of nested primers ensures that the second amplicon will be produced only if a sequence complementary to the inner primers has been amplified during the first round.[96,97]

Some other variations of PCR methods, such as *arbitrarily primed* (AP) PCR,[17,98] *enterobacterial repetitive intergenic consensus* (ERIC) PCR,[98,99] *randomly amplified polymeric DNA* (RAPD) PCR,[79,99,100] *repetitive extragenic palindromic* (REP) PCR,[14,99,101] and BOX PCR[101,102] generate multiple fragments of DNA that can be separated by electrophoresis to form unique fingerprint patterns. Closely related bacteria produce fingerprints that are similar. These methods may distinguish individual isolates of the same species, but they cannot be used to classify strains by serotype.[98]

Restriction fragment length polymorphism (RFLP) PCR[17,45] is a strategy for identification that uses a combination of PCR and restriction enzymes. In the first step the DNA is amplified using primers that have sequences identical to portions of a gene in the bacterium of interest. The amplicons then are digested into smaller fragments by a restriction enzyme and sorted by size, using electrophoresis on an agarose gel. A particular RFLP pattern is produced for each DNA sequence. Restriction fragments generated from known bacterial strains may be compared with fragments generated from unknown strains under the same conditions.[103]

Although PCR amplification begins at an exponential rate, it enters a stationary phase after approximately 30 cycles, and additional cycles do not increase the concentration of amplicons. For this reason it is not practical to use PCR for quantifying bacteria directly, and other methods, such as real-time PCR, are used for this purpose.

Real-time PCR is a quantitative method for measuring amplicons as they are produced by measuring the increase in fluorescence of a dye added to the reaction mixture.[12,104,105] Methods using fluorescent reporters, such as SYBR® Green,[104,106] TaqMan®,[107,108] or molecular beacons,[9] collect quantitative data at the time when DNA is in the exponential phase of amplification.

SYBR® Green binds to double-stranded DNA during real-time PCR and fluoresces after excitation. As the amount of amplicon increases, so does the amount of fluorescence. In contrast, TaqMan® and molecular beacons are hybridization probes that include both a fluorescent dye and a quenching dye. Excited fluorescent dye transfers energy to the nearby quenching dye, rendering the probe nonfluorescent. During PCR amplification the quenching dye is separated from the fluorescent dye by enzymatic cleavage (TaqMan®) or by hybridization to the target (molecular beacons). As PCR amplification proceeds, higher quantities of fluorescent dye are separated from the quenching dye, resulting in increased fluorescence that can be measured and analyzed automatically.

The instrumentation for real-time PCR includes a thermal cycler with a computer, a spectrophotometer for fluorescence detection, and software for acquisition and analysis of data.[105,109]

DNA microarray techniques involve hybridization of DNA on a solid support with fluorescent-labeled probes that have been produced by PCR.[79,110] The microarrays are used to screen simultaneously for a large number of genes, which may even be from different bacteria. Single-stranded synthetic oligonucleotides containing various target DNA sequences or PCR amplicons of different genes are spotted by a robotic spotter in a lattice pattern on a solid surface, usually a treated glass slide called an array chip. Fluorescent-labeled probes are produced by PCR amplification of the sample DNA and then hybridized to the array chip, using an automated hybridization station. Specific bacteria can be detected in a complex microbial population if a fluorescent-labeled DNA probe derived from this population hybridizes with the DNA for a particular bacterium on the array chip. Multiple bacteria can be identified in a mixture if the probes for different species have been labeled with different fluorescent dyes.[15,111,112]

1.5 OTHER METHODS USED FOR BACTERIAL CHARACTERIZATION

Phage typing of bacterial cultures uses bacteriophages, which are viruses that infect bacteria. This method has been useful for epidemiological studies and for the identification of *Mycobacterium*, *Salmonella*, and *Listeria* spp., even though most phages are not restricted to a particular genus or species.[11,48] Certain phages of fecal bacteria that persist in natural waters can be used to show when water has been polluted by sewage.[35]

Tissue culture of susceptible mammalian cells may be used to characterize some bacteria by the toxins they produce. For example, Shiga-like toxins (verotoxins) are produced by *E. coli* O157:H7.[4,8]

Several additional instrumental techniques have also been developed for bacterial characterization. *Capillary electrophoresis* of bacteria, which requires little sample preparation,[42] is possible because most bacteria act as colloidal particles in suspension and can be separated by their electrical charge. Capillary electrophoresis provides information that may be useful for identification. *Flow cytometry* also can be used to identify and separate individual cells in a mixture.[11,42] *Infrared spectroscopy* has been used to characterize bacteria caught on transparent filters.[113] Fourier-transform infrared (FTIR) spectroscopy, with linear discriminant analysis and artificial neural networks, has been adapted for identifying foodborne bacteria[25,113] and pathogenic bacteria in the blood.[5]

The total proteins synthesized by bacteria grown under standard conditions, when analyzed by *polyacrylamide gel electrophoresis* (PAGE), form patterns that can be compared to those of known strains by visual or computer-assisted methods.[25,38,39]

Some nucleic acid-based techniques that are useful for epidemiological studies are *amplified ribosomal DNA restriction analysis* (ARDRA),[17,79] *ribo-*

typing,[14,34,79] and *pulsed-field gel electrophoresis* (PFGE).[25,34,79] PFGE uses a restriction enzyme, such as *Sma*I, which cuts DNA only at specific locations where certain sequences of 4 to 6 nucleotides are present. This enzyme is used to digest the complete genome of a bacterial strain in an agarose plug; the products then are subjected to electrophoresis on gels, using a specialized apparatus with an alternating electrical field.[100,114]

1.6 CONCLUSIONS

A great many methods have been developed for the identification of genera, species, and serotypes of bacteria. These include cultural methods (biochemical tests and chemical analyses of pure cultures), serological methods (ELISA and immunofluorescence assays), and genetic methods (nucleic acid hybridization and PCR techniques). The usual methods for identifying bacteria in contaminated samples require obtaining pure cultures first by streaking or serial dilution. Microscopic observation and biochemical tests are followed by confirmation of the suspected identity with specific antibodies or molecular probes. PCR and commercial kits containing the antibodies or probes are sometimes used with mixed cultures to determine quickly whether the DNA and proteins of a specific bacterium are present, and DNA microarrays can be used to test for several species of bacteria at one time. Nevertheless, if none of the probes or antibodies on hand correspond to any of the bacteria in the sample, then other tests with pure cultures will be necessary. Each of the methods has advantages for identification of certain species; however, none has yet achieved all of the advantages of speed and accuracy with no serious drawbacks.

ACKNOWLEDGMENTS

We thank Carl E. Cerniglia, Robert H. Heflich, and Donald D. Paine for their useful comments on the manuscript. The views presented in this article do not necessarily reflect those of the Food and Drug Administration.

REFERENCES

1. Philippe, H.; Douady, C. J. Horizontal gene transfer and phylogenetics. *Curr. Opin. Microbiol.* 2003, **6**, 498–505.
2. Millar, B. C.; Jiru, X.; Walker, M. J.; Evans, J. P.; Moore, J. E. False identification of *Coccidioides immitis*: Do molecular methods always get it right? *J. Clin. Microbiol.* 2003, **41**, 5778–5780.
3. Woo, P. C. Y.; Ng, K. H. L.; Lau, S. K. P.; Yip, K. T.; Fung, A. M. Y.; Leung, K. W.; Tam, D. M. W.; Que, T. L.; Yuen, K. Y. Usefulness of the MicroSeq 500 16S ribo-

somal DNA-based bacterial identification system for identification of clinically significant bacterial isolates with ambiguous biochemical profiles. *J. Clin. Microbiol.* 2003, **41**, 1996–2001.

4. Bettelheim, K. A.; Beutin, L. Rapid laboratory identification and characterization of verocytotoxigenic (Shiga toxin producing) *Escherichia coli* (VTEC/STEC). *J. Appl. Microbiol.* 2003, **95**, 205–217.
5. Maquelin, K.; Kirschner, C.; Choo-Smith, L. P.; Ngo-Thi, N. A.; van Vreeswijk, T.; Stämmler, M.; Endtz, H. P.; Bruining, H. A.; Naumann, D.; Puppels, G. J. Prospective study of the performance of vibrational spectroscopies for rapid identification of bacterial and fungal pathogens recovered from blood cultures. *J. Clin. Microbiol.* 2003, **41**, 324–329.
6. Bergeron, M. G.; Ouellette, M. Preventing antibiotic resistance through rapid genotypic identification of bacteria and of their antibiotic resistance genes in the clinical microbiology laboratory. *J. Clin. Microbiol.* 1998, **36**, 2169–2172.
7. Salyers, A. A.; Amábile-Cuevas, C. F. Why are antibiotic resistance genes so resistant to elimination? *Antimicrob. Agents Chemother.* 1997, **41**, 2321–2325.
8. Easton, L. *Escherichia coli* O157: Occurrence, transmission and laboratory detection. *Br. J. Biomed. Sci.* 1997, **54**, 57–64.
9. Fortin, N. Y.; Mulchandani, A.; Chen, W. Use of real-time polymerase chain reaction and molecular beacons for the detection of *Escherichia coli* O157:H7. *Anal. Biochem.* 2001, **289**, 281–288.
10. Pyle, B. H.; Broadaway, S. C.; McFeters, G. A. Sensitive detection of *Escherichia coli* O157:H7 in food and water by immunomagnetic separation and solid-phase laser cytometry. *Appl. Environ. Microbiol.* 1999, **65**, 1966–1972.
11. Ivnitski, D.; Abdel-Hamid, I.; Atanasov, P.; Wilkins, E. Biosensors for detection of pathogenic bacteria. *Biosens. Bioelectron.* 1999, **14**, 599–624.
12. Makino, S. I.; Cheun, H. I. Application of the real-time PCR for the detection of airborne microbial pathogens in reference to the anthrax spores. *J. Microbiol. Meth.* 2003, **53**, 141–147.
13. Peruski, L. F.; Peruski, A. H. Rapid diagnostic assays in the genomic biology era: Detection and identification of infectious disease and biological weapon agents. *BioTechniques* 2003, **35**, 840–846.
14. Blackall, P. J.; Miflin, J. K. Identification and typing of *Pasteurella multocida*: A review. *Avian Pathol.* 2000, **29**, 271–287.
15. Al-Khaldi, S. F.; Martin, S. A.; Rasooly, A.; Evans, J. D. DNA microarray technology used for studying foodborne pathogens and microbial habitats: Minireview. *J. AOAC Int.* 2002, **85**, 906–910.
16. Bertolini, E.; Penyalver, R.; García, A.; Olmos, A.; Quesada, J. M.; Cambra, M.; López, M. M. Highly sensitive detection of *Pseudomonas savastanoi* pv. *savastanoi* in asymptomatic olive plants by nested-PCR in a single closed tube. *J. Microbiol. Meth.* 2003, **52**, 261–266.
17. Louws, F. J.; Rademaker, J. L. W.; de Bruijn, F. J. The three Ds of PCR-based genomic analysis of phytobacteria: Diversity, detection, and disease diagnosis. *Annu. Rev. Phytopathol.* 1999, **37**, 81–125.

18. Scheu, P. M.; Berghof, K.; Stahl, U. Detection of pathogenic and spoilage microorganisms in food with the polymerase chain reaction. *Food Microbiol.* 1998, **15**, 13–31.
19. Valentín-Bon, I. E.; Brackett, R. E.; Seo, K. H.; Hammack, T. S.; Andrews, W. H. Preenrichment versus direct selective agar plating for the detection of *Salmonella enteritidis* in shell eggs. *J. Food Prot.* 2003, **66**, 1670–1674.
20. Balaban, N.; Rasooly, A. Staphylococcal enterotoxins. *Int. J. Food Microbiol.* 2000, **61**, 1–10.
21. Fahr, A. M.; Eigner, U.; Armbrust, M.; Caganic, A.; Dettori, G.; Chezzi, C.; Bertoncini, L.; Benecchi, M.; Menozzi, M. G. Two-center collaborative evaluation of the performance of the BD Phoenix automated microbiology system for identification and antimicrobial susceptibility testing of *Enterococcus* spp. and *Staphylococcus* spp. *J. Clin. Microbiol.* 2003, **41**, 1135–1142.
22. Giraffa, G.; Neviani, E. Molecular identification and characterization of food-associated lactobacilli. *Ital. J. Food Sci.* 2000, **12**, 403–423.
23. Rijpens, N. P.; Herman, L. M. F. Molecular methods for identification and detection of bacterial food pathogens. *J. AOAC Int.* 2002, **85**, 984–995.
24. Charteris, W. P.; Kelly, P. M.; Morelli, L.; Collins, J. K. Selective detection, enumeration and identification of potentially probiotic *Lactobacillus* and *Bifidobacterium* species in mixed bacterial populations. *Int. J. Food Microbiol.* 1997, **35**, 1–27.
25. Coeuret, V.; Dubernet, S.; Bernardeau, M.; Gueguen, M.; Vernoux, J. P. Isolation, characterisation and identification of lactobacilli focusing mainly on cheeses and other dairy products. *Lait* 2003, **83**, 269–306.
26. Aoi, Y. In situ identification of microorganisms in biofilm communities. *J. Biosci. Bioeng.* 2002, **94**, 552–556.
27. Greene, E. A.; Voordouw, G. Analysis of environmental microbial communities by reverse sample genome probing. *J. Microbiol. Meth.* 2003, **53**, 211–219.
28. Preston-Mafham, J.; Boddy, L.; Randerson, P. F. Analysis of microbial community functional diversity using sole-carbon-source utilisation profiles—A critique. *FEMS Microbiol. Ecol.* 2002, **42**, 1–14.
29. Wellington, E. M. H.; Berry, A.; Krsek, M. Resolving functional diversity in relation to microbial community structure in soil: Exploiting genomics and stable isotope probing. *Curr. Opin. Microbiol.* 2003, **6**, 295–301.
30. Widada, J.; Nojiri, H.; Omori, T. Recent developments in molecular techniques for identification and monitoring of xenobiotic-degrading bacteria and their catabolic genes in bioremediation. *Appl. Microbiol. Biotechnol.* 2002, **60**, 45–59.
31. Bej, A. K.; Steffan, R. J.; DiCesare, J.; Haff, L.; Atlas, R. M. Detection of coliform bacteria in water by polymerase chain reaction and gene probes. *Appl. Environ. Microbiol.* 1990, **56**, 307–314.
32. Iqbal, S.; Robinson, J.; Deere, D.; Saunders, J. R.; Edwards, C.; Porter, J. Efficiency of the polymerase chain reaction amplification of the *uid* gene for detection of *Escherichia coli* in contaminated water. *Lett. Appl. Microbiol.* 1997, **24**, 498–502.
33. Tsen, H. Y.; Lin, C. K.; Chi, W. R. Development and use of 16S rRNA gene targeted PCR primers for the identification of *Escherichia coli* cells in water. *J. Appl. Microbiol.* 1998, **85**, 554–560.

34. Simpson, J. M.; Santo Domingo, J. W.; Reasoner, D. J. Microbial source tracking: State of the science. *Environ. Sci. Technol.* 2002, **36**, 5279–5288.
35. Sinton, L. W.; Finlay, R. K.; Hannah, D. J. Distinguishing human from animal faecal contamination in water: A review. *New Zealand J. Mar. Freshw. Res.* 1998, **32**, 323–348.
36. Eduard, W.; Heederik, D. Methods for quantitative assessment of airborne levels of noninfectious microorganisms in highly contaminated work environments. *Am. Ind. Hyg. Assoc. J.* 1998, **59**, 113–127.
37. Hackstadt, T. The diverse habitats of obligate intracellular parasites. *Curr. Opin. Microbiol.* 1998, **1**, 82–87.
38. Bossis, E.; Lemanceau, P.; Latour, X.; Gardan, L. The taxonomy of *Pseudomonas fluorescens* and *Pseudomonas putida*: Current status and need for revision. *Agronomie* 2000, **20**, 51–63.
39. On, S. L. W. Identification methods for campylobacters, helicobacters, and related organisms. *Clin. Microbiol. Rev.* 1996, **9**, 405–422.
40. Spring, S.; Schulze, R.; Overmann, J.; Schleifer, K. H. Identification and characterization of ecologically significant prokaryotes in the sediment of freshwater lakes: Molecular and cultivation studies. *FEMS Microbiol. Rev.* 2000, **24**, 573–590.
41. Muilenberg, M. L. Sampling devices. *Immunol. Allergy Clin. North Am.* 2003, **23**, 337–355.
42. Desai, M. J.; Armstrong, D. W. Separation, identification, and characterization of microorganisms by capillary electrophoresis. *Microbiol. Mol. Biol. Rev.* 2003, **67**, 38–51.
43. Berkeley, R. C. W.; Ali, N. Classification and identification of endospore-forming bacteria. *J. Appl. Bacteriol. Symp. Suppl.* 1994, **76**, 1S–8S.
44. Robbins, E. I. Bacteria and Archaea in acidic environments and a key to morphological identification. *Hydrobiologia* 2000, **433**, 61–89.
45. Busse, H. J.; Denner, E. B. M.; Lubitz, W. Classification and identification of bacteria: Current approaches to an old problem; overview of methods used in bacterial systematics. *J. Biotechnol.* 1996, **47**, 3–38.
46. Gavin, P. J.; Warren, J. R.; Obias, A. A.; Collins, S. M.; Peterson, L. R. Evaluation of the Vitek 2 system for rapid identification of clinical isolates of Gram-negative bacilli and members of the family *Streptococcaceae*. *Eur. J. Clin. Microbiol. Infect. Dis.* 2002, **21**, 869–874.
47. Colquhoun, J. A.; Heald, S. C.; Li, L.; Tamaoka, J.; Kato, C.; Horikoshi, K.; Bull, A. T. Taxonomy and biotransformation activities of some deep-sea actinomycetes. *Extremophiles* 1998, **2**, 269–277.
48. Williams, S. T.; Locci, R.; Beswick, A.; Kurtböke, D. I.; Kuznetsov, V. D.; Le Monnier, F. J.; Long, P. F.; Maycroft, K. A.; Palma, R. A.; Petrolini, B.; Quaroni, S.; Todd, J. I.; West, M. Detection and identification of novel actinomycetes. *Res. Microbiol.* 1993, **144**, 653–656.
49. Hübner, I.; Steinmetz, I.; Obst, U.; Giebel, D.; Bitter-Suermann, D. Rapid determination of members of the family *Enterobacteriaceae* in drinking water by an immunological assay using a monoclonal antibody against enterobacterial common antigen. *Appl. Environ. Microbiol.* 1992, **58**, 3187–3191.

50. Levasseur, S.; Husson, M. O.; Leitz, R.; Merlin, F.; Laurent, F.; Peladan, F.; Drocourt, J. L.; Leclerc, H.; Van Hoegaerden, M. Rapid detection of members of the family *Enterobacteriaceae* by a monoclonal antibody. *Appl. Environ. Microbiol.* 1992, **58**, 1524–1529.

51. Zaccone, R.; Crisafi, E.; Caruso, G. Evaluation of fecal pollution in coastal Italian waters by immunofluorescence. *Aquat. Microb. Ecol.* 1995, **9**, 79–85.

52. Faude, U. C.; Höfle, M. G. Development and application of monoclonal antibodies for in situ detection of indigenous bacterial strains in aquatic ecosystems. *Appl. Environ. Microbiol.* 1997, **63**, 4534–4542.

53. Slotved, H. C.; Elliott, J.; Thompson, T.; Konradsen, H. B. Latex assay for serotyping of group B *Streptococcus* isolates. *J. Clin. Microbiol.* 2003, **41**, 4445–4447.

54. van Griethuysen, A.; Bes, M.; Etienne, J.; Zbinden, R.; Kluytmans, J. International multicenter evaluation of latex agglutination tests for identification of *Staphylococcus aureus. J. Clin. Microbiol.* 2001, **39**, 86–89.

55. Crowther, J. R. *ELISA*: *Theory and Practice.* Totowa, NJ: Humana Press, 1995.

56. Tymoczko, J. L.; Berg, J. M.; Stryer, L. *Biochemistry*, 5th ed. New York: Freeman, 2002.

57. Mansfield, L. P.; Forsythe, S. J. The detection of *Salmonella* using a combined immunomagnetic separation and ELISA end-detection procedure. *Lett. Appl. Microbiol.* 2000, **31**, 279–283.

58. Valdivieso-Garcia, A.; Riche, E.; Abubakar, O.; Waddell, T. E.; Brooks, B. W. A double antibody sandwich enzyme-linked immunosorbent assay for the detection of *Salmonella* using biotinylated monoclonal antibodies. *J. Food Prot.* 2001, **64**, 1166–1171.

59. Iqbal, S. S.; Mayo, M. W.; Bruno, J. G.; Bronk, B. V.; Batt, C. A.; Chambers, J. P. A review of molecular recognition technologies for detection of biological threat agents. *Biosens. Bioelectron.* 2000, **15**, 549–578.

60. Rompré, A.; Servais, P.; Baudart, J.; de-Roubin, M. R.; Laurent, P. Detection and enumeration of coliforms in drinking water: Current methods and emerging approaches. *J. Microbiol. Meth.* 2002, **49**, 31–54.

61. Rockabrand, D.; Austin, T.; Kaiser, R.; Blum, P. Bacterial growth state distinguished by single-cell protein profiling: Does chlorination kill coliforms in municipal effluent? *Appl. Environ. Microbiol.* 1999, **65**, 4181–4188.

62. Hanai, K.; Satake, M.; Nakanishi, H.; Venkateswaran, K. Comparison of commercially available kits with standard methods for detection of *Salmonella* strains in foods. *Appl. Environ. Microbiol.* 1997, **63**, 775–778.

63. Zarda, B.; Hahn, D.; Chatzinotas, A.; Schönhuber, W.; Neef, A.; Amann, R. I.; Zeyer, J. Analysis of bacterial community structure in bulk soil by in situ hybridization. *Arch. Microbiol.* 1997, **168**, 185–192.

64. Lebaron, P.; Catala, P.; Fajon, C.; Joux, F.; Baudart, J.; Bernard, L. A new sensitive, whole-cell hybridization technique for detection of bacteria involving a biotinylated oligonucleotide probe targeting rRNA and tyramide signal amplification. *Appl. Environ. Microbiol.* 1997, **63**, 3274–3278.

65. Manz, W. In situ analysis of microbial biofilms by rRNA-targeted oligonucleotide probing. *Methods Enzymol.* 1999, **310**, 79–91.

66. Manz, W.; Szewzyk, U.; Ericsson, P.; Amann, R.; Schleifer, K. H.; Stenström, T. A. In situ identification of bacteria in drinking water and adjoining biofilms by hybridization with 16S and 23S rRNA-directed fluorescent oligonucleotide probes. *Appl. Environ. Microbiol.* 1993, **59**, 2293–2298.

67. Fuchs, B. M.; Wallner, G.; Beisker, W.; Schwippl, I.; Ludwig, W.; Amann, R. Flow cytometric analysis of the in situ accessibility of *Escherichia coli* 16S rRNA for fluorescently labeled oligonucleotide probes. *Appl. Environ. Microbiol.* 1998, **64**, 4973–4982.

68. Mittelman, M. W.; Habash, M.; Lacroix, J. M.; Khoury, A. E.; Krajden, M. Rapid detection of *Enterobacteriaceae* in urine by fluorescent 16S rRNA in situ hybridization on membrane filters. *J. Microbiol. Meth.* 1997, **30**, 153–160.

69. Amann, R. I.; Ludwig, W.; Schleifer, K. H. Phylogenetic identification and in situ detection of individual microbial cells without cultivation. *Microbiol. Rev.* 1995, **59**, 143–169.

70. Olsen, G. J.; Lane, D. J.; Giovannoni, S. J.; Pace, N. R.; Stahl, D. A. Microbial ecology and evolution: A ribosomal RNA approach. *Annu. Rev. Microbiol.* 1986, **40**, 337–365.

71. Patel, J. B. 16S rRNA gene sequencing for bacterial pathogen identification in the clinical laboratory. *Mol. Diagn.* 2001, **6**, 313–321.

72. Maidak, B. L.; Cole, J. R.; Lilburn, T. G.; Parker, C. T.; Saxman, P. R.; Farris, R. J.; Garrity, G. M.; Olsen, G. J.; Schmidt, T. M.; Tiedje, J. M. The RDP-II (Ribosomal Database Project). *Nucl. Acids Res.* 2001, **29**, 173–174.

73. Renz, M.; Kurz, C. A colorimetric method for DNA hybridization. *Nucl. Acids Res.* 1984, **12**, 3435–3444.

74. Ouverney, C. C.; Fuhrman, J. A. Combined microautoradiography-16S rRNA probe technique for determination of radioisotope uptake by specific microbial cell types in situ. *Appl. Environ. Microbiol.* 1999, **65**, 1746–1752.

75. Regnault, B.; Martin-Delautre, S.; Lejay-Collin, M.; Lefèvre, M.; Grimont, P. A. D. Oligonucleotide probe for the visualization of *Escherichia coli/Escherichia fergusonii* cells by in situ hybridization: Specificity and potential applications. *Res. Microbiol.* 2000, **151**, 521–533.

76. Van Poucke, S. O.; Nelis, H. J. Rapid detection of fluorescent and chemiluminescent total coliforms and *Escherichia coli* on membrane filters. *J. Microbiol. Meth.* 2000, **42**, 233–244.

77. DeLong, E. F.; Wickham, G. S.; Pace, N. R. Phylogenetic stains: Ribosomal RNA-based probes for the identification of single cells. *Science* 1989, **243**, 1360–1363.

78. Moter, A.; Göbel, U. B. Fluorescence in situ hybridization (FISH) for direct visualization of microorganisms. *J. Microbiol. Meth.* 2000, **41**, 85–112.

79. Satokari, R. M.; Vaughan, E. E.; Smidt, H.; Saarela, M.; Mättö, J.; de Vos, W. M. Molecular approaches for the detection and identification of bifidobacteria and lactobacilli in the human gastrointestinal tract. *Syst. Appl. Microbiol.* 2003, **26**, 572–584.

80. Wagner, M.; Horn, M.; Daims, H. Fluorescence in situ hybridisation for the identification and characterisation of prokaryotes. *Curr. Opin. Microbiol.* 2003, **6**, 302–309.

81. Levsky, J. M.; Singer, R. H. Fluorescence in situ hybridization: past, present and future. *J. Cell Sci.* 2003, **116**, 2833–2838.

82. Swiger, R. R.; Tucker, J. D. Fluorescence in situ hybridization: A brief review. *Environ. Mol. Mutagen.* 1996, **27**, 245–254.

83. Hill, W. E. The polymerase chain reaction: Applications for the detection of food-borne pathogens. *Crit. Rev. Food Sci. Nutr.* 1996, **36**, 123–173.

84. Tani, K.; Kurokawa, K.; Nasu, M. Development of a direct in situ PCR method for detection of specific bacteria in natural environments. *Appl. Environ. Microbiol.* 1998, **64**, 1536–1540.

85. Bej, A. K.; Mahbubani, M. H.; DiCesare, J. L.; Atlas, R. M. Polymerase chain reaction-gene probe detection of microorganisms by using filter-concentrated samples. *Appl. Environ. Microbiol.* 1991, **57**, 3529–3534.

86. Lekowska-Kochaniak, A.; Czajkowska, D.; Popowski, J. Detection of *Escherichia coli* O157:H7 in raw meat by immunomagnetic separation and multiplex PCR. *Acta Microbiol. Polon.* 2002, **51**, 327–337.

87. Spierings, G.; Ockhuijsen, C.; Hofstra, H.; Tommassen, J. Polymerase chain reaction for the specific detection of *Escherichia coli/Shigella. Res. Microbiol.* 1993, **144**, 557–564.

88. Burtscher, C.; Fall, P. A.; Wilderer, P. A.; Wuertz, S. Detection of *Salmonella* and *Listeria monocytogenes* in suspended organic waste by nucleic acid extraction and PCR. *Appl. Environ. Microbiol.* 1999, **65**, 2235–2237.

89. Colwell, R. R.; Grimes, D. J. (Eds.). *Nonculturable Microorganisms in the Environment.* Washington, DC: American Society for Microbiology, 2000.

90. Rafii, F.; Holland, M. A.; Hill, W. H.; Cerniglia, C. E. Survival of *Shigella flexneri* on vegetables and detection by polymerase chain reaction. *J. Food Prot.* 1995, **58**, 727–732.

91. Rafii, F.; Lunsford, P. Survival and detection of *Shigella flexneri* in vegetables and commercially prepared salads. *J. AOAC Int.* 1997, **80**, 1191–1197.

92. Sheridan, G. E. C.; Masters, C. I.; Shallcross, J. A.; Mackey, B. M. Detection of mRNA by reverse transcription-PCR as an indicator of viability in *Escherichia coli* cells. *Appl. Environ. Microbiol.* 1998, **64**, 1313–1318.

93. Bej, A. K.; McCarty, S. C.; Atlas, R. M. Detection of coliform bacteria and *Escherichia coli* by multiplex polymerase chain reaction: comparison with defined substrate and plating methods for water quality monitoring. *Appl. Environ. Microbiol.* 1991, **57**, 2429–2432.

94. Dunbar, S. A.; Vander Zee, C. A.; Oliver, K. G.; Karem, K. L.; Jacobson, J. W. Quantitative, multiplexed detection of bacterial pathogens: DNA and protein applications of the Luminex LabMAP system. *J. Microbiol. Meth.* 2003, **53**, 245–252.

95. Way, J. S.; Josephson, K. L.; Pillai, S. D.; Abbaszadegan, M.; Gerba, C. P.; Pepper, I. L. Specific detection of *Salmonella* spp. by multiplex polymerase chain reaction. *Appl. Environ. Microbiol.* 1993, **59**, 1473–1479.

96. Waage, A. S.; Vardund, T.; Lund, V.; Kapperud, G. Detection of low numbers of *Salmonella* in environmental water, sewage and food samples by a nested polymerase chain reaction assay. *J. Appl. Microbiol.* 1999, **87**, 418–428.

97. Waage, A. S.; Vardund, T.; Lund, V.; Kapperud, G. Detection of low numbers of pathogenic *Yersinia enterocolitica* in environmental water and sewage samples by nested polymerase chain reaction. *J. Appl. Microbiol.* 1999, **87**, 814–821.

98. Burr, M. D.; Josephson, K. L.; Pepper, I. L. An evaluation of ERIC PCR and AP PCR fingerprinting for discriminating *Salmonella* serotypes. *Lett. Appl. Microbiol.* 1998, **27**, 24–30.

99. Pooler, M. R.; Ritchie, D. F.; Hartung, J. S. Genetic relationships among strains of *Xanthomonas fragariae* based on random amplified polymorphic DNA PCR, repetitive extragenic palindromic PCR, and enterobacterial repetitive intergenic consensus PCR data and generation of multiplexed PCR primers useful for the identification of this phytopathogen. *Appl. Environ. Microbiol.* 1996, **62**, 3121–3127.

100. Gürtler, V.; Mayall, B. C. Genomic approaches to typing, taxonomy and evolution of bacterial isolates. *Int. J. Syst. Evol. Microbiol.* 2001, **51**, 3–16.

101. Masco, L.; Huys, G.; Gevers, D.; Verbrugghen, L.; Swings, J. Identification of *Bifidobacterium* species using rep-PCR fingerprinting. *Syst. Appl. Microbiol.* 2003, **26**, 557–563.

102. Seurinck, S.; Verstraete, W.; Siciliano, S. D. Use of 16S-23S rRNA intergenic spacer region PCR and repetitive extragenic palindromic PCR analyses of *Escherichia coli* isolates to identify nonpoint fecal sources. *Appl. Environ. Microbiol.* 2003, **69**, 4942–4950.

103. Sato, T.; Hu, J. P.; Ohki, K.; Yamaura, M.; Washio, J.; Matsuyama, J.; Takahashi, N. Identification of mutans streptococci by restriction fragment length polymorphism analysis of polymerase chain reaction-amplified 16S ribosomal RNA genes. *Oral Microbiol. Immunol.* 2003, **18**, 323–326.

104. Bellin, T.; Pulz, M.; Matussek, A.; Hempen, H.-G.; Gunzer, F. Rapid detection of enterohemorrhagic *Escherichia coli* by real-time PCR with fluorescent hybridization probes. *J. Clin. Microbiol.* 2001, **39**, 370–374.

105. Wittwer, C.; Hahn, M.; Kaul, K. (Eds.). *Rapid Cycle Real-Time PCR: Methods and Applications*. Springer-Verlag: Berlin, 2004.

106. Fukushima, H.; Tsunomori, Y.; Seki, R. Duplex real-time SYBR Green PCR assays for detection of 17 species of food- or waterborne pathogens in stools. *J. Clin. Microbiol.* 2003, **41**, 5134–5146.

107. Heid, C. A.; Stevens, J.; Livak, K. J.; Williams, P. M. Real time quantitative PCR. *Genome Res.* 1996, **6**, 986–994.

108. Kuboniwa, M.; Amano, A.; Kimura, K. R.; Sekine, S.; Kato, S.; Yamamoto, Y.; Okahashi, N.; Iida, T.; Shizukuishi, S. Quantitative detection of periodontal pathogens using real-time polymerase chain reaction with TaqMan probes. *Oral Microbiol. Immunol.* 2004, **19**, 168–176.

109. Ambion, Inc. Real-time PCR goes prime time. *TechNotes* 2001, **8**, 1–4.

110. Ye, R. W.; Wang, T.; Bedzyk, L.; Croker, K. M. Applications of DNA microarrays in microbial systems. *J. Microbiol. Meth.* 2001, **47**, 257–272.

111. Call, D. R.; Bakko, M. K.; Krug, M. J.; Roberts, M. C. Identifying antimicrobial resistance genes with DNA microarrays. *Antimicrob. Agents Chemother.* 2003, **47**, 3290–3295.

112. Call, D. R.; Borucki, M. K.; Loge, F. J. Detection of bacterial pathogens in environmental samples using DNA microarrays. *J. Microbiol. Meth.* 2003, **53**, 235–243.

113. Mossoba, M. M.; Al-Khaldi, S. F.; Jacobson, A.; Segarra Crowe, L. I.; Fry, F. S. Application of a disposable transparent filtration membrane to the infrared spectroscopic discrimination among bacterial species. *J. Microbiol. Meth.* 2003, **55**, 311–314.

114. Maule, J. Pulsed-field gel electrophoresis. *Mol. Biotechnol.* 1998, **9**, 107–126.

115. Ivnitski, D.; O'Neil, D. J.; Gattuso, A.; Schlicht, R.; Calidonna, M.; Fisher, R. Nucleic acid approaches for detection and identification of biological warfare and infectious disease agents. *BioTechniques* 2003, **35**, 862–869.

2

MASS SPECTROMETRY: IDENTIFICATION AND BIODETECTION, LESSONS LEARNED AND FUTURE DEVELOPMENTS

ALVIN FOX
Department of Pathology and Microbiology University of South Carolina, School of Medicine, Columbia, SC 26208

2.1 INTRODUCTION

In the 1990 book *Analytical Microbiology: Chromatography and Mass Spectrometry*[1] ways in which mass spectrometry had been or might be successfully used in microbiology were discussed. The areas of research delineated included (1) identification of bacteria at the species, or strain, level, after prior culture (2) biodetection without culture, and (3) characterization of isolated macromolecules. All have relevance to the diagnosis of human infection, as relates to natural disease as well as occurring after a biological attack. Indeed, the events of 2001 and 2002 in which the US postal service was attacked with anthrax, and through it the nation, we have all been all too well educated about the need for biological countermeasures and homeland security.

In the preface to the 1990 book a technological revolution was predicted that would dramatically change the diagnostics field. Although this revolution

Identification of Microorganisms by Mass Spectrometry, Edited by Charles L. Wilkins and Jackson O. Lay, Jr.

has occurred, it did not come in the fashion predicted. Since this time analytical and molecular biology technology has improved dramatically with the widespread application of a battery of powerful techniques newly introduced, including real-time polymerase chain amplification (PCR) and genomics (i.e., full-scale DNA sequencing) in the molecular biology arena, and soft ionization mass spectrometry approaches (matrix assisted laser desorption/ionization mass spectrometry [MALDI] and electrospray ionization [ESI]) in the analytical chemical field. These molecular and analytical techniques in turn built on an earlier generation of taxonomy technologies (including gel electrophoresis, classical PCR and small molecule analysis, e.g., fatty acids and sugars, by gas chromatography (GC) and gas chromatography-mass spectrometry [GC-MS]). Today, as will be discussed here and elsewhere in this book, small molecule analysis has reached maturity, and developments in MS analysis now focus on higher mass DNA and proteins. A new generation has now entered the field since 1990. Thus it might be useful to document where we have gone and where we are going. In this fashion we can learn from successes and hopefully not repeat mistakes. The informed reader might also consider reading two other books.[2,3]

2.2 ANALYSIS OF FATTY ACID AND SUGAR MONOMERS BY GC-FID, GC-MS, AND GC-MS-MS

Profiling of fatty acid monomers, released from membrane phospholipids of whole cells, using GC with flame ionization detection (FID) is still the most widely used analytical method for bacterial speciation (after isolation and growth of individual bacterial species). When necessary the identity of these fatty acids is readily confirmed by GC-MS. Indeed, modern MALDI-TOF MS analysis of proteins shares remarkable similarity, to fatty acid profiling, in the type of information that can be obtained as relates to bacterial identification. Prior to GC analysis, fatty acids are released by methanolysis and subsequently converted to methyl esters and are thus referred to as FAMES (fatty acid method esters). The procedure, although not complicated, still requires several hours of manual derivatization prior to GC analysis. The GC analysis itself, although automated, is also time-consuming. FAMES reached maturity in the early 1980s and was being widely employed as nicely described by Wayne Moss many years ago.[4] Although less widely used, carbohydrate profiling of whole cell hydrolysates using gas chromatography-mass spectrometry (GC-MS) provides complementary information to FAMES. Bacterial whole cell hydrolyzates are converted to alditol acetates for analysis of neutral and amino sugars.[5,6] The widespread use of fatty acid, but not carbohydrate profiling, illustrates a lesson that should be well learned. Techniques must provide useful information that is highly reproducible, but to be widely used such methodology must also be readily learned (i.e., simplicity is essential).

Procedures developed for GC-FID (1960s–1970s) were readily adapted for GC-MS analysis (in the 1980s). Mass spectra can also be used to identify each fatty acid component in a FAME (or carbohydrate in an alditol acetate) profile. In the selected ion monitoring (SIM) MS mode where specific *m/z* ratios are utilized, simple chromatograms free of background interferences from other components of the bacterial cell are generated. This helps greatly in trace analysis of fatty acids in complex environmental matrices.[7] Detection of fatty acids has been widely used to assess the bacterial population in environmental samples. Although it should be stressed that the presence of particular species is difficult to assess from the profiles, since individual fatty acids (unlike certain DNA or protein sequences) may be shared among bacteria of diverse origin.

Additionally other small molecules are used extensively to assess the microbial content of airborne and surface dust, largely in indoor settings. This has primarily come from the work of Lennart Larsson and co-workers. Hydroxy fatty acids (components of the lipid A region of lipopolysaccharides) are used to quantitate the levels of Gram-negative bacteria and muramic acid (a component of the glycan backbone of peptidoglycan) as a marker for total bacterial load (including Gram-positive and Gram-negative bacteria.[8–11] The GC-MS techniques do not discriminate between live and dead bacteria or released bacterial constituents.

High-resolution chromatographic separations (e.g., GC) coupled with selective cleanup steps are important in improving the specificity of the detection of chemical markers in complex matrices. However, chromatographic separation is not sufficient to eliminate extraneous peaks when nonselective detectors are employed. The use of the mass spectrometer as a selective GC detector (i.e., GC-MS analysis in SIM mode) aids greatly in diminishing background noise by focusing only on ion(s) that are present in the compound of interest. However, even when using SIM, it is common to find extraneous background peaks. The tandem mass spectrometer, as a GC detector, provides even greater specificity in detecting trace amounts of chemical markers in complex matrices when used in multiple reaction monitoring (MRM) mode. Tandem mass spectrometry has the added advantage of generating a total ion mass spectrum from a selected precursor ion (product ion spectrum). The resulting product ion spectrum can be used for a definitive identification of the compound of interest at trace levels.[8–12] Similar concepts could be applied for detection of proteins at trace levels in complex matrices (i.e., selective cleanup and improved instrumental analysis using MS-MS monitoring for specific chemical markers).

In bacterial cells, marker compounds are present at the part per hundred to part per thousand level. In environmental samples, which represent a complex mixture of components, such markers are often present at the part per ten thousand to part per hundred thousand level. In certain clinical samples, in some instances, these markers may be present as low as parts per billion or less. Absolute identification of these markers in certain clinical and environmental samples is an exacting analytical task requiring sophisticated instrumentation. GC-MS-MS has currently been used with the great sensitivity and specificity.

GC-MS-MS has demonstrated utility for determining the levels of bacterial contamination for both clinical and environmental analyses. In addition to muramic acid and hydroxy fatty acid levels (markers respectively for bacterial PG, Gram-negative bacterial LPS) ergosterol provides a marker for fungi and all serve as a useful measures of biopollution of indoor air.[11] GC-MS-MS is also a powerful tool for detection of bacteria or their constituents in mammalian body fluids and tissues, which are sterile in the absence of infection.[8,13] Muramic acid is not synthesized by mammalian enzyme systems. Unfortunately, hydroxy fatty acids are present in normal blood and tissues. The presence of a high hydroxy fatty acid "background" diminishes the utility of hydroxy fatty acids in clinical applications.[14] High-performance liquid chromatography (LC) analysis coupled with electrospray ionization (ESI) MS and MS/MS is usually performed without prior derivatization simplifying sample preparation. However, GC-MS and GC-MS-MS are often less demanding for routine analysis. Furthermore, for complex matrices, GC-MS-MS have proved to have much greater sensitivity than LC-MS-MS for trace detection of monomers.[15–17] This helps explain why LC-MS-MS has not been widely adopted for trace analysis in the microbiology field.

Phospholipids can be ionized as intact entities for MS or MS-MS analysis. On MS-MS analysis, in the negative ion mode, individual fatty acids (present in bacterial cells isolated by culture) are identified from the product ion spectra. The class of the phospholipid can be determined by summing the masses of individual fatty acids and subtracting the value from the mass of the parent (molecular) ion. MS analysis of intact microbial phospholipids had been demonstrated earlier using laser desorption[18] and fast atom bombardment MS analysis.[19] Dramatic improvements in the sensitivity of phospholipid analysis have been achieved using ESI MS and MS-MS.[20] However, phospholipids are sufficiently widely distributed in nature that species level detection in complex matrices is not afforded. This observation has largely contributed to the lack of widespread application of these methods for biodetection. However, it is unclear why phospholipid analysis has not been more popular as a complementary or alternative technique to fatty acid profiling of isolated bacterial species using GC-MS.

2.3 ANALYSIS OF PCR PRODUCTS BY PCR, PCR–MS, AND PCR-MS-MS

Classical PCR involves detection of a PCR product by electophoretic mobility on a gel, which is time consuming. Real-time PCR is distinct from classical PCR, in that electrophoresis is avoided and the PCR product is detected simply by an increase in fluorescence. For example, the PCR reaction mixture contains a dye that is not fluorescent, except when bound to double-stranded (ds) DNA. Thus as the amount of ds DNA increases on amplification, fluorescence also increases. In both cases by simply changing the design of the primer (a short sequence of synthetic DNA) that recognizes the target sequence, it is simple to set up either identification or biodetection. Technical developments achieved for one bacterial species are readily applied to others with minimal changes.

For the non–mass spectrometrist, it should be pointed out that nowadays the analysis of large molecules (e.g., PCR products or proteins) is primarily based on MALDI (matrix-assisted laser desorption ionization mass spectrometry) or ESI ionization. In the former case the sample is present within an energy-absorbing matrix on a surface. When hit with a laser beam, the matrix absorbs the energy, transferring it to the molecule of interest (e.g., DNA). Generally only a singly ionized species is produced having a single charge. In contrast, ESI MS is performed in solution, and the sample is introduced into the source of the mass spectrometer using a syringe pump. The solution is sprayed in the presence of an electrical field, and charges are transferred to the resulting droplets. As the droplets evaporate, the charge is transferred to molecules present within the droplet. Ions are produced that can have multiple charge states (e.g., for a 30 mer, anywhere from –1 to –30 changes). Since mass analyzers (the next stage after the source) generally separate by the mass to charge ratio, simple spectra are generated for MALDI MS but ESI spectra are considered to be complex. However, a simple computer-assisted operation (de-convolution) converts the complex plots (mass spectra) of mass–charge ratios versus abundance to molecular weight also versus abundance. When plotted in this fashion, ESI mass spectra resemble MALDI spectra and are readily interpreted. The longer the DNA sequence that can be analyzed the more information that can be obtained, but generally MALDI MS has been most successfully used for analysis of low-mass DNA (e.g., primer extension reactions) due to limited resolution for high mass PCR products.[21] Furthermore it is only in the past couple of years that MALDI-MS-MS has begun to become more widely available, so analysis of PCR products has been very limited to this point.

PCR is the most widely used technique, as the first step, in technology to analyze structure diversity in DNA. Minor variations in sequence (e.g., single nucleotide polymorphisms or the presence of tandem repeats [variable numbers of a short sequence present sequentially in a single stretch of DNA]) are detected. PCR employs specific sets of primers to amplify only a gene of interest. Femtogram to picogram amounts of DNA are used to generate high nanogram or even microgram amounts of the gene of interest. Classically in trace analysis, the analytical chemist uses liquid chromatography, or other separation techniques, to concentrate an analyte present in low concentration and to separate it from high-abundance background peaks that would otherwise confound the analysis. However, with PCR such separations are often unnecessary, even when performing trace analysis on complex clinical or environmental samples (e.g., detecting bacterial infection). As noted above, generally, PCR products are characterized by classical molecular biology techniques (e.g., arrays or electrophoresis). Alternatively, one of the most powerful techniques to obtain structural information is mass spectrometry. This is evidenced by the proteomics revolution where the engine driving this has been mass spectrometry. Mass spectrometry could have a similar impact in the DNA-based molecular diagnostics arena. However, there is the difficulty of analytical chemists using a technique that they are not often familiar with (PCR) or molecular biologists using mass spectrometers.

Over the past decade ESI mass spectrometry analysis of PCR products has become a mature technology for genetic discrimination. The more mainstream technique, real-time PCR, is also powerful in that it provides a simple yes or no answer, by generation of a fluorescent signal, as to the presence of a characteristic sequence. However, there may be too many false positives in certain situations (e.g., detection of a terrorist attack) to be acceptable. It is also difficult to use real-time PCR to detect minor sequence variations. The only difference between the stages in sample preparation, for real-time PCR or PCR–mass spectrometry, relate to an additional cleanup step for PCR-MS and PCR-MS-MS. However, a great deal of structural information is additionally obtained by mass spectrometric analysis. The primary steps are first sample collection. This can involve an air sampling pump and either an impinger (liquid collection) or an impactor (for collection onto a filter) for environmental samples. Clinical samples (e.g., body fluids) are, of course, more simply collected. The next step involves release of DNA from cells using organic solvents to lyse cells. For bacteria the cell membrane is encased in a tough cell wall, and breakage may require the additional use of shaking with glass beads or enzymatic digestion. The DNA is purified; to remove potential contaminants that otherwise will inhibit the PCR reaction. Next real-time PCR is performed to generate the gene of interest. In real-time PCR, the fluorescent signal is generated as the DNA is amplified; that is, amplification and detection occur together in a standard PCR instrument. However, additionally for MS analysis of PCR products purification of the PCR product must be performed prior to MS or MS-MS analysis. As noted below, this is primarily to eliminate low-mass constituents of the PCR reaction mixture that otherwise inhibit ionization in the mass spectrometer.

Currently PCR and mass spectrometry are performed by two separate instruments. However, there is no reason why PCR followed by simple automated cleanup and mass spectrometry cannot be incorporated into a single integrated instrument. Essentially every configuration of the modern ESI mass spectrometer has been used successfully for the analysis of PCR products, from the highest to the lowest resolution involving. Fourier transform ion cyclotron resonance (FTICR), triple quadrupole, quadrupole-time of flight (Q-TOF), and ion trap.[22–24] MS discriminates between two structurally related PCR products by MW difference. Mass accuracy is needed to differentiate the pairs of coding or non-coding strands (each pair may differ by 9–40 Da, as adenine to thymine [A to T] and guanine to cytosine [G to C] switch, respectively). Large PCR products over 200 nucleotides in length have been analyzed with the required precision.[25] Tandem mass spectrometry (MS-MS) allows for the differentiation of PCR products of closely related sequence, by fingerprint, even if they have the same nucleotide composition; fragmentation of the parent molecule occurs in the tandem mass spectrometer (analogous to a restriction digest pattern). A particular mass is selected that is collided with a collision gas (usually argon or helium). The DNA strand breaks into fragments, at each nucleotide position, by scission of phospho-diester bridges. Thus a series of fragments are generated, rather like a Sanger sequencing ladder, each differing from one another by a single nucleotide. Unfortunately, each piece can differ by charge and mass, and a further complication is that there are four possible breakage points within a phospho-diester bridge. Thus MS-MS spectra are more complex than MS spectra, and currently it is difficult to visually read sequence, although this is possible with computer-assisted software. Alternatively, MS-MS spectra can be compared to detect sequence variation without the need for sequencing.

Samples must be cleaned up prior to mass spectrometry analysis. Other components of the PCR reaction (e.g., metal ions and nucleotides) inhibit ionization in the mass spectrometer source or preferentially ionize. Furthermore metal ions can bind to the phosphate backbone of DNA, thus changing the mass and complicating mass spectral interpretation. When PCR-MS was first introduced, in the mid-1990s, there was a great deal of interest in developing a simple approach for the cleanup of PCR products for MS analysis. For ESI MS analysis a single approach, involving ethanol precipitation, was used reliably by a number of laboratories, and this rapidly became the standard approach.[22] Generally, cleanup is performed with ethanol precipitation in the presence of ammonium hydroxide. Ammonia replaces the metal ions and dissociates from the DNA in the mass spectrometer. Such cleanup is time-consuming, taking a several hours and, more commonly, requiring overnight precipitation. After precipitation, the pellet has to be rapidly washed with ethanol and ethanol:water. Thus this approach cannot accommodate cases where rapid analysis is necessary (e.g., for biodetection). In a recent development that shows promise, the PCR product is bound to a weak anionic exchange column (e.g., a Zip-Tip) and washed to remove low-mass materials (e.g., nucleotides and dideoxynucleotides) that do not bind to the column. The column is then washed with solutions containing high concentrations of ammonia to remove metal ions.[26]

Two variations on the analysis of PCR products by ESI mass spectrometry have emerged: (1) direct-injection MS and tandem mass spectrometry (MS-MS) and (2) liquid chromatography–mass spectrometry (LC-MS) and tandem mass spectrometry (LC-MS/MS). In the former approach, the sample is cleaned manually, and as noted above, the cleanup is performed as simply and rapidly as possible. In the latter approach, the cleanup is done automatically by the LC. The advantages of LC-MS and LC-MS-MS are that more complex samples can be analyzed, since the sample is separated into components by LC before entry into the mass spectrometer. Additionally the automation makes it possible to run large batches of samples. However, the LC analysis is much more complicated instrumentally than direct injection. The molecular biology community is not generally well versed in LC instrumentation and seems to prefer electrophoresis. Thus they are more likely to employ direct MS analysis of PCR products. It is worth noting that modern MS instruments (e.g., MALDI-TOF and ESI-ion traps) are run by Windows-based computer software. There is very little to learning how to introduce a DNA sample into either instrumental configuration. Indeed, when MS spectra are plotted as mass versus abundance, there is little data interpretation required. MS-MS spectra are quite complex, and further simplification of computer-assisted data handling is still needed. Although developments in MALDI MS-MS may help, simpler spectra than for ESI MS-MS should be generated.[27,28] These reviews also cover applications in cancer and human genomics analysis, which are not directly relevant to this microbiology review, but include a number of important methodological developments that can be readily applied to bacterial identification and biodetection.

Because of the nature of PCR the identifications can be made either directly from clinical or environmental samples or after culture. However, because environmental samples contain many bacterial species, among which bacilli are the most common, the design of primers is important to avoid complex mixtures of PCR products. Such mixtures are extremely difficult to analyze by direct MS analysis. Furthermore the sensitivity of the analysis may be compromised if the signal is spread among many components. Successful analysis directly from environmental samples therefore is still a topic for current research.

As an example of bacterial discrimination, in pure bacterial cultures the ribosomal RNA (rrn) operon consists of 16S and 23S rRNA genes flanking an interspace region (ISR). This operon is present in all bacteria, but the sequence varies among bacteria in a species-specific fashion. For example, variation in ISR size readily distinguishes the *B. cereus* group of organisms (which includes *B. anthracis*, *B. cereus*, and *B. thuringiensis*) from the closely related *B. subtilis* group (which includes *B. subtilis* and *B. atrophaeus*). With appropriate primers, 89bp products are generated for the *B. cereus* group and 111 to 119bp products for the *B. subtilis* group. Both subgroups *of B. subtilis* (W23 and 168-like, respectively) produce 114bp products that are readily distinguished from the 119bp products produced by the *B. atrophaeus* group.[29,30]

Based on sequencing of single operons, until recently it was felt that *B. anthracis* and *B. cereus/B. thuringiensis* were identical in all regions of the rrn operon. However, work from the Centers for Disease Control comparing sequences of the 11 operons of the 16S rRNA gene in the *B. anthracis* genome, at position 1146, revealed that 5 genes contain Ts whereas 6 contain As. For *B. cereus* and *B. thuringiensis* only Ts were observed at this position.[31] Mass spectrometry has the discriminating power to detect such fine genetic differences (an A to T switch). Synthetic oligonucleotides, having the same sequence as the bacterial genes, were used, so more work is needed to develop a routine procedure utilizing PCR products.

The ability to detect small genetic changes becomes more difficult as mass increases. There is further an upper mass range where analysis is impractical. For low-resolution instruments this limit is around a 100mer. Thus the mass has to be minimized or a high-resolution instrument employed. Alternatively, the smaller the piece of DNA analyzed, the more it chemically resembles a primer or nucleotide monomer; thus separation of the two during cleanup is difficult to do. If the primers and nucleotides are not removed, they can provide a massive background on MS analysis or inhibit ionization of the PCR product by preferential ionization. Thus for practical reasons it is extremely difficult to employ a PCR product below a 40 to 50mer for direct ESI MS or ESI MS-MS analysis.

2.4 ANALYSIS OF PROTEINS BY MALDI-TOF MS

MALDI-TOF MS analysis of bacterial whole cell proteins is now well established,[32–35] and more recent developments have been reviewed by Lay.[36] An interesting recent development has been growth of bacteria in isotopically enriched media prior to MALDI MS analysis; this greatly increases effective mass spectral resolution.[37] This approach has proved useful in bacterial identification and forensics, but because real-time biodetection requires analysis without prior culture, this approach does not have utility in this instance. The power of MALDI-TOF MS for protein profiling resides in the minimal sample preparation and rapid computer-assisted data handling, as is described in greater detail elsewhere in this book. Isolated bacterial colonies are simply dried on the MS source plate in the presence of the appropriate laser-absorbing matrix. In fact it should be emphasized that this approach is distinct from classical proteomics based approaches. These generally employ 2D gel electophoresis followed by tryptic digestion, and extraction/sample cleanup prior to MALDI-TOF MS analysis. Alternatively, after tryptic digestion, the sample are subjected to cleanup (e.g., with hydrophobic Zip-Tips) prior to online LC-ESI-MS-MS analysis. In either case cleanup and chromatography are important in reducing the complexity of samples that at any one time enter the mass spectrometer. As noted by Lay,[36] the number of peaks increases, and sensitivity of analysis increases, as the complexity of the sample is simplified by fractionation. Nevertheless, proteomics approaches are time-consuming and not suitable for real-time biodetection. Recognizing the importance of cleanup, efforts have been made to develop simpilified cleanup methods using affinity chromatography (including both lectins[38] or antibodies[39] prior to MS analysis. The difficulty here is how to apply such approaches for universal detection, since the appropriate affinity reagent must be selected for each organism of interest. Phages specific to the organism of interest have also been employed.[39] After bacterial infection the phages are amplified by intracellular growth generating high concentrations of phage-specific proteins.[39] However, unlike PCR (see above) where amplification reagents (primers) can be simply designed, chemically synthesized and purchased commercially, phages have to be individually selected (e.g., for specificity) and tested for each organism of interest. A simpler universal method remains to be developed. Vilaneuva et al.,[40] for example, have attempted a process of reversed phase cleanup columns for purification of human serum peptides for MS analysis. However, even in this case it proved necessary to remove major serum proteins, prior to peptide isolation. The recent utilization of MALDI ion trap MS-MS for protein analysis[41] is the most encouraging development. Along with appropriate cleanup, it should eventually allow universal real-time biodetection. Nevertheless, the difficulty of achieving this development is not to be under-estimated.

2.5 CHEMICAL MARKERS FOR PROTEIN-BASED IDENTIFICATION OR BIODETECTION

A chemical marker may be defined as a compound that is consistently present in one species or strain (e.g., under certain culture conditions) but absent or present in a modified form in a second. There is often a great diversity in the levels of certain cellular constituents, present or absents, between species and among strains. There is also considerable variation in profiles resulting from sample preparation (e.g., culture conditions) and the instrumental analysis.[42] So inconsistencies in an analysis due to the presence of an artifact should not be confused with natural variability resulting from real differences in the biochemical composition of microbes. In the analysis of pure bacteria cells, fingerprinting is particularly useful for species or strain differentiation. Elaborate data handling algorithms can also certainly help by automating decision-making.[43] However, this will not change the biochemical makeup of a sample. In complex mixtures certain smaller peaks become buried in larger ones of similar mass, or there may be preferential ionization. In either case peaks will not be observed. By focusing on one peak that is consistently present in the organism of interest (i.e., a chemical marker) and optimizing instrumental analysis (e.g., by using MS-MS or MS^n multiple reaction monitoring as noted in earlier sections) coupled with appropriate cleanup steps, it is possible to dramatically improve the reproducibility, specificity, and sensitivity of analysis.

This result is important to fully understanding the biochemical and ultrastructural origin of peaks and the physiological basis for variation. It not only helps in designing the analytical strategy (e.g., in selection of cleanup columns) but, more important, in making a decision on whether the marker should be used for strain or species identification or for biodetection. For example, there are a number of low–molecular weight peptides (1500–8000 kDa) present in

B. anthracis and related species.[41,44–46] Some of these peaks have been identified (e.g., as small acid soluble spore proteins and cyclic lipopeptides), but others remain uncharacterized. There is no agreement among different laboratories as to which markers are suitable for chemotaxonomic differentiatation of species (i.e., are consistently found in one species versus another) or for strain identification (i.e., are reproducibly found in one strain but not another). Further, although it might be anticipated that surface proteins can be preferentially ionized or extracted, the ultra-structural origin of some peptides within the cell is not always clear.

By comparison, the spore surface is characterized by an external exosporium consisting of a basal layer surrounded by a nap of hair-like projections. The first spore-surface protein, BclA (*Bacillus*, collagen-like protein) was recently discovered in *B. anthracis*. BclA is localized to the exosporium nap as demonstrated by monoclonal antibody labeling. It was subsequently recognized that the glycoprotein is the source of the spore-specific carbohydrates discovered many years earlier.[6] Indeed, the exosporium is primarily composed of carbohydrate[47] (Waller et al., 2004). The exosporium surrounds an underlying layer of coat proteins.[48] The poly-D-glutamic acid capsule is absent from the spores, so BclA constitutes the surface layer. The number of amino acids in the open reading frames of various strains range from 233 to 445. The *N*- and *C*- termini of BclA are conserved among *B. anthracis* and thus provide the species level. However, the peptide backbone varies in length among strains due to the presence of variable numbers of collagen-like tandem repeats (different numbers of GPT and $[GPT]_5$GDTGTT repeats in the protein). The number of GPT repeats range from 17 in the smallest protein to 91 in the largest, and the number of $[GPT]_5$GDTGTT] repeats range from 1 to 8 characterization of the variable region would thus help identify strains (e.g., for forensic purposes).[49] Since many of these amino acids are threonines (Ts) that bind sugars through the side chain hydroxyl moiety, it is likely that much of this repeating structure of the peptide backbone serves as a backbone for glycosylation (e.g., attachment of rhamnose-oligosacharides). The carbohydrates are different in *B. anthracis* versus *B. thuringiensis*/*B. cereus*. Thus the species identification is also provided by carbohydrate analysis.[6,50,51] To employ such proteins as biodetection markers, since there is variability in the mass of the intact proteins, it is necessary to focus either on the *N*- or *C*-terminus or on the carbohydrate side-chains rather than the intact protein.

2.6 CONCLUSIONS

Analysis of bacterial monomers is performed routinely to detect bacteria at trace levels in environmental samples using GC-MS-MS. PCR is also used routinely by the molecular biology community for detection of clinical infections. MS and MS-MS greatly improve the specificity of analysis of PCR products and provide additional structural information that may be important in applying these techniques for environmental analysis (e.g., for real-time detection of a terrorist attack). Protein profiling by MALDI-TOF MS has great promise for bacterial identification and forensics but still requires substantial work in order to become a simple biodetection technique. Nevertheless, recent progress is encouraging, and the future appears bright.

REFERENCES

1. Fox, A.; Morgan, S. L.; Larsson, L.; Odham, G. (Eds.). *Analytical Microbiology Methods: Chromatography and Mass Spectrometry*. New York: Plenum, 1990.
2. Fenselau, C. (Ed.). *Mass Spectrometry for the Characterization of Microorganisms.* Washington, DC: American Chemical Society, 1994.
3. Odham, G.; Larsson, L.; Mardh, P.-A. *Gas Chromatographty/Mass Spectrometry Applications in Microbiology.* New York: Pleunum, 1984.
4. Moss, C. W. In *Analytical Microbiology Methods: Chromatography and Mass Spectrometry*. Fox, A., Morgan, S. L., Larsson, L., Odham, G. (Eds.). New York: Plenum, 1990, pp. 59–70.
5. Fox, A. In *Carbohydrate Analysis by Modern Chromatography and Electrophoresis*. Rassi, Z. E. (Ed.). Amsterdam: Elsevier, 2002, pp. 829–843.
6. Fox, A.; Black, G.; Fox, K.; Rostovtseva, S. Determination of carbohydrate profiles of Bacillus anthracis and Bacillus cereus including identification of *O*-methyl methylpentoses using gas chromatography–mass spectrometry. *J. Clin. Microbiol.* 1993, **31**, 887–894.
7. Tunlid, A.; White, D. C. In *Analytical Microbiology Methods: Chromatography and Mass Spectrometry.* Fox, A., Morgan, S. L., Larsson, L., Odham, G. (Eds.). New York: Plenum, 1990, pp. 259–274.
8. Fox, A.; Fox, K.; Christensson, B.; Harrelson, D.; Krahmer, M. Absolute identification of muramic acid at trace levels in human septic fluids in vivo and absence in aseptic fluids. *Infect. Immun.* 1996, **64**, 3911–3955.
9. Fox, A.; Wright, L.; Fox, K. Gas chromatography–tandem mass spectrometry for trace detection of muramic acid, a peptidoglycan marker in organic dust. *J. Microbiol. Meth.* 1995, **22**, 11–26.
10. Larsson, L.; Saraf, A. Use of gas chromatography-ion trap mass spectrometry for the detection and characterization of microorganisms in complex samples. *Mol. Biotechnol.* 1997, **7**, 279–287.
11. Saraf, A.; Larsson, L. In *Advances in Mass Spectrometry*. Karjalainen, E. J., Hesso, A. E., Jalonen, J. E., Karjalainen, U. P. (Eds.). Amsterdam: Elsevier, 1998, Vol. 14, pp. 449–459.
12. Fox, A.; Krahmer, M.; Harrelson, D. Monitoring muramic acid in air (after alditol acetate derivatization) using a gas chromatography-ion trap tandem mass spectrometer. *J. Microbiol. Meth.* 1996, **27**, 129–138.
13. Kozar, M.; Krahmer, M.; Fox, A.; Gray, B. M. Failure to detect muramic acid in normal rat tissues but detection in cerebrospinal fluid from patients with pneumococcal meningitis. *Infect. Immun.* 2000, **68**, 4688–4698.

14. Szponar, B.; Krasnik, L.; Hryniewiecki, T.; Gamian, A.; Larsson, L. Distribution of 3-hydroxy fatty acids in tissues after intra-peritoneal injection of endotoxin. *Clin. Chem.* 2003, **49**, 1149–1153.
15. Conboy, J. J.; Henion, J. High performance anion exchange chromatography coupled with mass spectrometry for the determination of carbohydrates. *Biol. Mass Spectrom.* 1992, **21**, 397–407.
16. Shahgholi, M.; Ohorodnik, S.; Callahan, J. H.; Fox, A. Trace detection of underivatized muramic acid in environmental dust samples by microcolumn liquid chromatography electrospray–tandem mass spectrometry. *Anal. Chem.* 1997, **69**, 1956–1960.
17. Simpson, R. C.; Fenselau, C.; Hardy, M. R.; Townsend, R. R.; Lee, Y. C.; Cotter, R. J. Adaptation of a thermospray liquid chromatography/mass spectrometry interface for use with alkaline exchange liquid chromatography of carbohydrates. *Anal. Chem.* 1990, **62**, 248–252.
18. Platt, J.; Uy, O.; Heller, D.; Cotter, R. J.; Fenselau, C. Computer-based linear regression analysis of desorption mass spectra of microorganisms. *Anal. Chem.* 1988, **60**, 1415–1419.
19. Cole, M.; Enke, C. Direct determination of phospholipid structures in microorganisms by fast atom bombardment triple quadropole mass spectrometry. *Anal. Chem.* 1991, **63**, 1032–1038.
20. Black, G. E.; Snyder, A.; Heroux, K. Chemotaxonomic differentiation between the *Bacillus cereus* group and *Bacillus subtilis* by phospholipid extracts analyzed with electospray tandem mass spectrometry. *J. Microbiol. Meth.* 1997, **28**, 187–190.
21. Hurst, G.; Doktycz, M.; Vass, A.; Buchanan, M. Detection of bacterial DNA polymerase chain reaction products by matrix assisted laser desorption/ionization mass spectrometry. *Rapid. Commun. Mass Spectrom.* 1996, **10**, 377–382.
22. Krahmer, M. T.; Johnson, Y. A.; Walters, J. J.; Fox, K. F.; Fox, A.; Nagpal, M. Electrospray quadrupole mass spectrometry analysis of model oligonucleotides and polymerase chain reaction products: determination of base substitutions, nucleotide additions/deletions, and chemical modifications. *Anal. Chem.* 1999, **71**, 2893–2900.
23. Muddiman, D. C.; Wunschel, D. S.; Liu, C.; Pasa-Tolic, L.; Fox, K. F.; Fox, A.; Anderson, G. A.; Smith, R. D. Characterization of PCR products from bacilli using electrospray ionization FTICR mass spectrometry. *Anal. Chem.* 1996, **68**, 3705–3712.
24. Naito, Y.; Ishikawa, K.; Koga, Y.; Tsuneyoshi, T.; Terunuma, H.; Arakawa, R. Molecular mass measurement of polymerase chain reaction products amplified from human blood DNA by electrospray ionization mass spectrometry. *Rapid Commun. Mass Spectrom.* 1995, **9**, 1484–1486.
25. Wunschel, D. S.; Pasa-Tolic, L.; Feng, B. B.; Smith, R. D. Electrospray ionization Fourier transform ion cyclotron resonance analysis of large polymerase chain reaction products. *J. Am. Soc. Mass Spectrom.* 2000, **11**, 333–337.
26. Yiang, Y.; Hofstadler, S. A. A highly efficient and automated method of purifying and desalting PCR products by electrospray ionization mass spectrometry. *Anal. Biochem.* 2003, **316**, 50–57.

27. Oefner, P. J.; Huber, C. G. A decade of high-resolution liquid chromatography of nucleic acids on styrene-divinylbenzene co-polymers. *J. Chromatogr. B* 2002, **782**, 27–55.
28. Walters, J.; Muhammad, W.; Fox, K. F.; Fox, A.; Xi, R. D.; Creek, C.; Pirisi, L. Genotyping single nucleotide polymorphisms using intact PCR products by electrospray quadrupole mass spectrometry. *Rapid Comm. Mass Spectrom.* 2001, **15**, 1752–1759.
29. Fox, A.; Fox, K. F.; Castanha, E.; Muhammad, W. T. Mass spectrometry: polymerase chain reaction products. In *Encylopedia of Analytical Science*. Amsterdam: Elsevier, 2004.
30. Johnson, Y. A.; Nagpal, M.; Krahmer, M. T.; Fox, K. F.; Fox, A. Precise molecular weight determination of PCR products of the rRNA intergenic spacer region using electrospray quadrupole mass spectrometry for differentiation of *B. subtilis* and *B. atrophaeus*, closely related species of bacilli. *J. Microbiol. Meth.* 2000, **40**, 241–254.
31. Sacchi, C. T.; Whitney, A. M.; Mayer, L. W.; Morey, R.; Steigerwalt, A.; Boras, A.; Weyant, R. S.; Popovic, T. Sequencing of 16S rRNA gene: A rapid tool for identification of *Bacillus anthracis*. *Emerg. Infect. Dis.* 2002, **8**, 1117–1123.
32. Cain, T.; Lubman, D.; Weber, W. J. Differentiation of bacteria using protein profiles from matrix assisted laser desorption/ionization time of flight mass spectrometry. *Rapid Comm. Mass Spectrom.* 1994, **8**, 1026–1030.
33. Claydon, M.; Davey, S. N.; Edwards-Jones, V.; Gordon, D. The rapid identification of intact microorganisms using mass spectrometry. *Nature Biotechnol.* 1996, **14**, 1584–1586.
34. Holland, R. D.; Wilkes, J. G.; Sutherland, J. B.; Persons, C. C.; Voorhees, K. J.; Lay, J. O. Rapid identification of intact whole bacteria based on spectral patterns using matrix assisted laser desorption/ionization with time-of-flight mass spectrometry. *Rapid Comm. Mass Spectrom.* 1996, **10**, 1227–1232.
35. Krishnamurthy, T.; Ross, P.; Rajamani, U. D. Detection of pathogenic and non-pathogenic bacteria by matrix assisted laser desorption/ionization time-of-flight mass spectrometry. *Rapid Comm. Mass Spectrom.* 1996, **10**, 883–888.
36. Lay, J. O. MALDI-TOF mass spectrometry of bacteria. *Mass Spectrom. Rev.* 2001, **20**, 172–194.
37. Jones, J. J.; Stump, M. J.; Fleming, R. C.; Lay, J. O.; Wilkins, C. L. Investigation of MALDI-TOF and FTMS techniques for analysis of *Escherichia coli* whole cells. *Anal. Chem.* 2003, **75**, 1340–1347.
38. Bundy, J.; Fenselau, C. Lectin-based affinity capture for MALDI-MS analysis of bacteria. *Anal. Chem.* 1999, **71**, 1460–1463.
39. Madonna, A. J.; Van Cuyk, S.; Voorhees, K. J. Detection of *Escherichia coli* using immunomagnetic separation and bacteriophage amplification coupled with matrix-assisted laser desorption/ionization time-of-flight mass spectrometry. *Rapid Comm. Mass Spectrom.* 2003, **17**, 257–263.
40. Villanueva, J.; Phillip, J.; Enterberg, D.; Chaparro, C. A.; Tanwar, M. K.; Holland, E. C.; Tempst, P. Serum peptide profiling by magnetic particle assisted automated sample processing and MALDI-TOF mass spectrometry. *Anal. Chem.* 2004, **76**, 1560–1570.
41. Madonna, A. J.; Voorhees, K. J.; Tarenko, N. I.; Laiko, V. V.; Doroshenko, V. M. Detection of cyclic lipopeptide biomarkers from *Bacillus* species using atmospheric

pressure matrix-assisted laser desorption/ionization mass spectrometry. *Anal. Chem.* 2003, **75**, 1628–1637.

42. Saenz, A. J.; Petersen, C. E.; Valentine, N.; Gantt, S. L.; Karman, K. H.; Kingsley, M. T.; Wahl, K. L. Reproducibility of matrix-assisted laser desorption/ionisation time-of-flight mass spectrometry for replicate bacterial culture analysis. *Rapid Comm. Mass Spectrom.* 1999, **13**, 1585–1585.
43. Wahl, K. L.; Wunschel, S. C.; Jarmon, K. H.; Valentine, N. B.; Peterson, C. E.; Kingsley, M. T.; Zartolos, K. A.; Saenz, A. J. Analysis of microbial mixtures by matrix-assisted laser desorption/ionization time-of-flight mass spectrometry. *Anal. Chem.* 2002, **74**, 6191–6199.
44. Elhanany, E.; Barak, R.; Fisher, M.; Kobiler, D.; Altboum, Z. Detection of specific *Bacillus anthracis* biomarkers by matrix-assisted laser desorption/ionization time-of-flight mass sspectrometry. *Rapid Commun. Mass Spectrom.* 2001, **15**, 2110–2116.
45. Hathout, Y.; Demirev, P. A.; Ho, Y.-P.; Bundy, J. L.; Ryzhov, V.; Sapp, L.; Stutler, J.; Jackman, J.; Fenselau, C. Identification of *Bacillus* spores by matrix-assisted laser desorption ionization-mass spectrometry. *Appl. Environm. Microbiol.* 1999, **65**, 4313–4319.
46. Ryzhov, V.; Hathout, Y.; Fenselau, C. Rapid characterization of spores of *Bacillus cereus* group bacteria by matrix assisted laser desorption/ionization time-of-flight mass spectrometry. *Appl. Environ. Microbiol.* 2000, **66**, 3828–3834.
47. Waller, L. N.; Fox, N.; Fox, K. F.; Fox, A.; Price, R. L. Ruthenium red staining for ultrastructural visualization of a glycoprotein layer surrounding the spore of *Bacillus anthracis* and *Bacillus subtilis*. *J. Microbiol. Meth.* 2004, **58**, 23–30.
48. Sylvestre, P.; Couture-Tosi, E.; Mock, M. A collagen-like surface glycoprotein is a structural component of the *Bacillus anthracis* exosporium. *Mol. Microbiol.* 2002, **45**, 169–178.
49. Sylvestre, P.; Couture-Tosi, E.; Mock, M. Polymorphism in the collagen-like region of the *Bacillus anthracis* BclA protein leads to variation in exosporium filament length. *J. Bacteriol.* 2003, **185**, 1555–1563.
50. Fox, A.; Stewart, G. C.; Waller, L. N.; Fox, K.; Harley, W. M.; Price, L. R. Carbohydrates and glycoproteins of *Bacillus anthracis* and related bacteria. *J. Microbiol. Meth.* 2003, **54**, 143–152.
51. Wunschel, D. K.; Fox, K.; Black, G.; Fox, A. Discrimination among the *Bacillus cereus* group, in comparison to *B. subtilis*, by structural carbohydrate profiles and ribosomal RNA spacer region PCR System. *Appl. Microbiol.* 1994, **17**, 625–635.

3

AN INTRODUCTION TO MALDI-TOF MS

ROHANA LIYANAGE AND JACKSON O. LAY, JR.
University of Arkansas, Fayetteville, AR 72701

3.1 INTRODUCTION

The identification of bacteria by chemical approaches is sometimes referred to as "bacterial chemotaxonomy." Taxonomic classification can be made based on any number of chemical classifiers. The classification can be based on the presence or absence of taxonomically characteristic components or on different ratios of chemicals whose abundances provide a taxonomically distinct fingerprint. Bacterial chemotaxonomy has been reported in the scientific literature for at least 40 years. However, it is difficult to retrace the history of its development, and there is high likelihood of early contributions being missed. Therefore we will consider the developments in this field that can be reconstructed by searches of the scientific literature back to the 1960s. This is when the number of reports of the chemotaxonomic classification of bacteria starts to grow. The first we found to appear in this decade was in 1966 when Shioiri–Nakano reported the use of several analytical chemistry measurements for the classification of "streptococcal components".[1] In the following year, 1967, there were 11 reports of bacterial chemotaxonomy.[2–12] The chemical differences used for taxonomic differentiation ranged from changes in small molecules to differences in DNA. One of these early reports was a review of bacterial chemotaxonomy.[9] In 1968 Yamada reported the classification of acetic acid bacteria based on a mass spectrometry measurement.[13]

Identification of Microorganisms by Mass Spectrometry, Edited by Charles L. Wilkins and Jackson O. Lay, Jr.

Despite this report, very few additional mass spectrometry studies of bacterial taxonomy appeared before 1977. From 1977 to 1994 a number of mass spectrometry methods for the taxonomic classification of bacteria were reported.[14–47] The methods included GC/MS,[14,15,17,20–27,29–31,33,35–37] Pyrolysis MS,[18,19,38,39,41] and FAB/MS[28,32,40,42,43,45] analysis. Some of the most important developments in mass spectrometry based bacterial chemotaxonomy up to this time were reported in an excellent *ACS Symposium Series* book[46] edited by Catherine Fenselau, one of the pioneers in this field. However, except for the GC/MS analysis of fatty acid methyl esters, and a recent resurgence in interest in pyrolysis MS, most of the early mass spectrometry methods for characterizing bacteria are no longer used today, nor are they the focus of significant continuing research.

One reason for the decline in interest in these methods was the development of alternative approaches aimed at analysis of polar biomolecules by mass spectrometry. At about the same time these first mass spectrometry methods were being applied to bacterial taxonomy, a major new ionization method was introduced, matrix-assisted laser desorption ionization (MALDI),[48] which was well suited to the analysis of polar biomolecules, including peptides and proteins. MALDI was a sample ionization and vaporization (desorption) method that allowed fragile biomolecules to be introduced into the mass spectrometer as ions without significant decomposition caused by either the ionization or the desorption steps. Shortly after the MALDI methodology was developed, it was being routinely applied to the analysis of proteins and peptides. It is not surprising then that proteins from bacterial extracts were also analyzed by MALDI, including proteins that could be used for taxonomic classification of bacteria.[44,49] Perhaps more important was the observation that taxonomically characteristic proteins could be detected by MALDI mass spectrometry applied directly to whole cells[50,51] without any separation, fractionation, or cleanup. The analysis of cells directly by MALDI-TOF mass spectrometry was simple and rapid, and this approach has developed into an important method for the characterization of bacteria.[52–54] Because the technique has shown such promise, and also because it may be unfamiliar to some readers, a brief introduction to TOF mass spectrometers and MALDI are presented below in the context of bacterial chemotaxonomy. While this chapter only describes MALDI desorption/ionizaton and TOF mass analysis (MALDI-TOF), some of the principles involved and the issues associated with differentiation of bacteria by mass spectrometry should also be of use to the nonchemist as they consider the different mass spectrometry approaches described in some of the other chapters in this book.

3.2 MASS SPECTROMETRY AND TIME-OF-FLIGHT MS

Mass spectrometry is not a new technique. The separation of charged particles based on mass, charge, and flight path has been known since J. J.

Thomson's famous experiments in 1897 measuring the charge-to-mass ratio of the electron. Many attribute the construction of the first mass spectrometer, in 1919 to Aston, one of Thomson's students.[55] The three primary components needed to produce a mass spectrometer are relatively straightforward as shown in Figure 3.1. These have not changed since that time, and include an ion source to vaporize and ionize samples, a device called a mass analyzer to separate ions based on their mass (or more correctly their mass to charge ratio), and finally a detector to measure the separated ions. While conceptually very straightforward, the use of ion flight times (time of flight) for mass separation was very slow to develop. The first suggestion that a useful mass spectrometer might be developed using the time-of-flight principle is attributed to Stephens[56] in 1946, nearly 50 years after Thomson's experiments and more than 25 years after Ashton's first mass spectrometer. In 1948 Cameron and Eggars[57] reported experimental data obtained using a TOF MS instrument they called a velocitron. An attractive feature of both the early velocitron instrument and their modern linear TOF counterparts is the simplicity of the design (see Figure 3.2). The three main regions of these instruments are the ion source, the drift region (mass analyzer), and the detector. Samples are evaporated, ionized, and accelerated in the ion source. Traditionally evaporation has been accomplished by simple heating, and ionization has been accomplished by ionization using 70 eV electrons. Both methods can result in decomposition of the analyte, and modern mass spectrometry has developed "softer" methods for sample evaporation and ionization. Nonequilibrium energy transfer can eject (desorb) samples into the gas phase without heating, and low-energy proton transfer can be used for "soft" ionization. Modern desorption/ionization methods can even combine vaporization and ionization into single step, one example of which is discussed in detail below. The accelerated ions obtain a mass-dependent velocity and move from the ion source and into a "separation" region. Ions separate in the drift or "field-free" region where

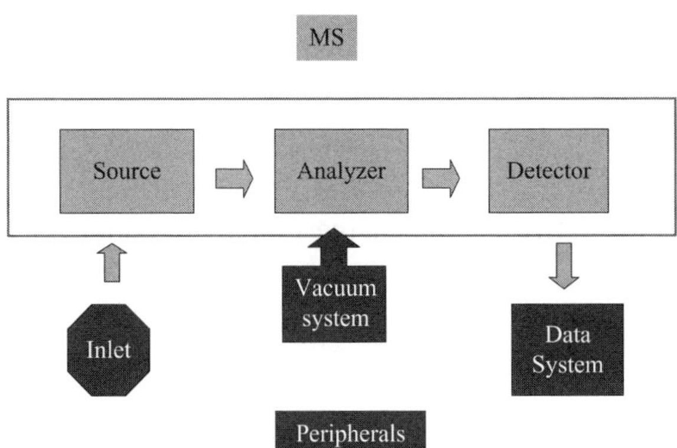

Figure 3.1 Components of a mass spectrometer.

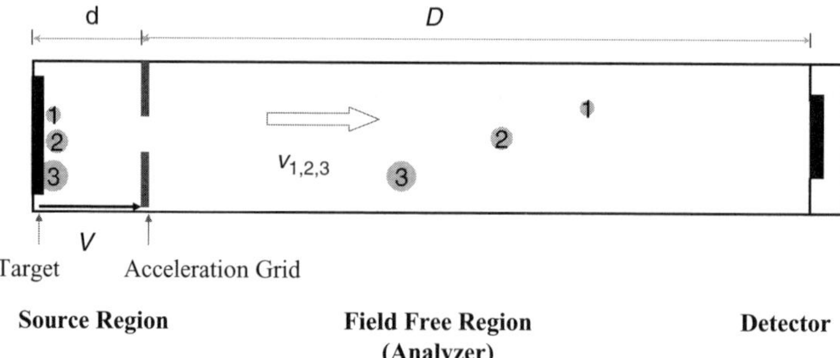

Figure 3.2 Principle of a time-of-flight mass spectrometer.

their different velocities (based on mass and charge) differentiate them as they "race" toward the detector. In the detector region the now mass-separated ions are counted, giving rise to a signal for each *m*/*z* value proportional to the number of ions present.

The TOF mass spectrometer operates in a pulsed mode. This means that each spectrum is produced from a discrete "event" where a packet of ions is formed and accelerated. All of these ions of the same sort (positively or negatively charged) formed in the original packet are separated in time (time of flight) based on their mass to charge ratio and detected. Since essentially all the ions from each pulse eventually reach the detector and are recorded a mass spectrum is produced with very high sensitivity. This is called "separation in time." The alternative approach is "separation in space." It is used in magnetic and quadrupole-based mass analyzers to differentiate one *m*/*z* value (in a certain location or "space") from all others by discarding all the nonselected values. This results in detection of only that small fraction of ions having the currently selected *m*/*z* value as the selected value changes in the scanning mode. The difference between separation in "time" and "space" is one of the major differences between TOF and many other mass spectrometers. With TOF mass spectrometers the entire process can be repeated very rapidly. Many spectra can be obtained in a few seconds, and typically several spectra are averaged to produce a final recorded mass spectrum with an enhanced signal-to-noise ratio.

In the absence of collisions as ions traverse the instrument, ion flight times can be directly related to mass or more correctly "mass to charge" (*m*/*z*) values. (In many experiments the charge value is 1 and the *m*/*z* value reflects mass.) However, ions need to be formed in a very specific region of the ion source. Ions formed in different regions would experience a different accelerating potential. With desorption/ionization coupled to TOF MS, this usually occurs on the surface of a stainless steel "probe" on which the sample has been deposited. This surface is labeled "target" in Figure 3.2. Ions formed on or near this surface in the ion source are accelerated across a potential difference (*V*) created within this region. The voltage difference results from the different potentials applied to the target and an acceleration grid at opposite ends of the source region. The voltage creates an electric field that imparts to the ions energy of motion equal to the product of the number of charges on each ion (z), the magnitude of the charge of a single electron (e) and the overall voltage difference over the region designated d on the Figure 3.2.

$$\text{Energy} = zeV \tag{3.1}$$

This energy of motion is a *kinetic energy (KE)* given by the equation

$$KE = \frac{mv^2}{2}. \tag{3.2}$$

Thus

$$\frac{mv^2}{2} = zeV. \tag{3.3}$$

In (3.3) the mass and velocity of the ions are m and v. Mass spectrometry data are usually plotted with ion abundance on the vertical axis and the mass-to-charge (m/z) ratio on the horizontal one. Solving for the mass-to-charge value (m/z), we obtain

$$\frac{m}{z} = \frac{2eV}{v^2}. \tag{3.4}$$

The value for the charge of the electron is a constant. In many experiments the voltage applied across the ion-source region is also a constant. Thus we can simplify (3.4) to give

$$\frac{m}{z} = \frac{k}{v^2}$$

or

$$v = \sqrt{\frac{kz}{m}} = \frac{k}{\sqrt{m/z}}.$$

Equation (3.4) shows that ions leave the source region with velocity values inversely related to the square root of their m/z values. So long as there are no collisions, the ions will maintain a constant velocity across the field-free region in route to the detector. The most convenient manner for discerning the mass (or m/z) values is to measure the velocity values indirectly. The detector of the ions is at a fixed location in the mass spectrometer at a distance D

from the ion source. Thus the time taken by ions to reach the detector can be related to velocity and is given by

$$t = \frac{D}{v}. \tag{3.5}$$

Substituting for v and recognizing the D is also a constant, we arrive at

$$t = \frac{D}{v} = \frac{D\sqrt{m}}{\sqrt{k'z}} = k'\sqrt{\frac{m}{z}}. \tag{3.6}$$

Equation (3.6) shows that ion arrival times at the detector can be related to the square root of their m/z values. To obtain the m/z values, we simply rearrange the equation again to give

$$\frac{m}{z} = k't^2. \tag{3.7}$$

While it is certainly possible to replace the constant with the appropriate values for constituent parameters ($2eV/D^2$), it is much more practical to determine the value empirically. A common calibration equation for TOF mass spectrometry is slightly more complicated (with the addition of an "offset" m/z value, b) and is given by

$$\frac{m}{z} = kt^2 + b. \tag{3.8}$$

The values of k (sometimes labeled a) and b are determined using spectra obtained with known mass standards.

The instrument and equations given above describe a liner TOF MS. For an instrument with typical values for the drift region length and for the ion acceleration voltage, the ion flight times are typically on the order of milliseconds. The typical resolving power of such an instrument would be around 300 to 400. Resolving power for a given signal in a mass spectrum is given by the mass value for the ion divided by the "mass width" at half the abundance for the same signal (see Figure 3.3). Thus a typical TOF MS signal at m/z 1000 that is 3 Da (m/z units) wide would have a resolving power of 333 ($m/\Delta m$). (Resolving power has also been referred to as resolution.) This simple linear TOF MS design can easily produce resolved mass spectra for small molecules and can also give "spectral fingerprints" for bacteria, but it cannot resolve unit mass differences (e.g., the masses of the carbon isotopes) much above mass 300 to 500. However, modifications to this simple design (described below) can increase resolving power by as much as an order of magnitude.

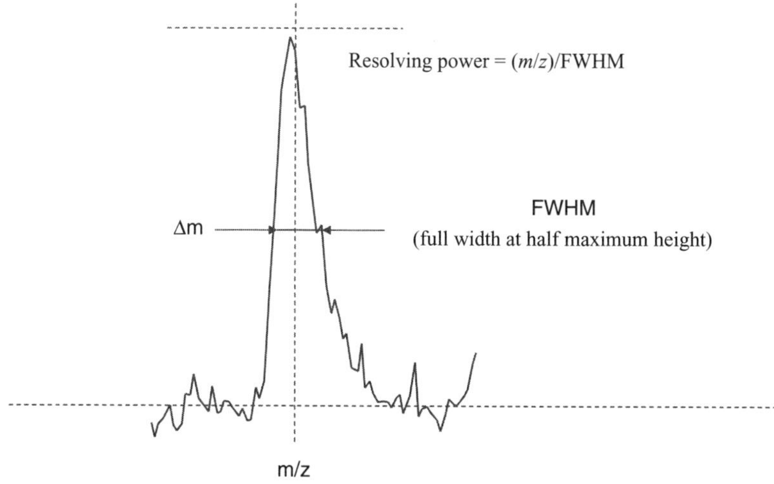

Figure 3.3 Mass spectral resolving power.

In practice, these equations do not *completely* describe the behavior observed with a simple linear TOF MS system, and in some circumstances the resolving power may be lower than expected. The initial conditions for the ions involving time, space, and kinetic energy distributions all contribute to uncertainty in the mass assignment because of a loss of mass (or time) resolution. For example, the time width of the initial pulse event (making the ion packet) defines the best time width (resolving power) possible in the TOF MS mass measurement. Thus it is an advantage to have a pulse event that has a very brief duration. It is also helpful (as noted above) that all ions are formed at the same location within the source. Finally it is important that all of the ions have velocities resulting only from the voltage applied to accelerate them. The velocity associated with random motion or the ionization event itself can contribute measurably to the overall ion velocity. One particular concern for TOF mass spectrometers is pressure and the resulting collisions.

The density of matter in the region around the plume produced in the laser desorption/ionization event can be relatively high for a very short time. The brief laser pulse desorbs both ions and neutral species at the same time, and there is sufficient material in this plume that collisions between the analyte ions and neutrals can occur. Since the ion velocities (and resulting flight times) are used to measure mass values, such collisions result in uncertainties in velocity and hence measured mass values. The magnitude of this uncertainty is sufficient to noticeably reduce the resolving power of the mass measurement. Ions having the same *m/z* value reach the detector at different times, giving rise to very broad peaks. As peak widths increase, the number of discrete peaks that can be measured in a spectrum also decreases. In the high vacuum of the TOF mass spectrometer (typically 10^{-7}–10^{-8} torr) the density in the region around the laser desorption plume is quickly reduced by expansion after the laser desorption event. Moreover heavy analyte species tend to be less mobile than the co-desorbed MALDI matrix and thus remains near the target region longer. If the ions are accelerated from the target out of the ion source after a very brief delay, collisions are much less likely, and the resolving power observed in the mass spectrum is greatly improved. This technique, of inserting a brief time delay between ion formation and ion acceleration, is called delayed extraction.[58] Delayed extraction has allowed MALDI ionization to be coupled with TOF mass spectrometers with retention of the inherent TOF resolution, but it has not significantly increased resolution.

A somewhat more complicated design for TOF mass spectrometers has increased their resolution significantly.[59] Within the same vacuum system it is possible to essentially double the field-free region by "reflecting" the ions back toward the source (field-free region 1 + field-free region 2 in Figure 3.4). In such a design ions travel from the source region toward the conventional detector, but are not detected. They are instead reflected by an opposing field, back toward the source region (V_2 in Figure 3.4). Effectively doubling the length of the field-free region increases the maximum achievable resolving power by a factor or about 2. The ions enter a region with a retarding potential that is actually greater than the acceleration potential in the ion source region. In this region the ions "turn around" and head back toward the detector. Because ions can be reflected at slightly less than 180°, it is possible to locate the second detector adjacent to the ion-source region. Such an arrangement produces a reflectron TOF mass spectrometer.[59] An important point is that the reflectron has an advantage beyond simply doubling the length of the field-free region. The retarding field may also act to somewhat refocus the ion beam. This is because ions having a small amount of excess velocity for a given mass penetrate slightly more into the field before they are reflected, whereas

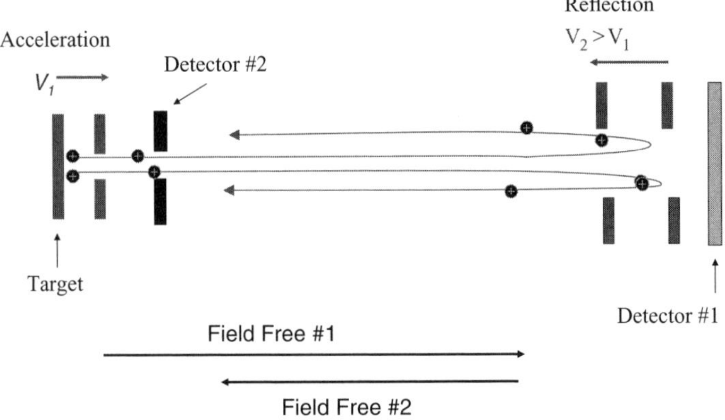

Figure 3.4 Principle of a reflectron mass spectrometer.

ions with slightly less velocity for that same mass do not penetrate much. The velocity imparted in the reverse direction (leaving the reflectron) depends on the depth of penetration of the ions at the exact moment they cease their forward motion (they stop briefly). Hence they are reflected with slightly different energies, and these differences compensate for (or cancel) the small differences in velocity they had entering the reflecting region. This refocusing capability allows mass resolving powers of greater than 5000 to be achieved when only TOF devices are used for mass separation.

Three main advantages of the TOF mass analyzer for the characterization of bacteria can be deduced from the treatment given above with a little additional information. These advantages are sensitivity, mass range, and speed. As noted above the TOF detects all ions of "like charge." Since traditional mass spectrometers (magnetic, quadrupole, and ion trap) differentiate one m/z value from others by simply discarding all ions with the nonselected m/z value, they throw away more ions than they detect. Hence they are much less sensitive. In the TOF the ions are separated in time (*a very short time*), but they are all eventually detected. Because the abundances of proteins in whole cells are not as great as other components, sensitivity is an important factor. The TOF is also capable of detecting high-mass ions. Indeed, in an excellent review of TOF MS in the *ACS Symposium Series*, R. J. Cotter notes that the technique has been used to detect ions over 300 kDa.[60] Traditional mass spectrometers have very limited capabilities for the detection of ions in the 5 to 25 kDa typically used for the differentiation of bacteria. Finally the TOF instrument can produce individual spectra in less than a second. Thus it is possible to obtain many spectra in a short period of time. The TOF instrument is also well suited for use with MALDI as both techniques work well in the pulsed mode.

3.3 MATRIX-ASSISTED LASER DESORPTION IONIZATION

Matrix-assisted laser desorption ionization (MALDI) is one of many vaporization and ionization methods that has been developed. MALDI actually accomplishes both vaporization and ionization in a single step. Because of its applicability to fragile biomolecules it has been one of the major achievements in mass spectrometry, and was the basis for part of the 2002 Nobel Prize in Chemistry awarded to Koichi Tanaka. Its advantages can most easily be explained by considering the first steps needed to obtain a mass spectrum. Analyte species must be converted from the solid phase (or liquid phase) to gas-phase ions. This process involves both vaporization and ionization. Traditionally vaporization was accomplished by evaporation of the sample with heating, if necessary. Many samples have sufficient vapor pressure in the high vacuum environment of the mass spectrometer (typically 10^{-6} torr) that they sublime directly into the ion source with little or no heating. However, many biological molecules, and proteins, will decompose prior to evaporation. For more than 70 years ionization in mass spectrometry was accomplished by bombarding the vaporized sample with electrons in a process called either electron bombardment (EB) or more commonly electron impact (EI) ionization.[61] The interaction between a 70 *eV* electron having a kinetic energy much higher than typical ionization potentials (10–15 *eV*) invariably results in ejection of a secondary electron and the formation of a radical cation. Often the newly formed radical cation (the molecular ion) is in a relatively energetic state and rapidly looses its excess energy by fragmentation to produce characteristic lower mass fragments. While the fingerprint produced by these fragments can be used to identify many organic molecules, it is less useful for very large molecules or for those molecules that fragment entirely to ions with very little diagnostic utility (i.e., *m*/*z* 43, 57, 71, . . . , or other very common ions). Alternative methods were developed, starting in the late 1960s, to produce more stable protonated molecules (not called protonated molecular ions!) by less energetic proton-transfer mechanisms.[62] The recently developed technique of MALDI has the advantage of providing a "soft" or nondestructive (at the molecular level) mechanism of converting solid phase analytes into gas-phase ions[63] with minimal fragmentation. Because MALDI accomplishes vaporization and ionization in the same process, it is referred to as a desorption/ionization technique.

The MALDI process is illustrated in Figure 3.5. The analyte is embedded in a very large excess of a matrix compound deposited on a solid surface called a target, usually made of a conducting metal. Typically about 1 ml of the analyte solution is mixed with the same volume of a saturated matrix solution, either before or during application to the target. The solvent, typically water, acetonitrile/water, or acetone/water, is allowed to dry completely on the target which is then placed into the mass spectrometer's ion source. This target is typically made of stainless steel, but it can be any rigid conducting material. The matrix compound is typically a weak acid that also absorbs light at the wavelength of the laser used in the experiment. Three typical matrix compounds are shown in Figure 3.6. They are all aromatic compounds, and each

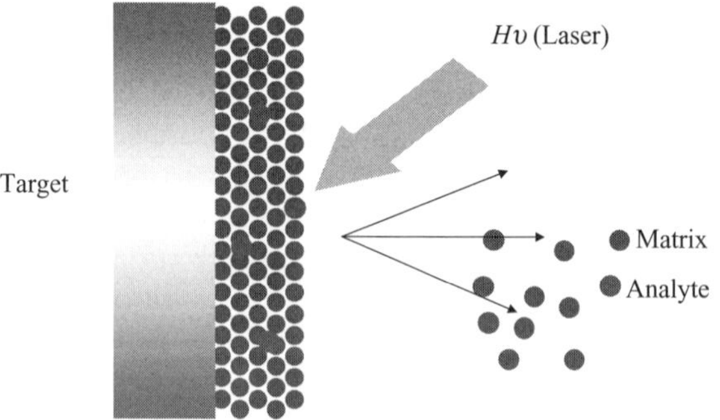

Figure 3.5 Principle of the MALDI process.

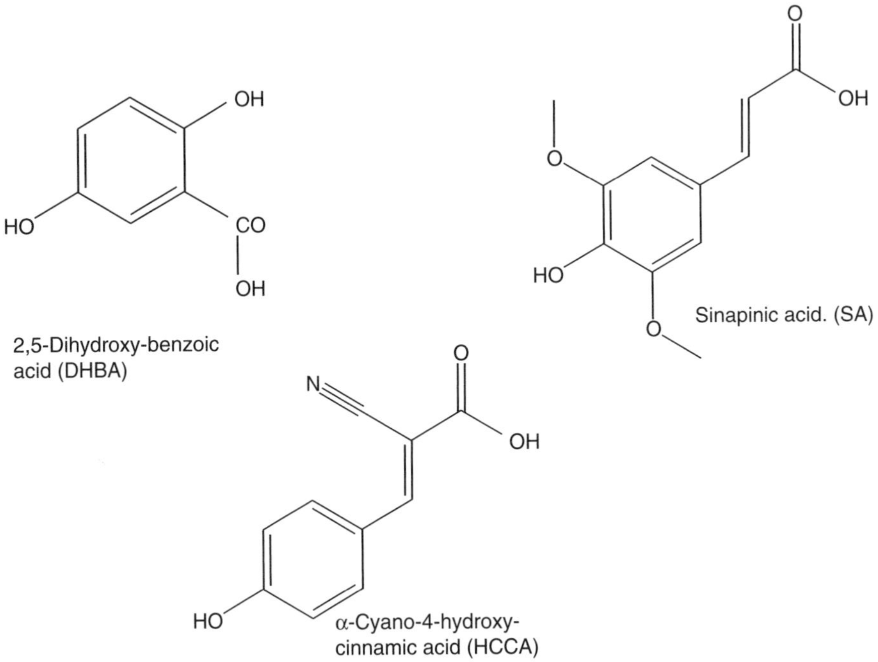

2,5-Dihydroxy-benzoic acid (DHBA)

Sinapinic acid. (SA)

α-Cyano-4-hydroxy-cinnamic acid (HCCA)

Figure 3.6 Structural formulae of three typical matrix compounds.

has a carboxylic acid moiety. Because the matrix absorbs strongly at the wavelength of the laser, there will be a strong interaction between the matrix/analyte solid layer regardless of the light-absorbing properties of the analyte. After a very brief laser pulse, the irradiated spot is rapidly heated and becomes vibrationally excited. Because the process does not involve an equilibrium step, but rather an explosive expansion into a high-vacuum region, the usual common-sense notion of heating does not apply. The laser energy is absorbed, and the vibrational energy results in expulsion of material in a plume just above the irradiated site. The emission is accompanied by minimal dissociative vibrational excitation of the analyte. The expanding plume of matrix compound physically "carries" the analyte into the vapor state with very little actually heating of the sample. For this reason intact proteins can be transferred into the gas phase without the decomposition that would be expected from normal heat-induced evaporation. This process results in the transfer of both ionic and neutral species into the vapor state. Because the matrix compound is also a weak acid, some of the polar analytes (like proteins) will attract protons from the carboxylic acid groups of the matrix in an acid-base equilibrium process. This can occur in the solid state or by gas-phase proton transfer in the region just above the site of desorption. In either case a gas-phase protonated analyte species can be produced. The mass spectrometer does not detect any of the abundant desorbed neutrals. Also the region corresponding to the matrix is typically much lower in *m/z* value than the analyte. The detection of the abundant protonated matrix species is avoided by not sampling ions in the *m/z* range; in other words, they are ignored, and not even detected. This MALDI approach is ideally suited for the analysis of proteins because they are good proton acceptors and are much higher in mass than protonated matrix, or even protonated matrix cluster ions. MALDI is also well suited to the analysis of mixtures of proteins (including enzymatic digests) because the technique typically produces one large ion (protonated molecule) per component. While fragmentation, and the presence of multi-charged ions are both possible as well, they are typically of such magnitude that they can be readily identified in spectra. Thus each component in a mixture can be associated with one significant signal. Figure 3.7 shows a typical MALDI mass spectrum obtained from cytochrome C. The large protonated molecule and a weaker diprotonated species are the only significant ions in the spectrum. The matrix peaks are all well below the 3500 Da lower *m/z* limit in this spectrum.

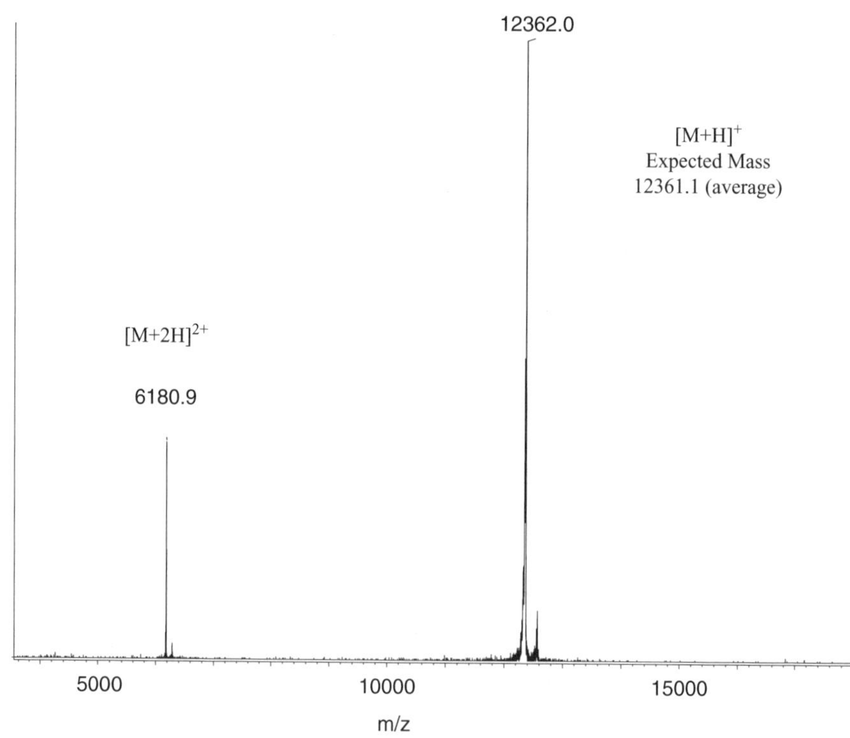

Figure 3.7 Time-of flight mass spectrum of cytochrome C.

3.4 MALDI-TOF MASS SPECTROMETRY

As noted above, TOF mass analysis is a pulse-based approach. A discrete event is needed to either produce or to accelerate the ions. Moreover a clear, reproducible initial time, or $t = 0$, is needed for proper calibration and mass assignment. Because the MALDI process involves the use of lasers, which are also pulsed, the laser pulse can be taken as the initiating event. For ion flight times in microseconds and laser pulse widths in nanoseconds it is clear that the ionization step is compatible with the TOF mass separation and does not unnecessarily broaden the peak widths. The TOF requires sufficiently high vacuum in the drift regions that collisions do not occur in flight. While the MALDI process does result in vaporization, the laser spot is in fact quite small, and very little material is actually evaporated into the MS. This and the low vapor pressure of the MALDI matrix compound allow relatively high vacuum (10^{-7}–10^{-8} torr) conditions to be easily maintained in the drift region. The integrating rather than scanning (all ions are detected) nature of the TOF is also well suited to MALDI. The very tiny amounts of material sampled by the laser are sufficient to produce mass spectra. It is not uncommon for MALDI mass spectra of proteins to be produced from picomolar levels of protein.

Figure 3.7 reflects the typical mass accuracy obtained by MALDI-TOF MS. The observed mass is usually within 1 Da of the expected mass. Problems arise, however, in the definition of the expected mass. For unresolved isotopes the expected mass would correspond to the average mass, whereas for resolved isotopes the expected mass would be the mono-isotopic mass or the mass of an isotope peak, depending on which is selected. Sometimes the mass spectrum has a mixture of isotopically resolved peaks, partially resolved peaks, and unresolved peaks over the mass range covered in a single spectrum. The resolving power in MALDI-TOF mass spectra ranges from a few hundred (simple linear TOF), to about 500 (delayed extraction, linear TOF) to as high as 5000 in the reflectron mode. While the reflectron mode results in higher resolving power, fewer ions are detected. This is because the linear TOF will detect signals for ions leaving the ion source, as if they were parent ions, even if they subsequently decompose. If they fragment in flight, both the neutral and the fragment ion retain the kinetic energy imparted leaving the ion source. On the other hand, the reflectron mode reflects only the ionic fragments, and this is based on their current mass, rather than the intial mass, resulting in loss of signal. The stable ions, surviving intact from source to detector are observed in the reflectron mode. The choice of the best operating mode depends on the signal level (sample amount) and resolution needed.

3.5 MALDI-TOF AND BACTERIAL IDENTIFICATION

The principle issue relating to the use of MALDI-TOF MS for bacterial characterization is resolving power. This factor has a profound effect in large measure because the characterization of whole bacteria by MALDI involves the analysis of a very complex mixture. This is done without any preliminary separation of components and also in a context where the ultimate objective is to produce a reproducible and taxonomically useful mass spectral fingerprint. The interplay between resolution and reproducibility in the context of this complex mixture analysis and application merits some specific discussion. It should also be noted that while MALDI-TOF MS provides a means for taxonomic identification of bacteria, it does not usually have much applicability to the detection of specific targeted proteins. The one exception is the analysis of greatly overexpressed proteins. Finally it should be mentioned that the dynamic range possible within the spectrum has important implications on the method.

A typical whole-cell MALDI-TOF mass spectrum is shown in Figure 3.8 for a strain of *Vibrio parahaemolyticus*. One of the key issues in bacterial characterization is the reproducibility of the spectral patterns produced. Replicate spectra of this organism would likely show mass values assigned to the most significant peaks change by up to 2 Da, even from the same culture. The intensity values for peaks could also vary considerably, even on different locations of the same sample spotting of a single sample and MALDI target. How then do we obtain, or even define reproducibility? Clearly in the earliest experiments spectra were sufficiently reproducible that blind coded samples were correctly identified. The short answer is that reproducibility is relative, and must be considered in the context of the application. In the

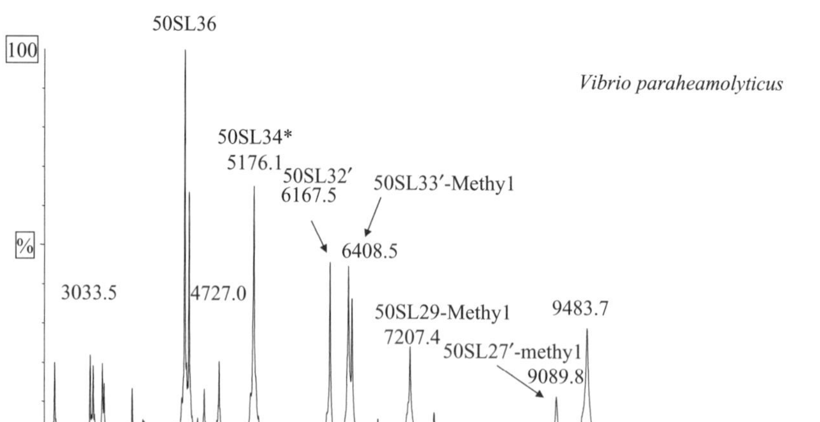

Figure 3.8 Whole cell time-of-flight mass spectrum of *Vibrio parahaemolyticus*.

context of whole-cell experiments reproducibility includes ion abundances to some extent, but because of inherent biological variability the mass assignment values are usually more important. In the best case the bacteria will produce (under similar experimental conditions) the same proteins in approximately the same abundances. However, even when the best efforts are made to control conditions, bacteria still respond to small uncontrollable changes in their environment, and protein abundances should be expected to vary somewhat. In this instance mass assignment (protein identity) is more important than peak (protein) abundance. In the worst case the bacteria will produce quite different profiles, and characterization will rely on the subset of conserved proteins (if any). The mass resolving power in these experiments plays an unexpected role in the mass assignments and the overall spectral patterns in a perhaps unexpected way because of the extreme complexity of these mixtures. This can be explained in two ways. The first involves changes in composition, and the other lack of homogeneity in the samples.

Changes in composition affect the mass spectra because very few of the peaks in any whole bacteria mass spectrum can be attributed to only one protein. Often the peaks contain contributions from multiple proteins and sometimes include overlap with salt clusters or fragments resulting from small molecule (water) loss. A careful look at Figure 3.8 will suggest that the peak widths are easily 100 Da wide at half the height. Thus they could contain additional proteins, fragments, and even contributions from sodium and potassium adducts. Thus small changes in the abundance of one or more of the overlapping peaks shifts the centroid (calculated center) of the overall peak and hence causes variation in the reported mass assignments. Under these conditions the apparent mass for a characteristic fingerprint-related peak might appear to change by 1 or 2 Da in replicate experiments, even if the peak envelope was measured with perfect mass accuracy. These changes in abundance correspond to changes in the composition of the peaks leading to the "weighted average" giving rise to the observed mass. For this reason it is very common, and should be considered "normal," for spectra from whole-cell bacteria to show significant spectral features that have mass assignments that are difficult to assign to a specific mass value. In contrast, for pure proteins the experimental average mass value (unresolved isotope cluster) is often is very close to the expected value even at m/z 10,000.

There is another explanation for the changes in spectra associated with changes in the composition of the underlying peaks, and thus the resulting observation of 1 to 2 Da mass shifts. Even if all cells from the same cell culture did contain identical chemical constituents and ratios, the chemical composition within the cell is certainly not homogeneous. Variations in sample deposition on the MALDI target are also well known. Even for a pure protein sample it is common that MALDI mass spectra vary somewhat across the sample target because of a lack of homogeneity across the target itself. Thus the usual practice is to seek the "sweet spot" on the MALDI target to obtain a mass spectrum, even with a pure compound. The homogeneity changes should also give rise to changes in composition, peak shapes, abundances, and mass assignments when components are only partially resolved.

For all these reasons incomplete resolution of variable chemical components (not just isotopes as for pure compounds) contributes to the uncertainty in the calculation of peak centers and thus an uncertainty in mass assignments of a few Daltons is expected in MALDI-TOF spectra from whole cells. This phenomenon also explains why the observed resolving power (and mass accuracy) in MALDI-TOF mass spectra from whole cells is typically lower than the observed resolving power (and accuracy) obtained using standard compounds on the same instrument. With our own instrument (Bruker Reflex III) we obtained near unit mass resolution on bovine insulin in the reflectron mode, and yet our resolving power at the same mass (about 5000 Da) was less than 700 when measured using whole cells. The overlapping peaks give rise to an enhanced peak width at half the height and lower measured resolving power.

Even so, for the identification of unknown proteins, the whole-cell MALDI-TOF methodology has applicability for chemotaxonomy because of the rapid detection of a unique fingerprint. This fingerprint is relatively stable and reproducible within the constraints outlined above. The characteristic signals have been attributed to the detection of the most basic, and abundant proteins within the cell. For many bacterial cell types this corresponds to the ribosomal proteins. While proteins may contribute 15% to the total weight of a protein, the ribosomal proteins may represent 20% of this portion or about 3% of the total cellular mass. On average, there are about 17,000 copies of the ribosomal proteins per cell. In addition to being very abundant, they are also typically very basic. Thus they are ideal candidates for detection. In 1998 Arnold and Reilly isolated ribosomes from *E. coli* and demonstrated significant taxonomic specificity for these proteins by differentiation of a large number of strains based on their profile.[64] Many of these ribosomal proteins have the same masses as the characteristic peaks in whole-cell spectra, and it is now widely accepted that these peaks correspond to detection of ribosomal proteins in whole-cell spectra. However, the easily detected ribosomal proteins typically mask most other proteins in the spectra from bacteria such as *E. coli*. The spectrum in Figure 3.8 from a *Vibrio* bacterium shows several major peaks. While these peaks cannot be assigned exactly based on these data, it is likely many of them are from ribosomal proteins. While it is not yet possible to predict all of the ribosomal proteins for this organism, and it has been shown that there exists taxonomically important strain-to-strain variation in these proteins, it is nevertheless interesting to consider how a set of known ribosomal proteins fit this spectrum. A search of the SWISSPROT and TREMBL databases using the sequence retrieval system from the Expasy Web site lists 16 ribosomal proteins, 1 hypothetical protein, and 1 ribosomal protein fragment. Five of the 50S ribosomal proteins (three with a reasonable modification) have a mass value that falls within 2 Da of the observed m/z values. Two of these masses correspond to monomethylation of the protein, and the third to the loss of the terminal MET group (both modifications are typical in bacteria). Another protein was within 5 Da of the expected value for a 50S ribosomal protein, but a mass error of 5 Da would be unusual. None of the peaks matched 30S or other ribosomal proteins. It is, of course, not possible to prove the identity of these proteins using only TOF mass spectrometry. In fact in other studies three different proteins have been assigned to the identity of a characteristic protein from *E. coli* in whole-cell MALDI experiments. The proteins cold shock A (CSP-A), B (CSP-B), and the 50S-L29 ribosomal proteins have masses that differ by only 3 Da. Because of the broad isotope profile at this mass and because of the mass shifts discussed above, these three proteins cannot be differentiated in whole-cell MALDI MS experiments using a TOF MS, and all three have been proposed by investigators. While the cold shock proteins are well known, readily induced, and often characterized using 2D gels, the primary protein observed at this mass in whole-cell MALDI has been confirmed to be the ribosomal protein in FTMS experiments.[65] This example demonstrates the importance of ribosomal proteins in whole-cell spectra, and how they can mask other biologically important (expected) proteins. In most cases it is simply unreasonable to expect to see a specific target protein directly from whole cells, and most mechanistic studies will continue to rely on some form of separation prior to analysis. There is, however, one application in which a specific target protein might be observed from whole cells that could represent an important application of the whole cell MALDI-TOF MS method, namely the analysis of recombinant proteins. Just as the rapidity of analysis is important in taxonomic applications, it may also be important for near–real–time monitoring of culture broths in experiments aimed at production of protein drugs. Currently there are over 90 FDA approved protein drugs in use, and many additional proteins are being tested. It has been estimated that it costs approximately $300,000,000 to bring each of these drugs to market. While recombination protein production represents only a fraction of the total cost, real-time monitoring of the culture broths might contribute to a reduction in the cost of making such protein drugs. The use of MALDI-TOF MS for monitoring recombinant expression from E. coli has already been reported.[66] Figure 3.9 shows a whole-cell MALDI TOF experiment where the soluble core domain protein of cytochrome B5 has been overexpressed in *E coli*. In addition to the abundant ribosomal proteins three new proteins are observed in this sample. The 11.2 kDa protein corresponds to the target protein, whereas the larger 11.5 and 9.7 kDa proteins represent unwanted overexpression products. While these proteins may not be as basic as the ribosomal proteins, they are present in such great excess that they are readily detected by MALDI-TOF MS.

3.6 CONCLUSIONS

MALDI ionization represents an ideal method for directly sampling the most abundant and basic proteins directly from whole cells. Typically these are ribosomal proteins. TOF mass spectrometry can be easily coupled to MALDI

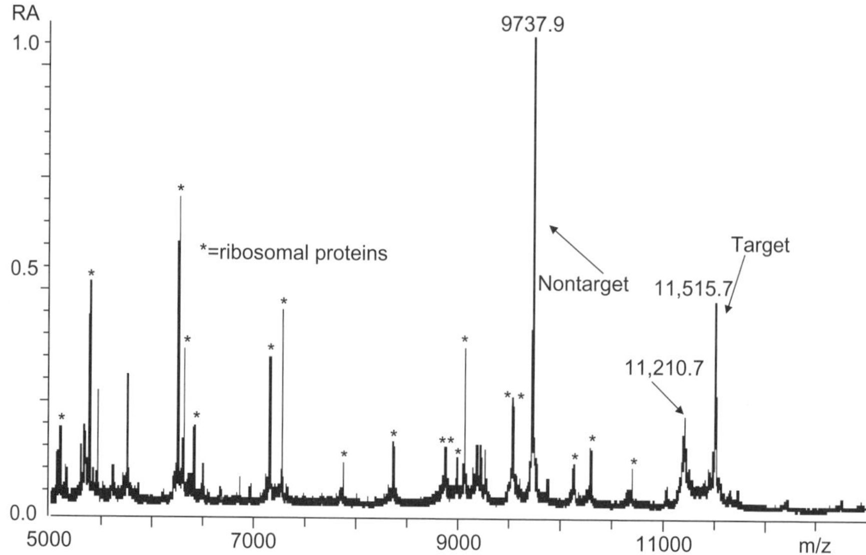

Figure 3.9 MALDI TOF mass spectrum showing over-expressed soluble core domain protein of cytochrome B5.

ionization sources, and MALDI-TOF MS represents a very rapid method for analyzing the proteins desorbed directly from whole cells. While the TOF MS is not a high-performance device, the whole-cell spectra produced by MALDI-TOF MS have taxonomically characteristic features that can be used to differentiate bacteria at the genus, species, and strain level, even thought only a small portion of the bacterial proteome can be detected by direct analysis. Because such a small number of proteins can be detected directly from whole cells, the method can normally not be used to look for target proteins outside the very small number of very basic proteins produced in large numbers within the cell. However, there is one possible and important exception to this generalization, the application of whole-cell MALDI-TOF MS to the characterization of recombinant proteins. Because both taxonomic identification and the analysis of overexpression products can both benefit form rapid and direct analysis, MALDI TOF is an important and useful method.

REFERENCES

1. Shioiri-Nakano, K.; Tadokoro, I.; Kudo, M. Instrumental analysis of streptococcal components. I. Carbohydrate and fatty acid constitution of hemolytic streptococci. *Jap. J. Exper. Med.* 1966, **36**, 563–576.
2. Boenicke, R.; Nolte, H. Diamine oxidases in mycobacteria. I. Possibilities of differentiation of several species of Mycobacterium by demonstrating diamine

oxidase. *Zentralbl. Bakteriol. Parasitenk. Infektionskr. Hyg., Abt. 1: Medizinisch-Hygienische Bakteriologie, Virusforschung und Parasitologie, Orig.* 1967, **202**, 479–487.

3. Bohacek, J.; Mraz, O. Deoxyribonucleic acid base composition of species *Pasteurella haemolytica, Actinobacillus lignieresii,* and *Actinobacillus equuli. Zentralbl. Bakteriol. Parasitenk. Infektionskr. Hyg., Abt. 1: Medizinisch-Hygienische Bakteriologie, Virusforschung und Parasitologie, Orig.* 1967, **202**, 468–478.
4. Heberlein, G. T.; De Ley, J.; Tijtgat, R. Deoxyribonucleic acid and taxonomy of Agrobacterium, Rhizobium, and Chromobacterium. *J. Bacteriol.* 1967, **94**, 116–124.
5. Henderson, A. Urease activity of *Acinetobacter lwoffi* and *A. anitratum. J. Gen. Microbiol.* 1967, **46**, 399–406.
6. Kandler, O. Chemical composition of bacterial cell walls as chemotaxonomic criterion. *Zentralbl. Bakteriol., Parasitenk., Infektionskr. Hyg., Abt. I. Orig.* 1967, **205**, 197–209.
7. Matteuzzi, D.; Crociani, F. Cellulosolytic sporogenous bacteria isolated from sheep rumen. *Annal. Microbiol. Enzimol.* 1967, **17**, 181–190.
8. Muftic, M. K. Application of chromogenic substrates to the determination of peptidases in mycobacteria. *Folia Microbiol.* (Prague, Czech Republic) 1967, **12**, 500–507.
9. Palleroni, N. J.; Solanes, R. E.; Rosell, P. F. Bacterial chemotaxonomy. *Rev. Facult. Ciencias Agrarias, Univ. Nac. Cuyo* 1967, **13**, 68–93.
10. Pine, L.; Boone, C. J. Comparative cell wall analyses of morphological forms within the genus. *Actinomyces. J. Bacteriol.* 1967, **94**, 875–883.
11. Rosypalova, A.; Rosypal, S. Physiological properties of violet pigmented micrococci: Their oxidation pattern and ability to grow in minimal synthetic medium. *Spisy Prirodov. Fakulty Univ. Brne* 1967, No. 479, 15–27.
12. Welker, N. E.; Campbell, L. L. Unrelatedness of *Bacillus amyloliquefaciens* and *Bacillus subtilis. J. Bacteriol.* 1967, **94**, 1124–1130.
13. Yamada, Y.; Aida, K.; Uemura, T. Distribution of ubiquinone 10 and 9 in acetic acid bacteria and its relation to the classification of genera Gluconobacter and Acetobacter, especially of so-called intermediate strains. *Agric. Biol. Chem.* 1968, **32**, 786–788.
14. Starr, M. P.; Jenkins, C. L.; Bussey, L. B.; Andrewes, A. G. Chemotaxonomic significance of the xanthomonadins, novel brominated aryl-polyene pigments produced by bacteria of the genus Xanthomonas. *Arch. Microbiol.* 1977, **113**, 1–9.
15. Yano, I.; Toriyama, S.; Ohno, Y.; Masui, M.; Kageyama, K.; Kusunose, M.; Kusunose, E.; Akimori, N. Comparative studies on molecular species of mycolic acids in nocardia and related bacteria by gas chromatography-mass spectrometry. *Iyo Masu Kenkyukai Koenshu* 1977, **2**, 35–40.
16. Collins, M. D.; Goodfellow, M.; Minnikin, D. E. Isoprenoid quinones in the classification of coryneform and related bacteria. *J. Gen. Microbiol.* 1979, **110**, 127–136.
17. Asselineau, C.; Clavel, S.; Clement, F.; Daffe, M.; David, H.; Laneelle, M. A.; Prome, J. C. Lipid constituents of *Mycobacterium leprae* isolated from experimentally infected armadillo. *Annal. Microbiol. (Paris)* 1981, **132A**, 19–30.

18. Eudy, L. W. Analytical pyrolysis and derivatization methods combined with gas chromatography-mass spectrometry for the characterization of bacteria and other nonvolatile materials. Univ. South Carolina, Columbia, SC, USA (1983), 197 pp. From: *Diss. Abstr. Int. B* 1984, **45**(1), 171.

19. Engman, H.; Mayfield, H. T.; Mar, T.; Bertsch, W. Classification of bacteria by pyrolysis-capillary column gas chromatography-mass spectrometry and pattern recognition. *J. Anal. Appl. Pyrolysis* 1984, **6**, 137–156.

20. Jantzen, E. Analysis of cellular components in bacterial classification and diagnosis. In *Gas Chromatogr./Mass Spectrom. Appl. Microbiol.* Odham, G.; Larsson, L.; Maardh, P.-A. (Eds.), New York: Plenum, 1984, 257–302.

21. Walla, M. D.; Lau, P. Y.; Morgan, S. L.; Fox, A.; Brown, A. Capillary gas chromatography-mass spectrometry of carbohydrate components of *Legionellae* and other bacteria. *J. Chromatogr.* 1984, **288**, 399–413.

22. Collins, M. D. Isoprenoid quinone analyses in bacterial classification and identification. *Soc. Appl. Bacteriol. Techn. Ser.* 1985, **20**, 267–287.

23. Eudy, L. W.; Walla, M. D.; Hudson, J. R.; Morgan, S. L.; Fox, A. Gas chromatography-mass spectrometry studies on the occurrence of acetamide, propionamide, and furfuryl alcohol in pyrolyzates of bacteria, bacterial fractions, and model compounds. *J. Anal. Appl. Pyrolysis* 1985, **7**, 231–247.

24. Yano, I. Analysis of bacterial metabolites and components by computerized GC/MS system—From shorter chain acids to very long-chain compounds up to C80. *Rapid Meth. Autom. Microbiol. Immunol. (4th Int. Symp.)* 1985, 239–247.

25. Brondz, I.; Olsen, I. Chemotaxonomy of selected species of the *Actinobacillus–Haemophilus–Pasteurella* group by means of gas chromatography, gas chromatography-mass spectrometry and bioenzymatic methods. *J. Chromatogr.* 1986, **380**, 1–17.

26. Carlier, J. P.; Sellier, N. Identification by gas chromatography-mass spectrometry of short-chain hydroxy acids produced by *Fusobacterium* species and *Clostridium inocuum*. *J. Chromatogr.* 1987, **420**, 121–128.

27. Gilbart, J.; Fox, A.; Morgan, S. L. Carbohydrate profiling of bacteria by gas chromatography-mass spectrometry: Chemical derivatization and analytical pyrolysis. *Eur. J. Clin. Microbiol.* 1987, **6**, 715–723.

28. Heller, D. N.; Cotter, R. J.; Fenselau, C.; Uy, O. M. Profiling of bacteria by fast atom bombardment mass spectrometry. *Anal. Chem.* 1987, **59**, 2806–2809.

29. Kaneda, K. Molecular species analysis of mycolic acids in acid-fast bacteria. *Iyo Masu Kenkyukai Koenshu* 1987, **12**, 63–70.

30. Suzuki, K. Application of mass spectrometry to bacterial taxonomy. Cellular fatty acids and isoprenoid quinones. *Iyo Masu Kenkyukai Koenshu* 1987, **12**, 45–53.

31. Zhou, F.; Zhu, H.; Tang, G.; Gao, S. Study on the discrimination of bacteria by gas chromatographic profiles of cellular fatty acids. *Weishengwu Xuebao* 1987, **27**, 95–104.

32. Heller, D. N.; Murphy, C. M.; Cotter, R. J.; Fenselau, C.; Uy, O. M. Constant neutral loss scanning for the characterization of bacterial phospholipids desorbed by fast atom bombardment. *Anal. Chem.* 1988, **60**, 2787–2791.

33. Morgan, S. L.; Fox, A. Chemotaxonomic characterization of microorganisms by capillary gas chromatography-mass spectrometry. Univ. South Carolina, Columbia,

SC, USA (1988), 8 pp. From: *Gov. Rep. Announce. Index* (U.S.) 1989, **89**(1), Abstr. No. 901,508

34. Platt, J. A.; Uy, O. M.; Heller, D. N.; Cotter, R. J.; Fenselau, C. Computer-based linear regression analysis of desorption mass spectra of microorganisms. *Anal. Chem.* 1988, **60**, 1415–1419.
35. Carlier, J. P.; Sellier, N. Gas chromatographic-mass spectral studies after methylation of metabolites produced by some anaerobic bacteria in spent media. *J. Chromatogr.* 1989, **493**, 257–273.
36. Drucker, D. B.; Jenkins, S. A. Applications of mass spectrometry including combined gas chromatography-mass spectrometry in taxonomic studies of bacteria. *Biochem. Soc. Trans.* 1989, **17**, 245–249.
37. Morgan, S. L.; Fox, A.; Gilbart, J. Profiling, structural characterization, and trace detection of chemical markers for microorganisms by gas chromatography-mass spectrometry. *J. Microbiol. Meth.* 1989, **9**, 57–69.
38. DeLuca, S.; Sarver, E. W.; Harrington, P. d. B.; Voorhees, K. J. Direct analysis of bacterial fatty acids by Curie-point pyrolysis tandem mass spectrometry. *Anal. Chem.* 1990, **62**, 1465–1472.
39. Dworzanski, J. P.; Berwald, L.; McClennen, W. H.; Meuzelaar, H. L. C. Mechanistic aspects of the pyrolytic methylation and transesterification of bacterial cell wall lipids. *J. Anal. Appl. Pyrolysis* 1991, **21**, 221–232.
40. Hotta-Hara, H.; Yabuuchi, E.; Mizuno, S.; Yano, I. Application of FAB/MS analysis of bacterial membrane lipids to the chemotaxonomy. *Nippon Iyo Masu Supekutoru Gakkai Koenshu.* 1991, **16**, 223–226.
41. Smith, P. B.; Snyder, A. P. Characterization of bacteria by quartz tube pyrolysis-gas chromatography/ion trap mass spectrometry. *J. Anal. Appl. Pyrolysis* 1992, **24**, 23–38.
42. Wait, R. The use of FAB MS of cellular lipids for the characterization of medically important bacteria. *NATO ASI Ser.*, 1992, **353**, 427–441.
43. Drucker, D. B.; Aluyi, H. S.; Boote, V.; Wilson, J. M.; Ling, Y. Polar lipids of strains of Prevotella, Bacteroides and Capnocytophaga analyzed by fast atom bombardment mass spectrometry. *Microbios* 1993, **75**, 45–56.
44. Cain, T. C.; Lubman, D. M.; Weber, W. J., Jr. Differentiation of bacteria using protein profiles from matrix-assisted laser desorption/ionization time-of-flight mass spectrometry. *Rapid Commun. Mass Spectrom.* 1994, **8**, 1026–1030.
45. Drucker, D. B. Fast atom bombardment mass spectrometry of phospholipids for bacterial chemotaxonomy. *ACS Symp. Ser.* 1994, **541**, 18–35.
46. Fenselau, C. (Ed.). *ACS Symp. Ser.: Mass Spectrometry for the Characterization of Microorganisms.* 1994, **541**.
47. Fenselau, C. Mass spectrometry for characterization of microorganisms: An overview. *ACS Symp. Ser.* 1994, **541**, 1–7.
48. Karas, M.; Bachmann, D.; Bahr, U.; Hillenkamp, F. Matrix-assisted ultraviolet laser desorption of non-volatile compounds. *Int. J. Mass Spectrom. Ion Process.* 1987, **78**, 53.
49. Krishnamurthy, T.; Ross, P. L. Rapid identification of bacteria by direct matrix-assisted laser desorption/ionization mass spectrometric analysis of whole cells. *Rapid Commun. Mass Spectrom.* 1996, **10**, 1992–1996.

50. Holland, R. D.; Wilkes, J. G.; Rafii, F.; Sutherland, J. B.; Persons, C. C.; Voorhees, K. J.; Lay, J. O., Jr. Rapid identification of intact whole bacteria based on spectral patterns using matrix-assisted laser desorption/ionization with time-of-flight mass spectrometry. *Rapid Commun. Mass Spectrom.* 1996, **10**, 1227–1232.
51. Claydon, M. A.; Davey, S. N.; Edwards-Jones, V.; Gordon, D. B. The rapid identification of intact microorganisms using mass spectrometry. *Nature Biotechnol.* 1996, **10**, 1992–1996.
52. Lay, J. O., Jr. MALDI-TOF mass spectrometry and bacterial taxonomy TrAC, *Trends Anal. Chem.* 2000, **19**, 507–516.
53. Lay, J. O., Jr. MALDI-TOF mass spectrometry of bacteria. *Mass Spectrom. Rev.* 2002, **20**, 172–194.
54. Fenselau, C.; Demirev, P. A. Characterization of intact microorganisms by MALDI mass spectrometry. *Mass Spectrom. Rev.* 2002, **20**, 157–171.
55. Aston, F. W. The mass-spectra of chemical elements. *Phil. Mag.* 1920, **39**, 611–625.
56. Stephens, W. E. A pulsed mass spectrometer with time dispersion. *(Proc. Am. Phys. Soc.) Phys. Rev.* 1946, **69**, 691.
57. Cameron, A. E.; Eggers, D. F. Ion "velocitron." *Rev. Sci. Instr.* 1948, **19**, 605–607.
58. Vestal, M. L.; Juhasz, P.; Martin, S. A. Delayed extraction matrix-assisted laser desorption time-of-flight mass spectrometry. *Rapid Comm. Mass Spectrom.* 1995, **9**, 1044–1050.
59. Mamyrin, B. A.; Karataev, V. I.; Shmikk, D. V.; Zagulin, V. A. Mass reflectron: New nonmagnetic time-of-flight high-resolution mass spectrometer. *Zh. Eksp. Teor. Fiz.* 1973, **64**, 82–89.
60. Cotter, R. J. In *ACS Symp. Ser.: Time-of-Flight Mass Spectrometry.* Cotter, R. J. (Ed.) 1994, **549**, 16–48.
61. Barton, H. A. The ionization of hydrochloric acid by electron impacts. *Phys. Rev.* 1927, **30**, 614–633.
62. Fales, H. M.; Milne, G. W.; Vestal, M. L. Chemical ionization mass spectrometry of complex molecules. *J. Am. Chem. Soc.* 1969, **91**, 3682–3685.
63. Hillenkamp, F.; Karas, M. Mass spectrometry of peptides and proteins by matrix-assisted ultraviolet laser desorption/ionization. *Meth. Enzymol.* 1990, **193** (Mass Spectrom.), 280–295.
64. Arnold, R. J.; Reilly, J. P. Fingerprint matching of *E. coli* strains with matrix-assisted laser desorption/ionization time-of-flight mass spectrometry of whole cells using a modified correlation approach. *Rapid Commun. Mass Spectrom.* 1998, **12**, 630–636.
65. Jones, J. J.; Stump, M. J.; Fleming, R. C.; Lay, J. O., Jr.; Wilkins, C. L. Investigation of MALDI-TOF and FT-MS techniques for analysis of *Escherichia coli* whole cells. *Anal. Chem.* 2003, **75**, 1340–1347.
66. Easterling, M. L.; Colangelo, C. M.; Scott, R. A.; Amster, I. J. Monitoring protein expression in whole bacterial cells with MALDI time-of-flight mass spectrometry. *Anal. Chem.* 1998, **70**, 2704–2709.

4

THE DEVELOPMENT OF THE BLOCK II CHEMICAL BIOLOGICAL MASS SPECTROMETER

Wayne H. Griest and Stephen A. Lammert
Oak Ridge National Laboratory, Chemical Sciences Division, Oak Ridge, TN 37831

4.1 INTRODUCTION

Since the use of mustard gas during World War I, chemical (and later biological) weapons pose an ever-increasing threat in today's military conflicts due to their role as a "force multiplier." Easier and cheaper to produce than nuclear weapons, chemical and biological weapons have the potential to inflict heavy casualties and can be delivered using conventional military munitions. In scenarios where an adversary is out-manned or at a technological disadvantage, the use or even the threat of use of these weapons can neutralize this advantage and multiply the effectiveness of a smaller enemy force. In addition to the obvious potential outcome of high soldier mortality, the aftermath of their use can consume significant medical resources to treat the surviving exposed troops. Additional resources would be required to decontaminate not only troops, but also their equipment. Even if decontamination measures are effective, key equipment will be rendered useless in the short-term due to contamination. Finally whether the actual use of these weapons is real or only threatened, troop effectiveness is significantly reduced if they are required to don *mission oriented protective posture* (MOPP) gear. To counter this threat, the US military has a stake in developing detection

Identification of Microorganisms by Mass Spectrometry, Edited by Charles L. Wilkins and Jackson O. Lay, Jr.

systems that are able to detect and identify the presence of chemical or biological agents before they can adversely affect the soldiers' health or influence the course of the battle.

In general, chemical and biological agents differ in the mechanism of their mode of action. Chemical agents[1] act like an acute poison while biological agents (except for toxins) are typically infections. Chemical and biological agents also differ in the time scale in which their symptoms develop. Chemical agents can be classified into four groups with examples (and their military identifiers in parenthesis) provided:

- *Nerve agents*, which interfere with nerve signaling such as Sarin (GB), Soman (GD), Tabun (GA), or methylphosphonothioic acid, *S*-[2-diethylamino)ethyl]*O*-2-methylpropyl ester (VX).
- *Blister agents*, which cause burns on the skin such as sulfur mustard gas, 1,1′-thiobis[2-chloroethane] (HD), and lewisite, (2-chloroethenyl) arsenous dichloride (L).
- *Riot control/incapacitants*, which cause extreme discomfort or mental confusion such as tear gas, chloroacetophenone (CN), or 3-quinuclidinylbenzilate (BZ).
- *Blood/pulmonary agents*, which interfere with metabolic functions such as hydrogen cyanide (AC) or phosgene (CG).

Chemical agents are typically fast-acting and symptoms develop within seconds to minutes of exposure. Even in the absence of protective detection systems, the use of chemical weapons will be immediately detected in the behavior of the troops on the field.

Biological agents[1,2] can be classified into four major groups also:

- *Bacterial agents*, such as *Bacillus anthracis* (the causative agent for anthrax), or *Yersinia pestis* (the causative agent for plague).
- *Viral agents*, such as the alpha virus that causes Venezuelan equine encephalitis (VEE) or *variola major* (the causative agent for smallpox).
- *Toxins* (typically high molecular weight proteins), such as botulinum toxin, ricin, or Staphyloccocal enterotoxin (SEB) or T-2 toxin (which actually is a small molecule).
- *Rickettsiae*, such as *Coxiella burnetti* and *Rickettsia rickettsii.*

Bacteria, viruses, and rickettsiae have similar symptom progressions in that exposure is followed by a period of reproductive growth (often nonsymptomatic) in the body. As their numbers increase, they often eventually overcome the immune system. Many produce toxins that interfere with bodily functions. Purified toxins such as botulinum toxin (produced by the *Clostridium botulinum* bacteria) act in a similar manner to chemical agents since, as complex chemical compounds, they do not reproduce but immediately interfere with bodily functions. However, most toxins are not absorbed through the skin, as are the nerve and blister CWA. The symptoms of exposure to biological agents such as *Bacillus anthracis* may take days to present and in some cases, by the time an infected person is symptomatic, death or serious illness is nearly inevitable. In the absence of detection systems, infectious biological agents can be deployed in aerosol clouds containing tiny particulates and delivered in stealth. However, if medical countermeasures are employed immediately after exposure, the casualty rate can be drastically reduced. The lack of immediate indicators of exposure and the delayed onset of symptoms makes the early detection of biological agents particularly important.

Thus there are two separate and distinct missions of any system designed to detect both chemical and biological weapons, namely *Detect to Warn* and *Detect to Treat*. The presence of a chemical agent must be detected *immediately* in order to warn troops to put on protective gear. Biological agents must be detected and identified promptly in order to minimize the number of exposed troops; however, immediate death or incapacitation is not likely. Instead, early detection allows early treatment and lower mortality rates.

From the perspective of an analytical chemist, the detection of threat chemicals or organisms must address two important issues, *detection limits* and *selectivity*. These directly relate, respectively, to the false negative and false positive rates. A false negative occurs when the detection system fails to alarm when the threat agent is present at concentrations above the minimum required level. A false positive is a situation where a detector system alarms when no threat agent is present. The detection limit of the system addresses the minimum amount of material that is needed to make a confident identification and as such directly affects the false negative rate. Selectivity is the ability of the detection method to distinguish between a targeted threat and other, nonpathogenic chemical or biological background species. Furthermore the method must be capable of detecting the targeted agents in the presence of sometimes large concentrations of chemicals (diesel fuel vapor, explosives, chemicals released by fires, etc.) or particulates (smoke, diesel exhaust, dust, and other particulate matter dispersed by moving vehicles or explosions). These background species will inevitably be a large part of a battlefield environment as a consequence of battle. They may even be intentionally introduced (e.g., using smoke as an obscurant or position marker) by friendly or opposition forces. The ability of the detection system to distinguish between the target agents and background and correctly identify the threat agent in the presence of background is the primary determinant in the false positive rate.

4.2 DEVELOPMENT HISTORY AND DESIGN PHILOSOPHY

Whether elegantly simple (e.g., the use of canaries as detectors) or technologically complex, a reliable detection system must attempt to address several, sometimes conflicting requirements. Trade-offs between size/weight/power considerations are often balanced against the requirements of speed of detection, high sensitivity, and confident identification. Reliability, support consumables, and maintenance schedules are other intangibles that enter into this consideration. Lacking an ideal approach, researchers have pursued many different technologies as platforms for chemical and biological agent detection systems. The latest version of the *NBC Product and Services Handbook*[3] lists 120 pages of military instrumentation designed to address "Contamination Avoidance" (i.e., detection) for chemical, biological, and nuclear threats. These systems range from handheld sensors to vehicle-mounted systems and cover a wide range of analytical techniques, including wet-chemical methods, surface acoustic wave sensors, infrared and ultraviolet spectroscopic detectors, ion mobility spectrometers, ionization detectors, manual and automated immunoassay techniques for biodetection, spectroscopic particle counters, and more. Many of the sensor approaches trade selectivity or sensitivity for portability. Included in this list is mass spectrometry,[4] a powerful analytical tool and often considered a "gold standard" in analytical identification. In addition to highly confident sample identification, mass spectrometers are also among the most sensitive forms of analytical instrumentation. Although traditionally considered a laboratory tool because of its historical size and weight, mass spectrometers have made significant strides in reducing the size, weight, and power requirements as the instrumentation has evolved. While not yet handheld, many current mass spectrometers are of a physical size and weight that allows portability. The combination of confident identification with high sensitivity, the ability to target both chemical and biological threat agents, and the inherent ability to identify unknowns makes mass spectrometers especially attractive as fieldable military threat detectors.

The Block II chemical biological mass spectrometer (CBMS II) is the most recent version in an evolution of fieldable mass spectrometer systems designed for military detection and identification of chemical and biological warfare agents (CWA and BWA, respectively). It builds on the experience and performance of previous versions and employs the latest advances in the components that comprise the system. Two of these predecessors in particular have made important contributions to this development, the mobile mass spectrometer (MM-1) and the Block I chemical biological mass spectrometer (CBMS I).

4.2.1 Mobile Mass Spectrometer, MM-1

The first mass spectrometer used for the battlefield detection of chemical weapons was the MM-1 mobile mass spectrometer[5] produced by Bruker-Franzen (Bremen, Germany). It was first deployed in the early 1980s as a component in the German military's Fuchs NBC (nuclear, biological, and chemical) reconnaissance vehicle. The role of the MM-1 was as a "persistent ground chemical agent detector" (chemical agent intentionally spread in liquid form on the ground in order to deny troop access to that area). The US military purchased 48 Fuchs systems in 1987 for evaluation and use in Europe. Later in 1990 the military contracted General Dynamics to build a US version of the vehicle system. When the Gulf War of 1991 broke out, the German government supplied the United States with 60 Fuchs NBC systems for use in the conflict. These systems became known as XM93 "Fox" NBC vehicles. In addition to the MM-1, the Fox NBC reconnaissance system employs other chemical agent detection systems such as the M43A1 chemical vapor detector (a system whose detection is based on ion mobility spectrometry) and the M256 Series Chemical Agent Detector Kit (which uses wet-chemical and reactive paper detection methods). Despite its "NBC" label the Fox system does not have any real-time biological agent detection capabilities but incorporates the ability to safely collect suspected biological agent samples for laboratory analysis. The Fox reconnaissance system monitors for chemical agent on the ground using a double wheel sampling system (DWSS). The DWSS employs two silicone sampling wheels that alternately sample the ground by rolling over the surface for a period of 20 seconds and then subsequently presenting the wheel to the head of a heated probe that protrudes through the vehicle's protective hull. When the silicone wheels contact the heated probe head, any adsorbed chemicals are thermally liberated and permeate through a membrane interface into a heated capillary transfer line that carries the sample to the MM-1 for analysis. Figure 4.1 gives a picture of the DWSS, with a wheel contacting the head of the ground probe. This figure actually is of the XM1135 Stryker nuclear, biological, chemical reconnaissance vehicle (NBCRV; see Section 4.4).

The MM-1 is a linear quadrupole-based mass spectrometer with electron impact (EI) ionization.[4] An ion pump provides the required vacuum. By today's standards, its size ($0.34\,m^3/12.2\,ft^3$) and weight (177 kg/390 lbs) are

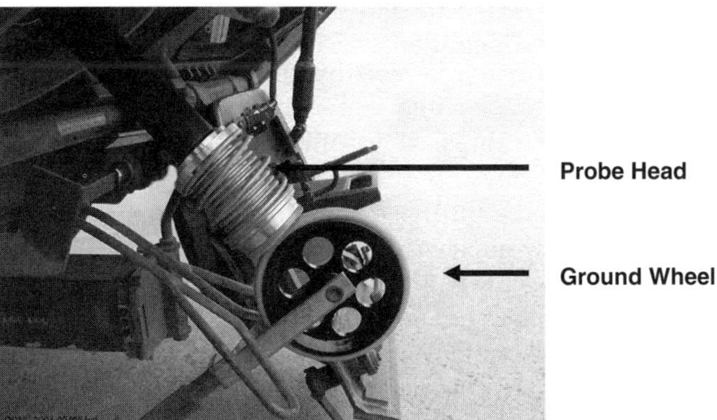

Figure 4.1 Double-wheel sampling system and head of chemical ground probe on the back of the XM1135 Stryker NBCRV Hull. (Image modified from original supplied by the US Army.)

considerable. As a linear quadrupole system, it is only capable of one stage of mass spectrometry—all ions generated in the EI spectrum (whether target ions or background ions) are summed into a single spectrum. This limits the ability to discriminate against background or other interferents, many of which may have some ions in common with the targeted agents. Identification is based on confirming the presence and relative ratios of four ions at predefined *m/z* values that represent the target agent's mass spectrum. The system's library contains the referenced entries for the targeted chemical agents and simulants, as well as entries that correspond to "fats, oils, and waxes" and represent the aliphatic hydrocarbon background often encountered in battlefield environments.

4.2.2 Chemical Biological Mass Spectrometer—Block I (CBMS I)

It was previously noted that the military NBC reconnaissance vehicles did not have the ability to detect and identify biological agents as part of their suite of detection systems. In 1987 the US military funded the development of a biological mass spectrometer-based detection system and awarded a contract to Teledyne (US Army ERDEC Contract DAAA15-87-C0008) for this work. Because of their experience with the MM-1, Bruker-Franzen again became an important player in this effort as Teledyne subcontracted the development of the mass spectrometer to them. This instrument, the *Chemical Biological Mass Spectrometer Block I* (CBMI),[6] is based on an ion-trap mass analyzer.[7] It was designed to be integrated into a suite of detector systems that collectively make up the *Biological Integrated Detection System* (BIDS).[8] The BIDS is a mobile biological point detection system that employs a laboratory shelter (containing two operators and the sampling/detection systems) and a towed power generator. The shelter is mounted on a high mobility multipurpose wheeled vehicle (HMMWV). Its operation would be as a stationary, point detection system for biological agents.

Ion traps had made a significant impact in the early 1980s when they were first introduced as mass spectrometers by Finnigan-MAT. As mass analyzers, ion traps have many characteristics that make them especially suited as fieldable instruments when compared to other forms of mass analysis. These include small size, lower power requirement, rugged, simple ion optics with no critical alignments, higher sensitivity, and, since ion traps can perform multiple stages of mass spectrometry (MS/MS or MS^n)[9] in a single mass analyzer, higher selectivity. Early in the development of ion traps, it was found that a buffer gas (helium is used in commercial laboratory instruments) was needed to improve the mass resolution of the system. Finnigan-MAT began to sell commercial versions of ion traps, and with their success came other competitors, including the Bruker-Franzen CBMS. What made the Bruker system exceptional was that unlike commercial ion traps of the time that used helium to kinetically cool the stored ion population prior to mass analysis, the CBMS employed a unique nonlinear ion ejection mechanism[10] that permitted the use of air as the buffer gas thus eliminating the necessity of compressed gas cylinders. Like its predecessor the MM-1, the CBMS uses an ion pump (4 liter/s) to provide the system vacuum and EI ionization. The vacuum interface is comprised of a permeable silicone membrane that is housed in a valve assembly that allows the vacuum system to be completely isolated from the atmosphere. Since the ion pump has only a single sealed inlet and no outlet, in the closed valve the analyzer compartment could remain under vacuum for weeks even though the system power was removed. In the event that the membrane interface was compromised, however, a oil-based mechanical pump had to be available to evacuate the system to moderate vacuum prior to restoring power to the ion pump. At 117 kg/258 lbs and 0.19 m^3/6.8 ft^3, the Block I CBMS made significant portability gains over its predecessor, the MM-1. The detection of biological agents was accomplished by measuring the relative proportions of the differing chain-length fatty acids in the phospholipids that comprise the cell wall of bacterial threats.[11–13] Other biomarkers used in the detection include ions derived from degradation of proteins (e.g., diketopiperazines) and other BWA components. This approach takes its origins from the FAMEs analysis technique (fatty acid methyl ester), a widely used, laboratory-based bacterial identification technique.[14] The FAME technique relies on the observation that bacterial cell walls differ in the relative proportions of fatty acids of various chain lengths, and this pattern can be used to distinguish bacteria from one another.

The Block I CBMS monitors for the presence of biological agents by sampling the air surrounding the BIDS unit. Particulates in the air are sampled and concentrated using a 1000 l/min XM-2 virtual impactor particle concentrator.[15] Particles in the 2 to 10 μm diameter range (typically considered easily respirable and therefore the most dangerous with respect to BWA infection) are selectively transported through the concentrator system concurrent with an elimination of particles outside of this range and a 1000 : 1 reduction in the airflow rate. The concentrated particles are then deposited onto a sintered quartz filter in a flow-through quartz tube. The fatty acids, are liberated from the cell wall phospholipids by thermolysis of the whole bacteria on the sintered quartz substrate at nominally 500°C and subsequently transferred through a fused silica transfer line to the mass spectrometer silicon membrane vacuum interface. After permeation through the interface, the fatty acids, and other biomarkers are profiled by the ion-trap mass spectrometer using electron impact ionization.

Despite the implication of its name the CBMS I does not have a chemical agent sampling capability and, as deployed in the BIDS system, is solely a bioagent detector. It should be noted that the CBMS I has been certified in military testing as a "classification detector" (i.e., a system that can determine if a detected threat is bacterial, viral, or toxin but not capable of identifying the actual threat agent; the CBMS I is more reliable as a classifier than as an identifier and so is used for the former). In addition the CBMS I was not designed for, nor did it undergo testing for radiation tolerance—a requirement for the reconnaissance mission but not for the BIDS program in which it is implemented.

4.2.3 Chemical Biological Mass Spectrometer—Block II (CBMS II)

In 1996 the US Army (Chemical Biological Defense Command, now the Office of the Joint Program Manager, Nuclear Biological Chemical Contamination Avoidance [JPMNBCCA]) began discussions with Oak Ridge National Laboratory (ORNL) to explore the possibilities of developing a Block II chemical biological mass spectrometer (CBMS II). After successful contract negotiations, the operational requirements were discussed at the program kickoff meeting in January 1997 at ORNL. Oak Ridge National Laboratory was chosen in part due to their experience on a program[16] to evaluate the Block I CBMS as a candidate instrument for on-site inspections under the Chemical Weapons Convention Treaty. As a part of this program, ORNL research staff took part in the regularly scheduled Block I CBMS users meetings. In addition ORNL had (and still has) extensive development experience on fieldable mass spectrometers for characterization and monitoring of chemical contamination at various Department of Energy (DOE) and Department of Defense (DOD) sites. The instrument and methods, sponsored under various funding agencies[17] are known as the direct sampling ion-trap mass spectrometer (DSITMS)[18] and enjoy widespread adoption and use of the instrumentation and techniques that have been developed. DSITMS is the basis of the US Environmental Protection Agency SW-846 test method 8265[19] for the analysis of volatile organic compounds in water, soil, and soil gases.

Several important partners were brought in by ORNL to strengthen the CBMS II team. Orbital Sciences Corporation (now Hamilton Sundstrand Sensor Systems [HSSS], Pomona, CA) was subcontracted as the Program Industrial Partner. Their primary responsibility was to ensure that the final design was producible at minimum cost and maximum reliability. HSSS also is collaborating in the CBMS II design and testing, and are responsible for manuals, training, integrated logistics support, and low-rate initial production (LRIP). MSP Corporation (Minneapolis, MN) provided extensive experience in the design and development of aerosol concentrators. Kent Voorhees' group at the Colorado School of Mines (Golden, CO) had long been active in using mass spectrometry to detect and classify bacteria.[12,13] Finally extensive contributions were made by military facilities, including Dugway Proving Ground (Dugway, UT), White Sands Missile Range (Las Cruces, NM), the Armed Forces Institute of Pathology (Washington, DC), and the sponsoring JPMNBCCA personnel (Edgewood, MD).

Figure 4.2 shows the complete CBMS II system. The main unit is comprised of three modules, the Biosampler Module, the Sample Introduction Module (SIM), and the Mass Spectrometer Module. The Biosampler Module houses the virtual impactor air particle concentrator and is only needed for the biological agent monitoring mode. The Sample Introduction Module contains the multiport sampling valve with its three input connections:

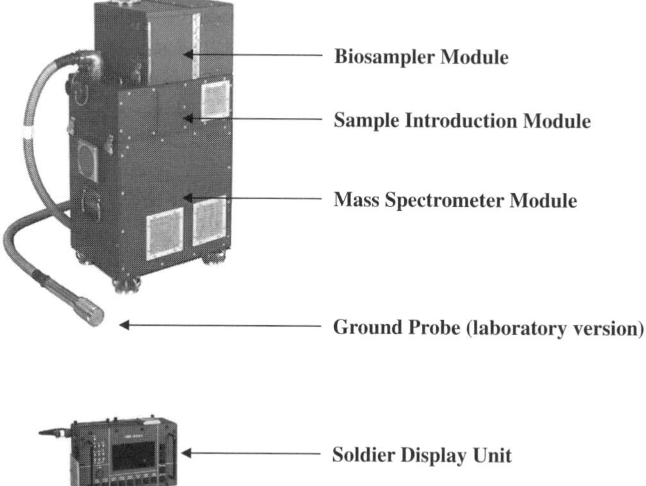

Figure 4.2 CBMS II and its modules. (Image supplied by Hamilton Sundstrand Sensor Systems.)

- The Biosampler Module for monitoring BWA in air.
- A heated air sampling line for monitoring CWA in air.
- A port to connect the government furnished equipment (GFE) ground probe transfer line from the DWSS system for monitoring persistent liquid CWA on the ground.

The sampling port is controlled by the operation software and can be set to continuously monitor a single one of the three inlets, or multiplexed between two (or all three although it is unlikely that a mission scenario will incorporate all three) of the modes (e.g., BWA in air and CWA in air). Also contained in the SIM is the pyrolyzer assembly, including the tetramethylammonium hydroxide (TMAH) solution delivery subsystem.

The Mass Spectrometer Module houses the vacuum system, capillary interface assembly, and ion-trap mass spectrometer in approximately half of the module. Also included are the reagent gas and calibration gas subassembly (a temperature-controlled housing that ensures consistent gas pressures). The other half contains the electronic printed circuit boards, power supplies, and instrument control computer.

The Soldier Display Unit (SDU) incorporates the data system computer used for identification and notification of threat detection as well as providing the means for soldier input (startup, mission selection, status and error monitoring). Contained in the Soldier Display Unit are two removable Personal Computer Memory Card International Association (PCMCIA) units. One contains the system high-level control software, including the data interpretation algorithm, and the other, logs significant system events (diagnostics, alarms, status/errors) and stores the spectral data acquired during detection events.

The philosophy of the CBMS II program was to build on the experience of both the MM-1 and Block I CBMS instruments and produce the first integrated battlefield detector capable of monitoring for both CWA and BWA. New capabilities would be added to increase the sensitivity and selectivity for both CWA and BWA. All instrumental advances depend on the capabilities and limitations of current state-of-the-art components. Nowhere is this more evident than in the vacuum system components. The primary enabling technology that allowed these advances was the development of commercially available turbomolecular vacuum pumps capable of operation in high vibration and shock environments. Experience on the DSITMS programs had indicated that these new turbomolecular pumps were both rugged and reliable in field instruments and later, extensive environmental testing as part of the CBMS II program would verify these initial findings. In fairness to the MM-1 and Block I CBMS programs, during the development of their programs in the 1970s and 1980s, only ion pumps were rugged enough to be considered for field use. Turbomolecular pump technology had not evolved sufficiently to allow operation outside of a controlled laboratory setting. The first rugged turbomolecular pumps became available in the early 1990s. No longer limited by low pumping speed ion pumps, the CBMS II program could use high-capacity (70 liter/s) pumps in the development with several positive outcomes. Since turbomolecular pumps can establish in minutes the high vacuum required for mass spectrometer operation, the requirement to maintain vacuum over long periods of time using ion pumps and valves was no longer a concern. Backup mechanical pumps were not needed as spares to accommodate the event of a membrane or other vacuum failure. Additionally the higher pumping capacity allows fused-silica capillary columns to be employed as the vacuum interface, increasing sample transmission and providing less discrimination against polar compounds than the membrane interfaces of its predecessors. Capillary interfaces also are more rugged than membrane interfaces and less prone to failure. Finally the higher pumping speed allows the use of chemical ionization (CI),[4] which produces spectra with less fragmentation of the molecular species. Careful selection of the CI reagent gas also allows the rejection of classes of potential interferents such as diesel fumes, fog oil, and other hydrocarbon backgrounds, resulting in higher selectivity for the targeted agents. Ethanol was selected as the chemical ionization reagent since its proton affinity is such that targeted chemical agents and simulants, as well as the biological target fatty acid biomarkers are efficiently ionized, primarily to their pseudomolecular ion at one atomic mass unit higher than their molecular weight. Most hydrocarbon species (the bulk of expected battlefield interferents), on the other hand, have proton affinities[20] below that of ethanol and as such are not ionized. As a liquid, ethanol does not require a compressed gas container, and a few milliliters of reagent are sufficient to last for many months.

While the Bruker-Franzen analyzer geometry and corresponding nonlinear scan function was retained, in the decade or so following the development of the CBMS I several advances in ion-trap mass spectrometry had occurred. Most notably was the application of broadband waveform ion manipulation[21] to the ion isolation and ion activation steps of MS/MS. This capability is important in ensuring that isolation and activation waveforms are independent of sample concentration. Ions of different mass-to-charge (*m/z*) values have storage trajectories of distinct and predictable frequencies. These frequencies can vary slightly depending on the ion population trapping conditions. High-sample concentration causes ion–ion repulsion (space charge), which leads to small changes in these trajectories and the subsequent ion frequency. Broadband techniques use computer-generated tailored waveform signals that are comprised of multiple discrete frequencies and ranges of frequencies. Waveforms can be fashioned that cover the entire range of frequencies that might be encountered in a range of sample concentrations. In this manner properly constructed broadband waveforms can ensure that the isolation and activation frequencies cover a wide range of possible sample concentrations. This capability is supplemented by an automatic sensitivity control, which adjusts EI ionization times and CI reaction times to prevent overloading the analyzer.

Automated and efficient sample introduction is typically one of the most challenging aspects of instrument development. Several new capabilities were developed in this area as part of the CBMS II program. The particle concentrator required to monitor for BWA in air was redesigned by MSP Corporation. A novel opposed jet virtual impactor[22] was developed that demonstrated improved efficiency and thereby allowed a lower sampled airflow (330 liter/min vs. 1000 liter/min for the XM-2) to be used. The result was a lower power requirement and considerably lower noise. In addition the opposed jet design also was more easily disassembled (without tools) for required cleaning in dusty environments. The pyrolysis chamber was redesigned to eliminate the flow-through characteristics of the CBMS I sintered quartz filter design that was prone to plugging in dusty environments. Instead, a reverse flow pyrolysis tube design is integrated with the flow from the virtual impactor to allow sampled particles to concentrate on the bottom of a quartz tube. An automated, in-situ derivatization subassembly delivers a precise quantity of a solution of TMAH in methanol onto the sampled particles prior to thermolysis. When the thermolysis heating cycle begins, the liberated fatty acid biomarkers are derivatized to their methyl esters and swept from the pyrolysis tube. As methyl esters, the biomarkers are considerably less polar than when in the free fatty acid form and as such are less likely to be lost due to adsorption in the subsequent transfer lines and analyzer components. The in-situ derivatization preparation step combined with the capillary interface greatly improves the efficiency of the fatty acid sample transport, increasing the method sensitivity. An outgrowth of the redesigned pyrolyzer tube assembly is that since the pyrolysis tube is anchored at only one point instead of two (as in a flow-through filter tube design), it is more rugged and less likely to break under high-vibration operation or during maintenance and replacement. In addition the tubes can be manufactured at a fraction of the cost of their predecessors. A unique carriage assembly provides for easy removal and replacement of the quartz pyrolysis tube, even by operators wearing the highest level MOPP IV gear.

Of special concern in the design is the requirement for the CBMS II to be radiation tolerant. The system must be capable of surviving a limited, tactical nuclear blast and subsequent exposure to radiation. After such an event the CBMS II must be back in full monitoring mode within 30 minutes. Radiation damage can occur to any component (vacuum seals, etc.) but of particular concern are electronic circuits, whose design requires a nontraditional approach. Damage to electronic circuits can occur from the initial electromagnetic pulse (EMP), prompt (initial blast) gamma radiation, neutron radiation, or accumulated or total gamma dose. Special grounding seals on the chassis are required to prevent exposure to EMP and electromagnetic interferences (EMI), and to prevent any possible EMI from the CBMS II itself from affecting other equipment in the vehicle. The high flux of prompt gamma radiation can induce large currents in powered circuits, which burn out fragile electronic junctions. To prevent this from occurring, fast protection circuits must be implemented that remove the power from the electronic circuits within a few hundred microseconds of detecting the gamma radiation front. Unpowered electronic circuits are not susceptible to damage from gamma radiation. Neutron damage can occur, however, whether or not the circuits are powered and inflict damage by direct collision and displacement of the tiny semiconductor switches contained in many microelectronics. As microprocessors become faster, these junctions become small enough to be completely destroyed by a single neutron passing through the junction. In order to prevent neutron damage, microprocessors must have larger feature sizes and as such must be older, slower speed versions. In order to regain calculation speed, several dedicated microprocessors must be employed for the various required functions of scan control, data acquisition, and data processing.

The very low storage temperature and low operating temperature specifications (see Section 4.3 below) also required dedicated design features because many of the components, such as the pumps and processors, cannot operate at very low temperatures. Because of this limitation there were added to several components cold start heaters controlled by snap switches. Further, an industrial temperature grade microprocessor was used in the SDU. With the snap switches the cold start heaters can come on when the power is applied at less than about 32°F/0°C. The snap switches cut off the cold start heaters and apply power to the full CBMS II system once their setpoint temperature is reached.

Last, it should not be surprising that the military sponsors who fund the development of next-generation instrumentation expect the final product to be smaller, lighter, less expensive to produce, less expensive to maintain, require fewer expendable parts and supplies, operate for longer periods with higher reliability, and employ measures that make the entire system easier to maintain, operate, repair and deploy.

4.3 REQUIREMENTS AND SPECIFICATIONS

4.3.1 General

The basic requirements for the CBMS II are to reliably detect and identify with sufficient sensitivity and selectivity both CWA and BWA in point detection and reconnaissance missions, in order to be deployable in wheeled reconnaissance vehicles and be operable by nontechnical personnel wearing, at the extreme, MOPP IV protective gear. Contrary to the usual practice for a military detector system, the CBMS II does not have its own requirements document. Instead, the requirements and specifications for the CBMS II are based on the detector requirements of the host platforms in which it will be deployed. These requirements are described in terms of performance, as opposed to the usual practice of being enumerated in volumes of detailed specifications. As is usual for a complex multiyear program, the requirements changed over the course of the CBMS II program as the requirements for the host platforms evolved.

4.3.2 Instrumental

Physical

The CBMS II is to be deployed aboard wheeled reconnaissance vehicles. The platforms include a light armored vehicle, the US Army XM1135 Stryker Nuclear, Biological, Chemical Reconnaissance Vehicle (NBCRV),[23] and two vehicles for the US Marine Corps' Joint Services Lightweight Nuclear, Biological, Chemical Reconnaissance System (JSLNBCRS):[24] the HMMWV and a wheeled, light armored vehicle very much like the XM1135 Stryker. Consequently the volume, mass, and power available to the instrument are limited. Table 4.1 lists the current physical specification and objective requirement for each parameter.

TABLE 4.1 CBMS II Physical Specifications

Parameter, Units	Current[a]	Objective
Volume, ft^3 (m^3)	5.7 (0.16)	4.5 (0.13)
Mass, lbs (kg)	193 (88)	130 (59)
Average power, W	600[b]	500
Peak power, W	1200[b]	1000

[a] Data listed for CBMS II PPUs. Data not yet available for LRIP units.

[b] Includes GFE chemical ground probe of around 200 W.

The current data for CBMS II preproduction units (PPU) show that the volume and mass are greater than the objective requirements, but they are within limits for reconnaissance vehicle deployment. LRIP unit data are not expected to be substantially different from those for PPUs. Further design work to reduce the mass will be carried out for future LRIP runs. The data for power draw is for the configuration including the chemical ground probe (a GFE sampling system), which is powered by the CBMS II. The ground probe draws around 200 W, which suggests that the objective requirements (which do not include the ground probe) are met. The peak power of 1200 W during startup is not a continuous power draw but rather intermittent peaks because the power management software apportions the power among the heated zones to bring critical zones to setpoint first.

The design of equipment to be simply and easily operated (manprint) is an important physical requirement. The CBMS II operator may have to operate the CBMS II and perform maintenance on it while the vehicle is vibrating and bouncing around in cross-country travel, without requiring special tools, and while wearing MOPP IV gear. The latter includes bulky rubber gloves that severely restrict manual dexterity and a mask that limits peripheral vision. Simple maintenance operations in the laboratory become very difficult under these conditions in the field, and manprint is critical to ensuring the optimum performance of the CBMS II and the success of a mission. Two modules in the CBMS II may require simple maintenance in the field and have been extensively engineered for manprint. One is the opposed-jet virtual impactor in the biosampler and the second is the pyrotube holder in the Sample Introduction Module (see Figure 4.2). The influence of manprint on the design of these parts was described previously in this chapter.

Monitoring Modes

The requirement of the CBMS II to detect and identify both CWA and BWA is the first time that multiple classes of agents have been specified for a single military detection/identification system. CWA are to be monitored by the CBMS II using two different sampling systems. The CBMS II is required to be compatible with the GFE chemical ground probe and DWSS, which picks up liquid CWA from the ground and transfers it to the CBMS II. The CBMS II is to monitor vapor phase CWA using a heated capillary line, which connects to a port on the vehicle hull. BWA in the air are to be monitored using a sampling stack. The classes of target agents to be monitored are described in a later section of this chapter. Such a wide range of target agents and monitoring modes places great demand on the applicability and flexibility of the sampling and analytical systems and the software controlling those systems and interpreting the analytical data.

Environmental

To be a deployed military detection system, the CBMS II must survive storage under severe conditions and reliably start up and operate under a wide range of field conditions. Achieving laboratory instrument performance under such constraints is a great challenge to selection of materials and components, instrument design and layout, and overall system function. The specifications are summarized in Table 4.2.

Military detector systems will experience wide extremes in temperature and humidity in storage and operation, both in shipping containers and installed inside reconnaissance vehicles deployed anywhere in the world. The lower extreme of the storage temperature range (–60°C/–51°F) limits the selection of components. For example, colorful, relatively high-resolution liquid crystal displays cannot survive such low temperatures. This constraint led to the selection of a monochrome plasma display with relatively low numbers of character spaces in which to present messages on the Soldier Display Unit (SDU; see Figure 4.2). It also required the use of an industrial-grade processor in the SDU. The nuclear radiation tolerance requirement greatly limited the computing power; older, slower (ca. 200 MHz) processors are incorporated in the design because their relatively large electronic features are less liable to radiation damage. Radiation tolerance also required adding circumvention circuits to power down critical components and circuits on detection of radiation.

Resistance to physical shocks and vibration required careful attention to selection of rugged components and to securing electrical and vacuum systems, wiring, connectors, components, and boards. Chemical ionization (CI) was used for the first time in a fieldable military detector because of the advent of rugged turbomolecular pumps capable of handling the gas load from the CI reagent.

Automated Operation

The CBMS II must be highly automated in all aspects of operation because the soldiers using the unit are not academically trained in mass spectrometry.

TABLE 4.2 Environmental and Other Requirements

Parameter	Requirement
Survive storage temperatures, °F (°C)	–60°F (–51°C) to +160°F (+71°C)
Operate in temperature range, °F (°C)	–25°F (–32°C) to +120°F (+49°C)
Startup time, from cold start at –25°F (–32°C), with 28 V power and vehicle interior temp. rises to above 32°F (0°C) in 10 min.	30 min
Operate in relative humidity range	5 to 95 %RH
Survive physical shock for vehicle	JSLNBCRS HMMWV
Survive vibration profile for vehicle	JSLNBCRS HMMWV
Survive radiation tolerance tests	(Classified)
EMF/EMP/EMI	MIL-STD-461D, -462D, and HEMP requirements to QSTAG 244 for JSLNBCRS
Mean time between hardware mission failures	801 h

Startup must be fully automatic once the main power switch is turned on. Instrument warmup is monitored, and once critical zones are within temperature limits, analyzer mass and frequency calibration and collection of background spectra are conducted automatically. During startup, faults and other problems are located and diagnosed by a built in test (BIT) that alerts the operator to faults and errors and records them in an electronic log. Critical errors under which the instrument cannot operate (e.g., failure to calibrate) result in instrument shutdown, while less serious problems that do not prevent the instrument from performing its mission (e.g., a heated zone not used in the current monitoring mode not reaching its setpoint) result only in a message. Once calibration and background collection are complete, the CBMS II starts monitoring in the default monitoring mode, which currently is for liquid CWA on the ground.

The CBMS II must start up and be ready for monitoring within 30 minutes, to fit within the requirements of a reconnaissance mission in which the vehicle and sensors are started up at a depot, company area, or area of bivouac and leave on a mission shortly afterward. The mission departure point may be close to a bivouac area.

All operations with the CBMS II must be easy to enable. The SDU that is used by the soldier to operate the CBMS II features a screen and pushbutton input similar to that found on automated teller machines (ATMs). The soldier pushes buttons to make choices among simple commands such as changing the CBMS II monitoring modes, recollecting the background spectrum, or reviewing the system electronics log. An example of the display page for changing monitoring modes is shown in Figure 4.3*a*. Two monitoring modes are shown. The monitoring mode is selected by pressing the button to the right side of the display, next to the desired monitoring mode.

The CBMS II must automatically detect, identify, and alarm when one of its sampling systems encounters a target agent. For example, when the system is monitoring for chemical agents, the spectra are continuously examined by the software for key agent-derived ions in CI and EI full-scan and in MS/MS modes, and when they significantly exceed the rolling average background, an alarm is displayed on the SDU (see Figure 4.3*b*), the alarm warning light and warning tone are activated, and the alarm message is sent to the central data processing unit on the vehicle and also is stored in the electronic log on the flash card in the SDU. During operation the condition of the CBMS II is continuously monitored by the runtime BIT, which is similar to the startup BIT discussed previously.

4.3.3 Performance

Agents and Monitoring Modes

As was noted previously, the CBMS II is required to detect and identify both CWA and BWA in three different monitoring modes. The list of CWA target agents is not available for public release, but it includes the common nerve,

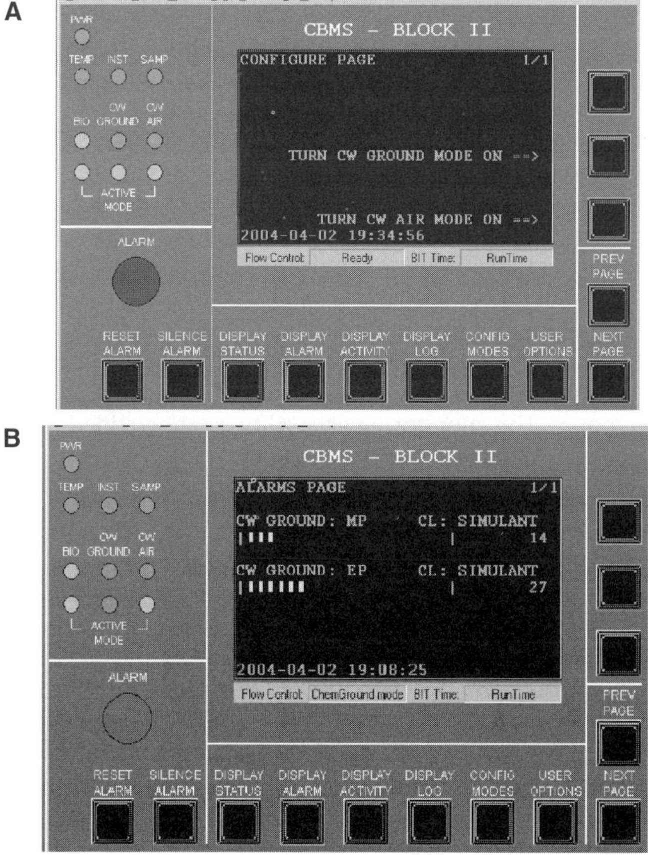

Figure 4.3 SDU and screen pages showing (*a*) choice of monitoring modes and (*b*) an alarm. (Images prepared by Kevin Hart.)

blister, riot, blood, and incapacitating agents, and CWA simulants, such as methyl salicylate (MES) and diethyl malonate (DEM). CWA are monitored in modes that sample the surface of the ground for liquid agents (termed the Chemical Ground monitoring mode) and the air for vapor phase agents (called the Chemical Air mode). The CWA list for the latter is shorter than for the former, since some of the CWA do not have appreciable vapor pressures under ambient conditions. There are two additional requirements for the CWA monitoring modes. One is that the CBMS II recognizes preprogrammed battlefield interferents, such as JP8, fog oil, and DF2, similar to the "fats, oils, and waxes" recognition capability of the MM1. The CBMS II does not alarm on these interferents but notifies the operator that the interferent has been identified. The second requirement for CWA monitoring is the ability to alarm on encountering unknown agents. An unknown agent is defined as a substance the CBMS II encounters that is not identified as either a target agent or a preprogrammed interferent. If the unknown agent option has been turned on, the CBMS II assigns the unknown agent a unique number, sounds an alarm, and stores the information that will allow it to recognize and alarm on that same substance when it encounters it again.

The BWAs monitored in the air (Bio Air mode) are those listed in the classified (secret) International Task Force (ITF)-6A list (1990). The list includes bacteria, rickettsiae, toxins, and viruses. The unknown agent option is not available for the Bio Air monitoring mode.

Later in the program a requirement for toxic industrial chemicals and toxic industrial materials (TICs/TIMs) was added. The targeted TICs/TIMs currently are 42 highly toxic or otherwise hazardous compounds taken from the ITF-40 list.[25] They are monitored as vapors and as liquid chemicals in separate sampling methods. The unknown agents option is not used in the TICs/TIMs modes. This monitoring capability is being added in 2005. Finally the nontraditional agents (NTA) are a classified list of agents added later in the Program. They are based on intelligence estimates of future CWA threats. The NTA consist mainly of very low volatility nerve agents. This monitoring capability is being added in future years.

Sensitivity and Response Time

Sensitivity and response time are important requirements. They ensure that alarms are sounded at agent concentrations that are not immediately dangerous to the soldiers, and that sufficient warning is given to allow effective protective measures to be taken, such as taking cover in a protective shelter or donning a mask. The agent detection limits of the CBMS II are required to be approximately equal to or lower than those of current CWA and BWA detectors, such as the MM1 (liquid CWA), the Automatic Chemical Agent Detection Alarm (ACADA, CWA vapors), and the Joint Biological Point Detection System (JBPDS, BWA aerosols). The detection limit specifications for those detectors, where available, are listed in Table 4.3. For liquid CWA on the ground, no specifications were located for the MM1. However, the detection limits of the CBMS II for liquid CWA, expressed as the mass of CWA samples applied to the DWSS wheel, are required to be lower than physiological effects levels. For example, the detection limits must be lower than miosis levels for nerve agents (typically, low mg masses). The response time specification is that the correct alarm must be sounded within 45 seconds of the time the CWA is applied to the probe head.

The specifications for CWA vapors are an agent concentration (mass of agent per volume of air) at the inlet to the CWA vapor sampling line and a response time. The specifications for the ACADA include two response times. A longer response time before an alarm requires a lower detection limit in order to offset the delay and greater potential exposure of the soldiers, versus the specification for the shorter response time. The specifications range from tens down to hundredths of a mg/m^3, and depend on the toxicity of the CWA. A response time of 10 seconds or less has been requested for CWA in air.

TABLE 4.3 Agent Detection Limit Specifications for the ACADA and JBPDS

Detector	Monitoring Mode	Agent	Specification
ACADA	CWA vapor	G-Series nerve (GA/GB/GD)	1 mg/m^3 in <10 s 0.1 mg/m^3 in <30 s
		VX	1 mg/m^3 in <10 s 0.04 mg/m^3 in <90 s
		HD and L	50 mg/m^3 in <10 s 2.0 mg/m^3 in <120 s
JBPDS	BWA in air	Bacteria, viruses, and toxins	25 ACPLA (threshold) 1 ACPLA (objective)

For BWA the detection limit is expressed in terms of the numbers of agent-containing particles per liter of air (ACPLA), and refers to BWA air concentrations at the BWA air sampling stack. The BWA are bacteria, rickettsiae, viruses, and toxins. Although an ACPLA is a convenient unified means of expressing BWA air concentrations for all four classes of BWA, the term is imprecise on such important factors as the particle size distribution and the actual number of bacteria/rickettsiae cells or virus strands, or mass of toxin molecules per particle or per liter of air. The objective specification is the ultimate, desired specification which may not be feasible with current technology; certainly BWA identification at 1 ACPLA is beyond current fielded technology. Therefore a less stringent threshold specification (25 ACPLA) is listed that must be met. The response time requirement for the CBMS II is a correct alarm within 3 minutes (objective) of the start of air sampling and 5 minutes (threshold).

Selectivity

This requirement is important to ensuring that the CBMS II alarms for the correct agent. The requirement is for the CBMS II to produce substantially fewer false positive and false negative alarms and misidentifications than the current CWA and BWA detectors. The design of the CBMS II incorporates several features that improve selectivity over that of previous detectors. These include CI and MS/MS, and for some target agents, MS/MS/MS.

4.4 PERFORMANCE TESTING

The ability of a system to meet its requirements and specifications must be demonstrated and documented before it can be accepted and fielded as a military system. Performance tests of a military detector system typically are performed by the contractor (Contractor Tests) and by the government (Government Tests). The former often are witnessed by government personnel who independently report on the methodology and results. Contractor testing is conducted at contractor facilities and at government facilities (but the instrument is run by contractor personnel). The Government Tests are performed at government facilities (Dugway Proving Ground, UT, etc.) by government personnel with the system installed both in a laboratory, in vehicles, or in shelters for field tests. Laboratory evaluations include both physical/environmental testing and CWA and BWA target agent testing with the actual CWA and BWA. The laboratory tests with agents utilize field sampling systems under controlled conditions and the actual agents that normally cannot be dispersed in the field. (Live pathogenic BWA normally are not tested because of the extreme protective measures required to aerosolize them. Instead, killed pathogens are usually tested.) Interferents also are run with and without the CWA and BWA. These tests include determining limits of detection, resistance to false negative and false positive alarms, and instrument recovery times. Field tests usually are conducted using CWA and BWA simulants and sometimes using vaccine (nonpathogenic) strains of bacteria. Interferents in field tests are normally those emitted from the vehicles themselves (e.g., engine exhaust), the natural background and substances (e.g., oil spills and road tars) on the surfaces of paved and gravel roads, and vegetation in the cross-country landscape.

4.4.1 Liquid CWA Performance Testing in the Laboratory

Laboratory tests to determine liquid CWA performance were performed with the same chemical ground probe that is used in the reconnaissance vehicles. Liquid samples containing known volumes and masses of CWA and/or interferents were applied to both glass plates (for a measure of the absolute sensitivity at the ground probe head) and ground wheel sections (to estimate performance of the wheel and probe DWSS sampling system), which were then held to the ground probe head for the same time period (5 s) and force (5 lbs-force) as the ground wheel is held to the probe head by the DWSS mechanism on the reconnaissance vehicles. CWA were first analyzed separately (without interferences) to determine the limits of detection and the ability of the CBMS II to correctly identify the agents. Next pure samples of common battlefield interferents were analyzed separately to test for false positive alarms due to the interferent itself and the speed with which the instrument clears itself of an excess of the interferent. Interferents included JP8 (turbine fuel), fog oil (used to generate obscuring smokes), seawater (coastal or off-shore use), and *N*,*N*-diethyl-*m*-toluamide (DEET, the active ingredient in insect repellants). Decontamination reagents also were examined, and included decontamination solution DS-2 (an alkaline solution of diethylenetriamine, ethylene glycol monomethyl ether, and sodium hydroxide), supertropical bleach, a skin decontamination lotion, and Sandia foam (a decontaminating foam). Finally the CWA were tested with varying interferent to CWA ratios to determine the ability of the CBMS II to correctly identify the agent as well as to avoid both false positive and false negative alarms.

Contractor testing of liquid CWA performance was conducted in 2002, followed by Government Tests later that year. The latter test confirmed the good sensitivity, response time, and cleardown time reported in the former test, but also revealed an unacceptable rate of false positive alarms with larger (microliter) volumes of interferents than were run in the Contractor Test. For this reason, the chemical agent detection/identification software was redesigned for much greater specificity. Basically the isolation and excitation of multiple CWA parent ions in a scan function was replaced with a single-ion isolation and excitation scan function for each agent. Ion isolation windows also were narrowed considerably, and scanning time was minimized by trimming scan ranges and using faster scan rates. These changes placed greater demands on, and required improvements in, the precision of the mass and frequency calibrations.

Both Contractor[26] and Government Tests[27] were repeated, and the improved selectivity and resistance to false positive and false negative alarms were verified. The actual CWA performance data are not available for public release, so only results for simulants can be discussed. Both the Contractor and Government Tests confirmed the ability of the CBMS II to correctly identify two commonly used CWA simulants, MES and DEM, applied to glass plates and wheels as part of the sampling process with the chemical ground probe. Two to 2.8 μg of MES (Contractor and Government Test results, respectively) and 5.0 to 5.1 μg of DEM were reliably identified on ground wheels and approximately one-tenth that mass on glass plates. The higher detection limits determined on the ground wheels was due to the fact that the ground wheels absorb and retain some agent in the sampling process. The CBMS II correctly identified MES at the detection limit in the presence of a 585-fold volume excess of common battlefield interferents such as JP8 and fog oil. DEM was identified in a 218-fold volume excess of these interferents. Seawater and DEET also were tested, with seawater giving similar results to fog oil and JP8; DEET yielded a slightly lower resistance to false negatives. The false positive rate was very low for tests with CWA mixed with the interferents and was zero for tests with volumes of interferents (only) up to 10 μl applied to the ground wheel. One of the decontamination reagents gave a single false alarm (alarmed for the simulant DEM) in tests of 2 μl volumes. This resistance to false positives and negatives is very hard to achieve, and particularly with interferents like JP8 and fog oil, which exhibit a peak at every m/z across the spectrum. By way of comparison, the MM1 tended to have slightly better absolute sensitivity but a far worse rate of false positive and false negative alarms. Furthermore the CBMS II cleared down much faster (a couple of minutes) from exposure to large volumes of interferents (e.g., 3 μl of neat fog oil applied on a ground wheel) and was ready much sooner to resume agent monitoring than was the MM1 (a couple of hours).

4.4.2 Field Tests of Liquid CWA Performance

Several field tests have been carried out in 2000 and in 2003 to 2004 with the CBMS II and the DWSS installed in both the JSLNBCRS HMMWV and Stryker NBCRV platforms. These tests focused primarily on most aspects of the vehicle performance, such as roadability, in addition to tests of the detector suite. In the DWSS-CBMS II detector system tests, the vehicle was driven over various types of terrain that would be encountered in a mission (e.g., primary and secondary roads and cross-country, off-road travel). Forty and 111 m strips of the terrain were sprayed with MES and DEM at 0.5 g/m^2 surface concentrations, and the ability was determined for the DWSS-CBMS II system to correctly alarm on the simulant while the vehicle was driven over the strips at various speeds. Interpretation of the test results was difficult because in the 2000 JSLNBCRS tests, the DWSS-CBMS II system gave a 100% correct alarm rate, but in the 2003 tests, the alarm rate varied from 19% to 100%, depending on the platform and terrain. It was observed that the DWSS failed to operate properly in many of the 2003 tests (e.g., the wheels were not properly contacting the probe head), which prevented the CBMS II from receiving any sample and alarming. It is important to note that after every field no-detect, the CBMS II was tested with a "confidence" (quality assurance) MES sample that was manually applied by the operator directly to the ground probe head, and that the alarm rate for these samples was 100%. This result indicates that the CBMS II was capable of detecting/identifying. A follow-up test was performed in the summer of 2004, after improvements were made to the DWSS.

Any problems in the setup, startup, and operation of the CBMS also were recorded in the course of the field tests. These observations are important to improving the system before it goes into production. The field tests revealed three problems. The protective screens over electronics cooling air inlets and outlets on the CBMS II housing were damaged by operator activities in the vehicles and required reinforcement. The ground probe head, which protrudes outside the vehicle hull, required more power to maintain the correct temperature under colder or wetter weather conditions. Finally the automated mass and frequency calibration procedure was not reliable in the field and required modification. These problems have been corrected and are being incorporated in the LRIP units.

4.4.3 BWA Laboratory and Field Performance Tests

As described in more detail elsewhere,[28–30] the CBMS II discriminates among and identifies BWA using biomarker ions derived from thermolysis/methylation of their cell walls and other components, such as membrane fatty acids ranging from 10 to 24 carbons in length. The ability of the latter to provide a means of differentiating BWA is illustrated by the spectra of four pathogens in Figure 4.4. (The identities of the pathogens cannot be revealed because they are classified.) The major ions at *m/z* of 243 and greater (but also including *m/z* 187) are mainly membrane fatty acids, and are clearly different among the different pathogens.

This performance was demonstrated in a proof-of-principle BWA contractor performance test conducted at ORNL early in 2002, and witnessed and reported[31] by staff from the Dugway Proving Ground. This test was conducted

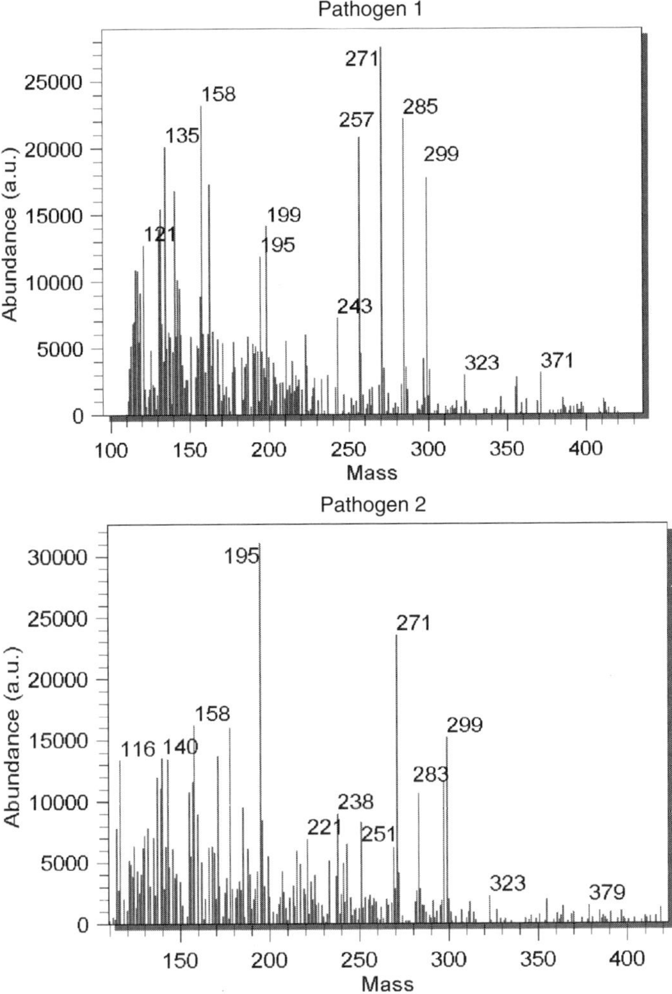

Figure 4.4 Full-scan CI mass spectra from the thermolysis/methylation of four pathogens.

in two parts. The ability of the CBMS II to discriminate among and correctly identify BWA target agents was tested by liquid injections made directly into the pyrolyzer tube. The biosampler was bypassed. The known numbers of bacteria or rickettsiae organisms and masses of crude virus samples (because viruses are not aerosolized in the pure form) and toxins were injected into the pyrotube, and the pyrocycle was run normally. Blank runs were made between sample sets and between individual samples to test for carryover and false positive alarms. There was only enough time to test one BWA simulant aerosol in the formal test, but a small number of gamma-killed bacteria also were

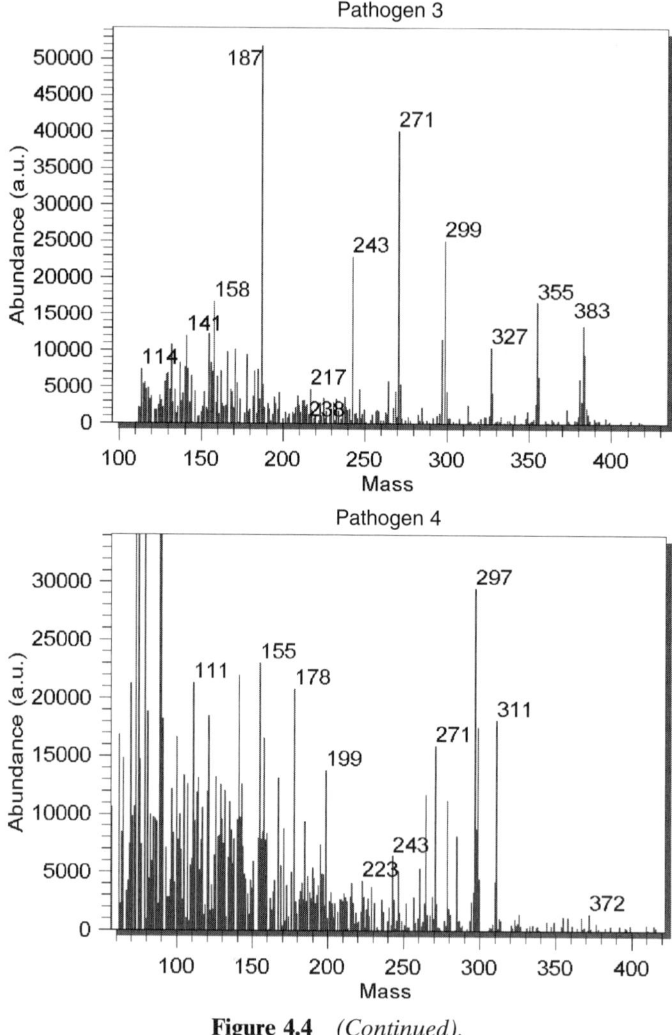

Figure 4.4 *(Continued).*

tested as aerosols after the formal test. The CBMS II was successful in distinguishing among and correctly identifying the BWA on the classified ITF6A (1990) list, plus a large number of variants, yielding between 20 and 30 BWA variants. The limited aerosol tests demonstrated that the 25 ACPLA specification can be met.

Since the proof-of-principle test, considerable improvements to the BWA monitoring method hardware and software were made by Hamilton Sundstrand Sensor Systems, and they are now being incorporated into LRIP units for the final phase of BWA method work. The BWA monitoring method and algorithm are being refined and the database expanded for final performance testing in the near future.

Although the CBMS II has not yet been subjected to formal field testing of BWA performance, units have participated in field tests. An early design PPU without an automated bioidentification algorithm was used in the Joint Field Trials-6, at the Defense Research Establishment Suffield, Alberta, Canada, in 2000. Although instrument reliability problems constrained the performance for much of the field trials, useful spectra were obtained from sampling disseminations of *Bacillus subtilis* var. niger (also called *Bacillus globigii*, BG) and *Erwinia herbicola* (EH) simulants at 20 to 30 ACPLA. Much was learned from the fielding experience, and that led to improvements in instrument reliability and stability. In 2003 the HSSS CBMS II unit with their internally funded improvements "piggybacked" on the JBPDS MOT&E at Eglin Air Force Base, Florida. The unit's reliability was excellent, with 100% availability for monitoring. It achieved a 50% probability of detection/identification for BG and EH at 15 ACPLA. Other simulants, ovalbumin (simulant for toxins) and the male specific coliphage, MS2 (simulant for pathogenic viruses), also were correctly identified. Two false positive alarms and two misidentifications were observed during highly concentrated agent disseminations, and are being examined for bioidentification algorithm improvements. Overall, these results are very good for a fieldable BWA identifier, and they indicate that the prospect is very good for successful completion of the BWA monitoring capability.

4.4.4 Nuclear Radiation Tests

Early in the program, critical components (e.g., the turbomolecular pump) and circuit boards were tested for their ability to survive neutron and gamma irradiation rates and doses similar to those that would be received from exposure to the detonation of a tactical nuclear device. All components were powered up at the start of the gamma irradiation tests but not during the neutron irradiation tests. Circuit boards were protected by circumvention circuits that powered down critical circuits in 10 to 100 μs upon detecting radiation. All components survived the nuclear radiation tests. This unusual performance was noted with positive commendations by the staff at the White Sands Missile Range, where the tests were performed. Tests of the fully integrated CBMS II system, installed in a reconnaissance vehicle, will be conducted in the future.

4.4.5 Environmental and Physical Tests

The CBMS II was exposed to the full temperature and supply voltage specification range in chamber tests and also to vibration tests several times during the development program. In extensive tests at HSSS in 2002, the CBMS II was equilibrated in chambers at ambient (+68°F/+20°C), cold (+32°F/0°C), and the upper operating temperature extreme (+120°/+49°C) with a low (20V DC), normal (28V DC), and high (31V DC) supply voltage. After correction of some problems encountered in the first testing, such as a failed rough pump cold start heater, a failed SDU power supply board, and a leaking elastomeric seal, the unit successfully started up and alarmed on a MES test pen in each test. It also successfully started up and alarmed after being equilibrated at the lower operating temperature extreme (−25°F/−32°C). The spectra of the calibration gas was confirmed as stable in tests over the range +32°F/0°C to +104°F/+40°C. The CBMS was turned off for the storage temperature extreme (−60°F/−51°C and +160°F/+71°C) exposures, following which the temperatures were raised or lowered to a point within the operating temperature range and the CBMS II was started up and tested using a MES test pen. The low-temperature tests revealed a formating problem with the extended temperature flash disks that was solved by using commercial temperature range disks from a different supplier. The startups were successful and the CBMS II alarmed on the MES. Three-axis vibration tests at ambient temperature also were run at HSSS. The vibration test profiles used mean power spectral densities derived from the FOX and JSLNBCRS HMMWV platforms operated on roads. No structural failures were observed, but the fastening of some cables, filters, and pneumatic hoses was found to be inadequate and was improved.

In the tests at the Edgewood Chemical Biological Center (ECBC), the CBMS II was exposed to vibration tests at ambient and reduced temperature (+39°F/+4°C) temperature. No damage was observed in ambient temperature/transverse axis of vibration tests. However, some damage to the shock mounts was noted after the ambient temperature/longitudinal axis of vibration test using the FOX vehicle vibration profile. After replacing that vibration profile with a newer vibration profile based on the Stryker vehicle and running tests at reduced temperature, no damage was observed. Tests of EMP and EMI at other government labs revealed a problem with external cable connections that is being investigated; so far it appears to be only a simple error in how those particular cables were fabricated. Finally maintenance and repair data for laboratory and fielded units indicate that the mean time between hardware mission failures specification will be met.

4.5 CONCLUSIONS

The CBMS II is in LRIP by HSSS, and will be fielded in about two years. At that point the CBMS II should be fully certified by the government as a detector/identifier for liquid and vapor CWA, TICS/TIMs, NTA, and BWA. No other integrated system has or will have such a wide range of detection/identification capabilities. It will greatly improve the protection of the military against weapons of mass destruction as well as industrial contaminants. It has obvious applications to homeland security as well.

ACKNOWLEDGMENTS

The authors gratefully acknowledge the dedication and hard work of their colleagues in the CBMS II Program Team at ORNL, HSSS, the Colorado School of Mines, and MSP Corporation, and the support and encouragement of the Army program sponsors at the Office of the Joint Program Manager, Nuclear, Biological, and Chemical Contamination Avoidance at the ECBC. This research was sponsored by the US Army Office of the Joint Program Manager, Nuclear, Biological, and Chemical Contamination Avoidance, DOE No. 2182-K011-A1, Department of Energy, under contract DE-AC05-00OR2275 with Oak Ridge National Laboratory, managed and operated by UT-Battelle, LLC.

REFERENCES

1. Sidell, F. R.; Takafuji, E. T. (Eds.) *Medical Aspects of Chemical and Biological-Warfare.* Falls Church, VA: Office of the Surgeon General, Department of the Army, 1997.
2. *Medical Management of Biological Casualties Handbook.* Fort Detrick, Frederick, MD: US Army Medical Research Institute of Infectious Diseases, 1996.
3. Braddock Smith Group, Inc., *NBC Products and Services Handbook.* Great Falls, VA, May 2004. Available online at http://www.nbcindustrygroup.com/handbook/inex08.htm.
4. Watson, J. T. *Introduction to Mass Spectrometry.* New York: Raven Press, 1976.
5. Rostker, B., Special Assistant for Gulf War Illnesses. *Information Paper. The FOX NBC Reconnaissance Vehicle.* Department of Defense. Available at www.gulflink.osd.mil/foxnbc/index.html. Also, *PM Recon M93A1 Block I Modification Fox Nucear, Biological, Chemical Reconnaissance System (NBCRX).* Joint Program Executive Office, Chemical Biological Defense. Currently available at http://www.jpeocbd.osd.mil/ca fox1.html.
6. *Chemical Biological Mass Spectrometer.* At Ref. 5, p. 72.
7. Todd, J. F. K.; March, R. E. (Eds.) *Practical Aspects of Ion Trap Mass Spectrometry.* Boca Raton, FL: CRC Press, 1995, Vols. 1–3.
8. *PM Bio Det Biological Integrated Detection System.* Joint Program Joint Program Executive Office, Chemical Biological Defense. Currently available at http://www.jpeocbd.osd.mil/ca_bids.htm.
9. Busch, K. L.; Gishe, G. L.; McLuckey, S. A. *Mass Spectrometry/Mass Spectrometry: Techniques and Applications of Tandem Mass Spectrometry.* New York: VCH Publishers, 1988.
10. Franzen, J.; Gabling, R. H.; Schubert, M.; Wang, Y. Nonlinear ion traps, Chap. 3. In *Practical Aspects of Ion Trap Mass Spectrometry*, Todd, J. F. J.; March, R. E. (Eds.), Boca Raton, FL: CRC Press, 1995, pp. 49–168.
11. First, J. P.; Dworzanski, L.; Bernwald, W. H.; McClennen, W. Mechanistic aspects of the pyrolytic methylation and transesterification of bacterial cell wall lipids. *J. Anal. Appl. Pyrolysis* 1991, **21**, 221–232.

12. Basile, F.; Voorhees, K. J.; Hadfield, T. L. Microorganism Gram-type differentiation based on pyrolysis mass-spectrometry of bacterial fatty-acid methyl-ester extracts. *Appl. Environ. Microbiol.* 1995, **61**, 1534–1539.
13. Basile, F.; Beverly, M. B.; Abbas-Hawks, C.; Mowry, C. D.; Voorhees, K. J.; Hadfield, T. L. Direct mass spectrometric analysis of in situ thermally hydrolyzed and methylated lipids from whole bacterial cells. *Anal. Chem.* 1998, **70**, 1555–1562.
14. Sasser, M. *Identification of Bacteria by Gas Chromatography of Cellular Fatty Acids.* Technical Note 101, MIDI, Inc., 2001. Available at http://www.midilabs.com.
15. The sampler is shown in Ref. 7.
16. Palausky, M. A.; Lammert, S. A.; Merriweather, R.; Sarver, M. B.; Sarver, E. W. A field portable ion trap for the detection of chemical weapons compunds. *Proc. 43rd ASMS Conf. on Mass Spectrometry and Allied Topics*, Atlanta, GA, 1995.
17. The primary funding agencies for the DSITMS development were the US Department of Energy, Office of Technology Development, and the US Army Environmental Center (DOE No. 1769-F054-A1).
18. Wise, M. B.; Guerin, M. R. Direct sampling MS for environmental screening. *Anal. Chem.* 1997, **60**, 26A–32A.
19. EPA SW-846 Method 8265, *Volatile Organic Compounds in Water, Soil, Soil Gas, and Air by Direct Sampling Ion Trap Mass Spectrometry*. Electronic version of EPA SW-846 Manual available at http://www.epa.gov/epaoswer/hazwaste/test/new-meth.htm#8265.
20. Hunter, E. P. L.; Lias, S. Evaluated gas phase basicities and proton affinities of molecules: An update. *J. Phys. Chem. Ref. Data* 1998, **27**, 413–457.
21. Kelley, P. E., US Patent 5,134,286 (1992).
22. Romay, F. J.; Roberts, D. L.; Marple, V. A.; Liu, B. Y. H.; Olsen, B. A. A high-performance aerosol concentrator for biological agent detection aerosol. *Sci. Technol.* 2002, **36**, 217–226.
23. *Styker 8-Wheel Drive Armored Combat Vehicle.* Defense Industries, Industry Projects. Available at http//www.army.mil/fact_files_site/stryker/. *PM Recond, Sensor Suite for the XM1135 Nuclear Biological, Chemical Reconnaissance Vehicle (NBSRV) Stryker.* Joint Program Executive Office, Chemical Biological Defense. Currently available at http://www.jpeocbd.osd.mil/ca nbcrv.htm.
24. *PM Recon, Joint Services Lightweight NBC Reconnaissance System (JSLNBCRS).* Joint Program Executive Office Chemical Biological Defense. Currently available at http://www.jpeocbd/osd.mil/ca jslnbcrs.htm.
25. *Industrial Chemical Prioritization and Determination of Critical Hazards of Concern.* Technical Annex and Supporting Documents for International Task Force (ITF)-40, Industrial Chemical Hazards: Medical and Operational Concerns. UASCHPPM Report 47-EM-6154-03. US Army Center for Health Promotion and Preventive Medicine, Aberdeen Proving Ground, MD, 2003.
26. Luo, S.; Andersen, D.; Mohr, A. J. *Abbreviated Test Report (Part II) for the Chemical/Biological Mass Spectrometer (CBMS) Block II Chemical Test.* DPG Document No. WDTC-TR-03-081, West Desert Test Center, US Army Dugway Proving Ground, Dugway, UT, 2003.
27. Woolery, D. O.; Thornton, K.; Lian, N. D.; Hunter, S. A. *Abbreviated Test Report for the Chemical Performance Retest of the Chemical Biological Mass*

Spectrometer Block II (Phases 2 and 3) Liquids. WDTC Document No. WDTC-TR-04-002, West Desert Test Center, Army Dugway Proving Ground, Dugway, UT, 2004.

28. Griest, W. H.; Wise, M. B.; Hart, K. J.; Lammert, S. A.; Thompson, C. V.; Vass, A. A. Biological agent detection and identification by the Block II chemical biological mass spectrometer. *Field Anal. Chem. Technol.* 2001, **5**, 177–184.
29. Hart, K. J.; Wise, M. B.; Griest, W. H.; Lammert, S. A. Design, development, and performance of a fieldable chemical and biological agent detector. *Field Anal. Chem. Technol.* 2000, **4**, 93–110.
30. Barshick, S. A.; Wolf, D. A.; Vass, A. A. Differentiation of microorganisms based on pyrolysis ion trap mass spectrometry using chemical ionization. *Anal. Chem.* 1999, **71**, 633–641.
31. Luo, S.; Andersen, D.; Mohr, A. J.; *Abbreviated Test Report for the Chemical Biological Mass Spectrometer Block II Biological Agent Performance Test*, DPG Document No. WDTC-TR-01-145, West Desert Test Center, U.S. Army Dugway Proving Ground, Dugway, UT, February 2002.

5

METHOD REPRODUCIBILITY AND SPECTRAL LIBRARY ASSEMBLY FOR RAPID BACTERIAL CHARACTERIZATION BY METASTABLE ATOM BOMBARDMENT PYROLYSIS MASS SPECTROMETRY

Jon G. Wilkes, Gary Miertschin, Todd Eschler, Les Hosey, Fatemeh Rafii, Larry Rushing, Dan A. Buzatu, and Michel J. Bertrand

National Center for Toxicological Research, Food and Drug Administration, Jefferson, AR 72079

5.1 INTRODUCTION

Rapid sub-typing of bacteria is needed for protection of public health and in civil-, criminal-, or terror-related forensics. Distinction of microbiological sub-types can signal important differences that affect the health risk from microbial infection and treatment strategies for disease. It can also be used to monitor the emergence of mutant strains.[1] In cases of nosocomial (hospital-incurred) infections and outbreaks, sub-typing capability could be used as an alternative for identifying the route by which infection spreads. Many studies

Identification of Microorganisms by Mass Spectrometry, Edited by Charles L. Wilkins and Jackson O. Lay, Jr.

from the early 1990s used pyrolysis mass spectral "fingerprint" patterns for that purpose.[2–11] Rapid methods for sub-typing analysis are crucial in clinical contexts, particularly to distinguish strains with respect to antibiotic resistance, acid resistance, toxin production capability, and other qualities that confer pathogenic character. During the last third of the past century, analytical chemists have periodically visited or revisited the potential use of mass spectrometric methods for rapid characterization of microbiological samples.[12–20] These studies and others have investigated a variety of sample introduction and ionization approaches. Still others have also surveyed a variety of data analysis techniques, including multilinear and nonlinear pattern recognition methods.[21–24] Results typically showed that over a limited time, a mass spectrometric instrumental and data analysis system could assess samples of whole cells or cell extracts and detect subtle differences among similar strains. Before a recent innovation these efforts did not yield a database of "standard" spectra that could be employed for routine analysis in an instrument-independent fashion.

Waters Corporation has now introduced the MicrobeLynx™, a MALDI-TOF MS system developed for "fingerprint bacterial ID." The system and its applications are described thus:

> The networked MALDI mass analyzer coupled with MicrobeLynx bioinformatics technology form the basis of this automated bacterial 'mass-fingerprinting' approach to speciation and typing of microorganisms. Macromolecules expressed on the surface of bacteria are sampled and characterized by molecular weight. The resulting mass spectrum provides a unique fingerprint for the species tested. Bacterial mass fingerprints of unknowns can be reliably matched against a database of quality-controlled reference spectra. Suggested applications include medical microbiology, and the food, water, pharmaceutical and biotechnology industries. (http://www.micromass.co.uk)

Using a different MALDI-TOF MS instrument, we have confirmed that species-level distinctions are possible. Recent discussions with Dr. Diane Dare of Waters Corporation indicated that speciation, rather than more subtle distinctions, remains the goal for the MicrobeLynx system.[25] We continue to investigate MALDI-TOF MS, as well as pyrolysis MS, for rapid characterization of bioterror agents. For such an application, or for susceptibility-informed antibiotic treatment of infections, there is a need for specificity below the species level. Bacteria of the same species can have very different pathogenic potentials and drug susceptibilities. The present state of the art for MALDI-TOF MS characterization appears not to support this degree of specificity. So we are continuing to investigate the MS "mass fingerprinting" concept using pyrolysis sample introduction, a variation of the same approach that proved useful for infraspecific strain comparisons during the 1990s.

The major issues or questions affecting the practical applicability of mass spectrometric approaches can be listed:

1. *Inherent MS information content* (ionization mode dependent).
2. *Spectral pattern reproducibility* (ionization mode dependent).
3. Ability to *explain the biological meaning* of mass spectral features responsible for class distinctions (ionization mode dependent).
4. *Computational methods* flexible enough to reflect non-linear relationships in the data structures, rugged enough for reliable classification of unknowns, and usable by technically trained, nonresearch personnel.
5. *Sample preparation*, because traditional culture methods require days of resuscitation, selection, and isolation prior to analysis, and so compromise the advantage of MS rapidity.

The Waters Corporation MALDI MS system successfully addresses several of these issues but appears to be weaker than PyMS in respect to spectral information content and pattern reproducibility, respectively issues 1 and 2. The MALDI spectra do not consistently provide unique biomarkers for distinguishing similar strains. Relative ion intensities by MALDI MS, which reflect differences in biomarker expression levels, have not proved reproducible and, except for the Waters system, have been typically omitted as a part of the pattern. Jarman et al. even advocated ignoring not only the intensity dimension of all ions and also excluding certain biomarker ions in the "fingerprint" if for any reason those ions were not consistently picked in replicate analyses.[26] They define sample identity by matching the combination of observations for each species, using only those markers identified as reliable by statistical comparisons among the replicate spectra. While this does give reliable certainty for grouping spectra, it cannot, by definition, be used to distinguish strains in which the difference is only protein expression level.

This chapter presents an integrated instrumental and computational system capable of addressing most issues listed above but with only partial success for criterion 3, the ability to correlate biological meaning with mass spectral features. We assume that microbiologists and clinicians would be willing to utilize infraspecific inferences from a rapid, reliable fingerprinting system, even without identification of all biochemical constituents giving rise to distinctive spectral features.

Analysis via the integrated system involves eight major steps:

Step 1. Automated, accelerated bacterial pre-enrichment, enrichment, and selective enrichment.

Step 2. Instrumental isolation out of liquid culture media of individual cells.

Step 3. Automated cell washing and cell suspension normalization to standard optical density.

Step 4. Automated sampling and application of cell suspension to a pyrolysis probe tip.

Step 5. Automated pyrolysis MS using a novel, highly reproducible ionization mode and a TOF mass analyzer.

Step 6. Spectra averaged across the entire program peak.

Step 7. Spectra normalized with respect to total ion intensity and any remaining elements of instrumental or sample preparation drift.

Step 8. Spectra classified using an artificial neural network pattern recognition program. (This program is enabled on a parallel-distributed network of several personal computers [PCs] that facilitates optimization of neural network architecture).

The following instrumentation and specialized equipment comprise the integrated system:

- A robot that dispenses culture media, solvents, cell suspensions, antimicrobial agents, or other liquid reagents into 96-well microtiter plates.
- A centrifuge equipped with a rotor to hold microtiter plates.
- A UV-Vis spectrometer that reads optical density in individual wells of a microtiter plate.
- A small microbiology incubator.

The preceding four instruments are used for steps 1 and 3.

- A flow cytometer equipped with forward-scattering and side-scattering detectors and a sorting option that can distinguish cell size-and-shape, sorting specified cells of 1 to 3 μM length into small volumes of culture broth in individual plate wells. (This instrument is used for step 2 and in another mode may contribute to step 1.)
- An autosampler using tapered mini-sample vials in a 54-well format. (Used for step 4)
- An autoprobe for sample introduction of pyrolysis probe into vacuum. (Used for the first half of step 5)
- A reflectron TOF MS equipped with a metastable atom bombardment ion source. (For steps 5 and 6)
- Software designed to normalize spectra to database standards. (Step 7)
- Pattern recognition software operating, in some cases, by distributed calculations over a network of high-performance PCs. (Step 8)

Together these subsystems can give rapid distinction of bacteria at near strain level after a single working day's operation, meaning less than eight hours from the time a sample arrives at the lab. The equipment is expensive. However, the cost per analysis is low due to automation and rapid operation. Enormous benefits could obtain from the availability of such integrated, automated systems for routine analyses in clinical or public health laboratories.

5.2 SAMPLE PREPARATION FOR RAPID, REPRODUCIBLE CELL CULTURE

Operation of the instrumental ensemble for rapid sample preparation is summarized as a flowchart in Figure 5.1.

5.2.1 Liquid Robotics

A food sample presented for routine microbiological screening assay would be blended or rinsed to provide suspended cells in liquid. The cells would then be inoculated into a liquid resuscitation, pre-enrichment or enrichment medium, with optional shaking to maintain aerobic conditions during incubation. The analogous process using the Biomek liquid robot (Beckman Coulter, Fullerton, CA) involves sorting such suspensions or washes into 100 μl of the same culture medium inside one well of a sterile 96-well microtiter plate. This is followed by a short period of incubation until some of the cells have begun to multiply (beyond lag phase).

Standard culture procedure normally requires 24 to 48 hours for sporulation and bacterial growth so that dormant or distressed-but-viable cells have passed out of lag phase. For the automated microculture system we describe to as few as two hours (this time can be significantly reduced), because the minimum number of growing cells required for further manipulation is small, the liquid volumes are small, and the optimum suspension concentration needed for single cell sorting (Figure 5.1, blue arrow flow path) is very dilute. We estimate only 6000 growing CFUs per well are necessary, assuming 100 μl of media sampled at 1 μl/min to give a cytometer "event" count of 1/s. Because of the shallow depth of media in each well, aerobes have ready access to air, so no shaking is required during incubation. Analysis of a sample for anaerobic species would use a separate microtiter plate containing appropriate media and incubated under standard anaerobic conditions.

The Biomek can be used in the same physical configuration (but using different selection parameters from the sorting operations described in Section 5.2.3) to accomplish several tasks (pre-enrichment, enrichment, selective enrichment, or inhibitor-based back-selection), each differing with respect to liquid growth media formulation, inclusion of antimicrobial agents, and so forth. These may be accomplished in parallel by cell selection and addition to wells filled with the different formulations, indicated in Figure 5.1 by varying colors on the microtiter plate symbols. Alternatively, a series of such operations can be conducted iteratively as indicated by the red arrow portion of the diagram. To protect personnel from sample exposure and the sample's integrity from environmental contamination, the Biomek apparatus is located inside a BSL-2 safety cabinet.

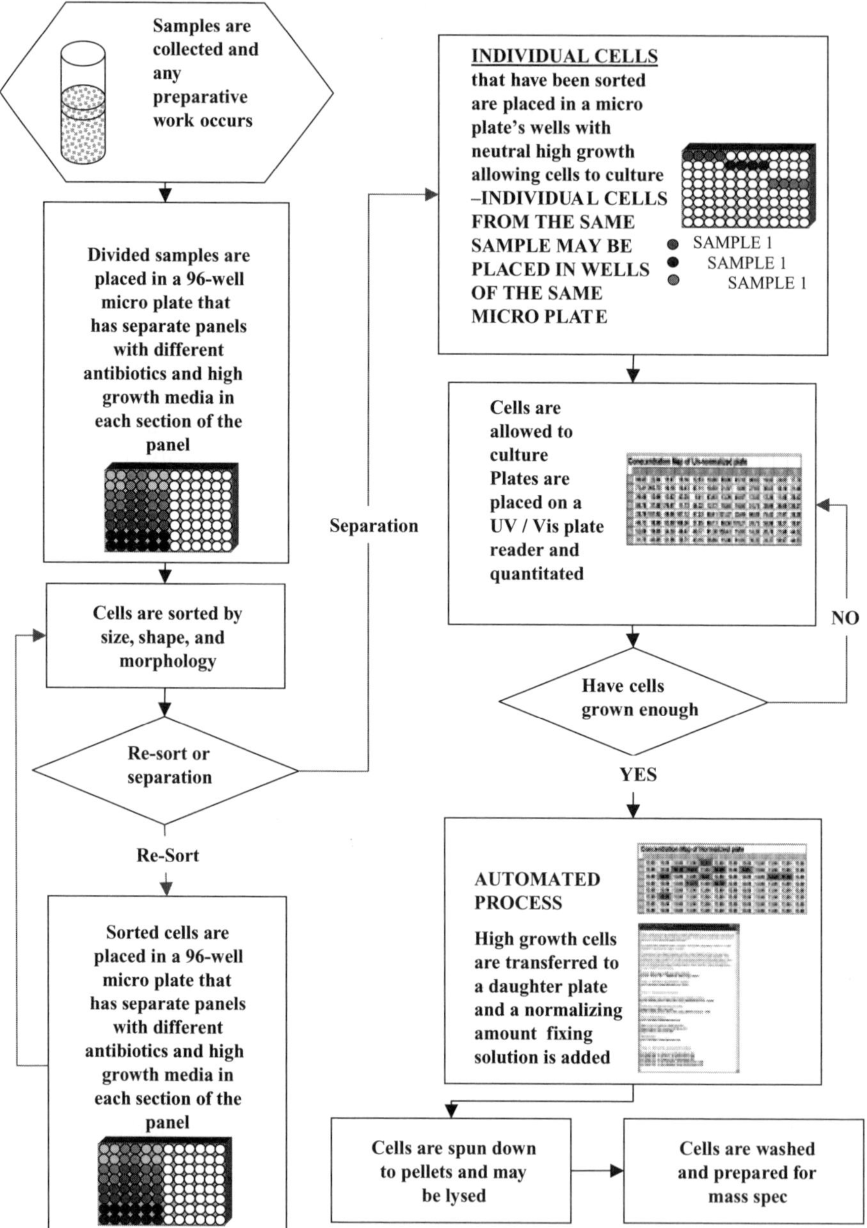

Figure 5.1 Flow diagram of the automated sample preparation and instrumental cell isolation equipment.

5.2.2 Large Batch Centrifugation

Comprehensive investigation of the microflora in a single environmental, food, or complex clinical sample (e.g., a stool sample) could generate a large number of diverse mixed cultures, each possibly requiring additional growth steps before isolation. Cultures in microtiter (96-well) plates rather than flasks have been specified so that standard operating procedures developed for this instrumental suite are compatible with automated, production-scale analysis. During the transfer of cells between different media for re-culturing, it is necessary to separate out the cells of interest from the liquid. In most cases, the best way to re-culture is by serial centrifugation, decanting, and resuspension. The process involves repeatedly transferring 96-well microtiter plates between the liquid handling robot, flow cytometer, incubators, and centrifuge. If records are transferred automatically, sample identity and cell culture history can be tracked efficiently through this process and passed on to the subsequent mass spectrometric and pattern recognition subsystems. Synchronized batch mode operations and record keeping are compatible with the goal of production analysis. These performance criteria are met using the computer interfaces of the Biomek, the Altra flow cytometer, the microtiter plate optical density reader, and those in the other subsystems.

In the final step before MS analysis, cells are transferred from standard growth medium to a fixing suspension. Batch mode cell washing and fixing also involves serial centrifugation and decanting. All of the liquid exchange processes are enabled with an Allegra 6 Tabletop Centrifuge (Beckman-Coulter, Fullerton, CA) fitted with a rotor designed to hold up to 6 microtiter plates. Other than the microtiter plate rotor, the application does not require specialty features of the centrifuge. To ensure sample integrity, sterilized microtiter plates are used, and these are always handled by aseptic technique and covered during transfers between the centrifuge, scanner, or other instruments.

5.2.3 Forward and Side Scatter Flow Cytometry

The use of flow cytometry for manipulation of viable bacteria depends on the technical capacity to sort cells by their shape, size, and "granularity" (meaning intracellular structure). Bacterial cells have much smaller average size than the mammalian cells for which flow cytometry was first developed. Fluorescent or other dyes could be added to enhance sensitivity for detection of small cells. However, it is preferable that this strategy not be used because there is the possibility that such a chemical label might either poison the cells (stressing them and decreasing their viability) or else might not have equal affinity for all bacteria. Currently flow cytometry is used with high-velocity sorting of *Mycobacterium tuberculosis* at the NIH's Vaccine Research Center, NAID, Bethesda, MD.[27]

The following general descriptions reference the particular instrumental and methodological features of the Beckman-Coulter EPICS Altra Flow Cytometer. Other manufacturers' products have analogous features, though some of the cell sorting and sample labeling characteristics of the EPICS Altra are particularly suited for this purpose.

Flow cytometers operate by passing cells in liquid suspension through a capillary channel in an optically clear quartz flow cell through which laser light is passed, reflecting off of the entrained cells. The quartz flow cell is vibrated up and down, which breaks the liquid stream into fine droplets after it exits into the air. With proper calibration, it is possible to determine exactly which downstream droplet contains a biological cell flagged by its scattered light signature while passing through the flow cell. Upon identification, the droplet with its single cell can be electrically charged and steered (using deflection plates at ~6000 V) out of the stream of nonselected droplets into a designated well in a 96-well microtiter plate.

Because single cell sorting is a crucial process in our method, the following description details the sorting mechanism. It is now a validated procedure. A designated droplet can be charged via the electrolyte stream so long as it retains even the slimmest physical connection. Timing and component locations are adjusted so that, after a calibrated delay, the last connected droplet contains the previously detected, single bacterial cell of interest and also is located directly opposite a "charge collar." When sort criteria are met, the system waits the designated delay time—for the cell inside the droplet to reach this location. That droplet is then electrically charged by a low-voltage pulse on the charge collar so that the downstream high-voltage deflection plates will steer it away from the stream of noncharged droplets (which go on to waste) and along a path that propels it toward a target. The particular well occupying the target location is software/actuator controlled using a stepper-motor-controlled X–Y table. Sample identity is stored in relation to the sort criteria that defined the droplet's selection, the well's media content, and all other experimental information.

Figure 5.2 shows a typical droplet stream with the last connected droplet indicated by a thin, horizontal cursor line and the stream charge collar shown as dark rectangles on each side of the droplet stream. In this view the stream flows downward and the quartz flow cell and laser optics are located out of the frame, above.

The newest flow cytometers have an optional forward-scattering detector that exhibits sensitivity for detection of very small suspended particles without depending on attachment of fluorescent markers, as in earlier technology. They can detect individual bacterial cells suspended in the stream. With an option known as pulse-pileup (PPU, discussed below in more detail), it is possible to differentiate an incipient droplet (an aliquot of the liquid flow in the cell that will become a single droplet) containing only one suspended cell from a droplet containing two or more cells. This second capability allows one to sort a single viable cell into a designated micro-titer plate well containing liquid

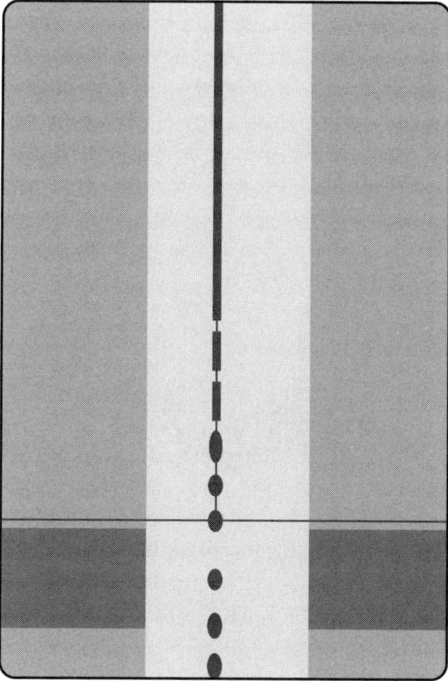

Figure 5.2 Synchronized, strobe-illuminated photograph showing the flow cytometer's liquid stream breaking into droplets that separate just as the stream enters a "charge collar."

culture medium to grow it out. For cocci and other cells that tend to agglomerate in suspension, this capability assures that the resulting culture is pure.

The shape and granularity of cells is reflected in the intensity profile of side-scattered light. If the cells are spheroidal, as are cocci, all signal intensity appears in a relatively narrow band reflecting the average size and granularity of the spheres. If the cells are elongated, as are rods, the side scatter profile reflects the probability distributions for size and granularity as they depend on cell orientation at the instant laser light scatters off of it. This phenomenon explains the much broader side-scatter peak for rods compared to similar sized cells having a spheroidal shape. This general rule governs the plot appearance for bacterial cells of all sizes and shapes.

The flow cytometer, fitted with both forward and side scatter detectors, generates a 2D plot indicating the distribution of light intensity, forward scatter (FS) versus side scatter (SS), and showing the physical profile of the particle responsible for the scatter. Figure 5.3 is such a plot for a mixture of five rod-shaped bacteria from our laboratory. The different strains appear in partly separated clusters (indicated by square boxes and dot color) along the side-scatter axis in the lower part of the plot.

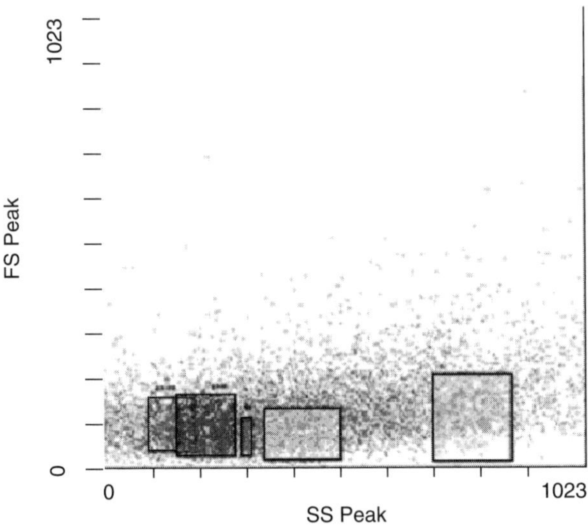

Figure 5.3 2D FS versus SS plot for a mixture of five rod-shaped bacteria.

From left to right the five strains are *Salmonella enterica* serovar Typhimurium, *Mycobacterium frederiksbergense*, *Corynebacterium* sp., *Flavobacterium* sp., and *Mycobacterium* [sp.] *PYR-100*. The *Flavobacterium* and *Salmonella* species are Gram-negative rods; the others are Gram positive. The *Mycobacterium frederiksbergense* are very small, short rods that produce reduced side-scatter range (data not shown) compared to the other Gram-positive cells. The location of these boxes was determined in the same session by running an authentic standard suspension of each cell type used to make up the test mixture.

Any signals appearing in the upper 75% of the plot would arise from larger nonbacterial particles or cell agglomerates, neither of which are of interest for cell isolation. The portion of the total particle population to be sorted is determined by placing the identification "box" in the desired place and specifying a desired number of "counts" within that box. Note that the region designated by the "box" can be made very small and located to discriminate against other unwanted cell populations. Thus an individual cell (count equals 1 with PPU option selected) selected in a high-purity sort (very small box) will itself have been derived from an elite subset of the detected particulate population.

For cell isolation purposes an important cytometer sort feature is "pulse-pileup" also referred to as "peak-pileup" or PPU. This recognizes "split" peak intensities arising from two or more cells in the same droplet that are strung together in chains or that coincidentally partially eclipse the laser beam. The two split peaks would each have the same forward- and side-scatter intensity as a single cell, but the hypothesized situation would not represent a desired droplet. The two cells might not be identical, so a mixture rather than an isolate would be obtained by culturing from that droplet. When split peaks occur, the cytometer's cell sorter, using the PPU option, can be set to reject the signal so the particular droplet is not diverted into a well for further growth.

It is possible to distinguish vegetative cells from spores of identical size and shape based on difference in granularity (side-scatter intensity). The 2D FS versus SS plot can therefore be used to confirm the absence of spores in a cell suspension, an important factor in method development to assure the mass spectrometers never become contaminated with live pathogens. Use of vegetative cells also eliminates the lag phase associated with spore germination and decreases subsequent incubation time for the culture to attain required optical density.

In cases where a target cell type's light-scatter distribution partially overlaps that of background strains, a pure culture of the strain can be obtained by sorting based on criteria at the edge of the target region farthest from background signals. This way the resulting isolate is most likely to comprise cells of interest. The boxes shown in the 2D plot in this case do not represent the total distribution of each cell type but a portion reflecting selection criteria for each strain in the presence of the other four mixture components. Several of the distributions overlap, especially the two on the left. The aforementioned calibration plot of pure suspension (each containing only one of these rod-shaped species) would have shown a horizontal, rectangular distribution (because of the rod shape). This distribution reflects in the plot rectangle an aspect ratio that corresponds qualitatively to the aspect ratio of the rod-shaped cells: long, narrow rods that have a much broader SS than FS distribution.

The fortuitous location in the 2D plot of *Salmonella typhimurium* to the far left of the other four cell types suggests the possibility of a cytometry-based sequential sample handling protocol, first involving instrumental pre-enrichment, then a short period of resuscitation or selective enrichment, and then one-cell-per-well sorting followed by rapid growth for strain determination. Infectious outbreaks have occurred from mixtures of a few *Salmonella* sp. cells (e.g., 10^2 cfu/ml) among a much larger population (10^8 cfu/ml) of competing background microflora.[28] The strategy outlined above might permit a major improvement in the early detection and characterization of pathogenic bacteria from contaminated food.

Sometimes no target cell type can be defined in advance, perhaps in a case of bioterror attack where analysts examining rinses from a salad bar lack a normal ecological context. In this scenario the pathogenic strains probably will be found in very high concentrations, so the challenge is not that of finding a needle in a haystack but of recognizing an unanticipated haystack. Very general sort criteria can be established that encompass the range of cell morphology (a wide, narrow box in the bottom 10% of the FS versus SS plot, thus excluding nonbacterial particles) and a one-cell-per-droplet count with the PPU specification retained. To address the intermediate situation of culturing isolates from a pathogen/nonpathogen mixture, the operator could fill multiple microtiter plates, one cell per well, with a rich nutrient medium that supports rapid growth of many known pathogens. This would improve the probability of growing and subsequently detecting an unanticipated pathogenic agent.

Several publications support aspects of the proposed sorting procedure.[29–37] These and many other papers involving bacteria and flow cytometry have concentrated on the use of fluorescent dyes to assess the proportion of viable cells or to provide other population data. Experiments with cell sorting have typically involved high-speed work aimed at enrichment of a particular strain relative to others in a mixture. In that case the absolute purity of the result was of lesser priority than rapid enrichment. However, Darvey and Kell's 1997 review[30] pointed out that recent advances in reliability and operational ruggedness now enabled technicians to operate flow cytometers for single cell characterization studies. These facts support the rationale for such sophisticated instruments applied to the challenge of rapid bacterial culture and isolation.

Since mass spectrometers are used for cell characterization, a cluster of safety and maintenance issues must be addressed. The MS vacuum systems generate aerosols that in counter-bioterror or other applications might become contaminated with viable spores of pathogens. Safety and maintenance issues might not obviously relate to performance (sample preparation and characterization) questions, but they actually do in this case. Performance can be effectively addressed by reproducible automated culture methods. For spore-forming bacteria, the requirement for MS fingerprint reproducibility implies the analysis of either spores or vegetative cells, but not of an undefinable mixture of the two. Safety and maintenance issues can both be addressed if the sample preparation is terminated by flow cytometry, either for isolation/assay or for rapid sorting, because the cytometer can distinguish spores from vegetative cells, sorting out the latter from the former so that spores are never introduced into the mass spectrometers.

The last step in preparation of spore-formers before analysis would be introduction of a counted number of vegetative cells into a small volume of 70% ethanol. This solution does not kill spores but does inactivate vegetative cells, rendering them safe for introduction into the mass spectrometer. At 2000 counts/s, in 4 minutes the EPICS Altra can sort any spores to waste and also dispense enough vegetative cells for Py-MAB/TOF MS analysis into 50 μl of the ethanolic fixing solution. This step of quality assurance/safety may not be necessary once rapid culture methods are fully developed because flow cytometry can also determine whether a significant number of spores form after only five or six hours of iterative division in a rapid growth, liquid culture medium.

5.3 ANALYTICAL INSTRUMENTATION FOR SENSITIVE DETECTION AND SPECTRAL REPRODUCIBILITY

5.3.1 Autosampler and Autoprobe

The analytical instrumentation requires automated sample handling and introduction not only for efficient batch operation and safety but also for achieving reproducible results acceptable for building or consulting a spectral library. The first element of this subsystem involves an LC•PAL autosampler (CTC Analytics, Zwinger, Switzerland) conjoined with the second, an Autoprobe (Scientific Instrument Services, Ringoes, NJ 08551) engineered for unattended sample introduction, and programmed pyrolysis inside the MAB/TOF mass spectrometer (Dephy Technologies, Montreal, Canada).

The LC•PAL autosampler collects liquid solutions from sealed vials in a 9 × 6, 54-vial layout. It can be programmed to stir the sample prior to uptake, a critical feature for this application. The autosampler stirs by rapidly and repeatedly sampling from and ejecting back into the sample vial a user-definable volume. This mixing volume is limited at the upper end by the volume of the sampling syringe, typically no more than 10 μl to ensure accurate dispensing of 0.5 μl during analysis.

Bacterial suspensions of interest from the Biomek or Altra (50 μl each) are transferred into sealed 2 ml vials for safe transport out of BSL-2 containment to either storage or the analytical system. The vials are fitted with 150 μl conical inserts. These are placed in numbered wells of the 54-vial plate, which is then positioned on the autosampler stage. Awaiting analysis, cells settle out of suspension to the bottom. Five rapid 101 μl uptake/ejection cycles (100 μl/s) are conducted within the 50 μl total sample volume. During this operation the tip of the autosampler syringe is located near the bottom of the conical insert, and this turbulent stirring effects re-suspension of cells. Immediately thereafter but slowly (at 2 μl/s), the syringe is filled with 0.5 μl of the suspension. This volume is translated to the pyrolysis probe and even more slowly (200 nl/s) injected onto the autoprobe's coiled platinum wire tip.

The SIS Autoprobe begins the process of sample insertion through a vacuum lock. After the first stage of lock penetration, the tip is warmed (100 mA) for 60 seconds to accelerate drying of the 0.5 μl sample suspension. After drying, rough vacuum is drawn on the probe tip until the inner port of the lock can be safely opened. Then the probe is inserted directly into the MS ion source, programmed pyrolysis is initiated, and MS data acquisition begins. After pyrolysis is complete, the autoprobe retracts the tip from the ion source. Before exiting high vacuum, the tip is heated to high current (1400 mA) for 10 seconds to evaporate semi-volatile sample residues and prime the wire surface for the next sample. The process continues with probe withdrawal back through the vacuum lock and finishes with the wire tip in sample loading position. The programming code for all these steps is written into and controlled for batch operations by the Dephy MAB-TOF MS data acquisition software.

5.3.2 Pyrolysis and Metastable Atom Bombardment (MAB) Ionization

Upon sample insertion into the ion source, pyrolysis is initiated by passing current through the probe tip, a linear ramp of 20 mA/s from an initial current of 500 mA to a maximum of 1000 mA, which is then sustained for a period of 50 seconds to complete vaporization of semi-volatile constituents. Sample deposition as a thin layer on a low-heat-capacity wire tip allows rapid, consistent heat transfer to the analyte. This produces a sharp and sudden outgassing of pyrolysates from the wire surface giving a single, narrow, tall peak in the total ion pyrogram. Since vapor is evolved directly inside the ion source, without a band-broadening transfer line, the favorable profile is retained. It is favorable because it increases detection sensitivity: the majority of components evolve and are detected within a few seconds. Increased sensitivity allows routine use of small, dilute suspensions and this reduces contamination in the MS ion optics. Dilute cell suspensions deposit by evaporation more uniformly on the pyrolysis wire, so heat transfer during pyrolysis is more consistent. This improves spectral reproducibility. There are other advantages and one potential disadvantage.

The other advantages relate to biomarker information content. Rapid heating immediately before ionization minimizes thermal decomposition. Use of platinum rather than rhenium wire reduces catalytic degradation effects. Consequently, if exposed to a soft ionization process, biomolecular pyrolysates can form heavier ions. These are more easily associated with the original constituents in the cells than smaller thermal fragments would be. The more informative ions might include dephosphorolated diglycerides in the 400 to 600 m/z range. (Unlike MALDI MS, whole proteins or even small polypeptides cannot be observed intact by pyrolysis MS because the amino acid polymer units dimerize upon heating to form six-member rings).

The potential disadvantage would occur if a conventional quadrupole or other scanning mass spectrometer were used for detection. The pyrolysis peak is so narrow that irreproducibility between replicate acquisitions would arise from small variations in the timing of the ion scan relative to the pyrolysis program. This consideration suggested using a time-of-flight (nonscanning) mass spectrometer. Features of the Dephy MAB-TOF MS, are described in Section 5.3.3.

Metastable atom bombardment (MAB) is a novel ionization method for mass spectrometry invented by Michel Bertrand's group at the University of Montreal, Quebec, Canada, and described by Faubert et al.[38] For the identification of bacteria by MS, MAB has a number of significant advantages relative to more familiar ionization techniques. Electron ionization (EI) imparts so much excess energy that labile biomolecules break into very small fragments, from which the diagnostic information content is limited since all biomolecules in the pyrolysate are producing fragments in a narrow mass range. Other "soft" (low-energy) ionization techniques typically have other problems. Low-energy EI, less reproducible for technical reasons than 70 *eV* EI, is also very inefficient and gives very poor sensitivity. Chemical ionization (CI) involves bi-molecular mechanisms highly dependent of temperature, source geometry, trace contaminants in CI reagent gases, and other variables, so the theoretical advantage of greater information content inherent in the observable ions is lost due to spectral irreproducibility. This represents an unacceptable compromise for a fingerprinting technique. In contrast, MAB offers excellent ionization efficiency, highly reproducible fragmentation patterns (due to a narrow energy distribution), and user-selectable ionization energy.

In MAB ionization, gas molecules are passed through a high-voltage-induced plasma and form ions (M^+) as well as metastables (M^*, meaning electrically neutral atoms with internal energy elevated relative to ground state). An electrode deflects the ions so that they do not contribute to analyte ionization. The electrically neutral metastables proceed without deflection into the ion source. When a metastable atom touches an organic molecule (e.g., one of the cell pyrolysates), it transfers excess energy to the molecule (causing the latter's ionization) as the metstable relaxes to ground state. Since only a few energy levels are available for atoms (metastable or ground state species) and their relative energies are quantum mechanically determined, the amount of energy transferred during ionization is consistent, essentially independent of or insensitive to other experimental factors.

The energetics involved in MAB ionization can be represented by an equation:

$$E_{\mathrm{int}} = E^* - \mathrm{IE} + E_k.$$

Here E_{int} represents the excess internal energy of an ionized molecule resulting from its close encounter with a metastable atom having an energy E^* above-ground state and impacting with kinetic energy E_k. The organic molecule has an ionization potential or ionization energy, IE. In MAB the metastables are not accelerated, so their E_k values are thermal, relatively close to zero. It is possible to choose, among several noble gases, one in which the available E^* value exceeds typical organic compound IEs by a small amount. If such an atom is used for MAB, the internal energy of the organic ions produced can be almost zero. Those ions have little excess energy and therefore the mass spectrum of the organic compound shows very little fragmentation. Conversely, use of a high-energy metastable will produce extensive fragmentation, but with even greater reproducibility than 70 *eV* EI. Finally use of a MAB gas that yields intermediate energy metastables can produce a pyrolysate spectrum "blind" to those sample contaminants with high IEs (which do not ionize at all upon contact with the metastable). Table 5.1 lists relevant energy states for noble gases (and N_2) used in MAB. Note that under N_2 MAB the low

TABLE 5.1 Excitation Energy of Metastable Species and Internal Energy of a Typical M^+

Gas	Helium		Neon		Argon		Krypton		Xenon		N_2			
Metastable	$^1S_0$	$^3S_1$	$^3P_0$	$^3P_2$	$^3P_0$	$^3P_2$	$^3P_0$	$^3P_2$	$^3P_0$	$^3P_2$	E $^3\Sigma_g^+$	w $^1\Delta_u$	a $^1\Pi_g$	**a′ $^1\Sigma_u^-$**
Energy (eV)	20.61	19.82	16.72	16.62	11.72	11.55	10.56	9.92	9.45	8.32	11.88	9.02	8.67	**8.52**
Proportion (%)	10	**90**	20	80	16	84	3	97	3	97	14		85	
E_{int} M^+(IE = 0eV)		**11.82**		8.62		3.55		1.92		0.32				**0.52**

internal energy (0.52eV) of an M^+ having IE = 8eV would result in minimal fragmentation.

After a number of preliminary experiments, we have begun to create a Py-MAB-TOF MS spectral library of clinical and environmental isolates from the USFDA's Office of Regulatory Affairs collection. We are using Ar* (11.72eV or 11.55eV). Argon is inexpensive and gives efficient, reasonably soft ionization without contaminating the MAB plasma "gun" as N_2 does or increasing costs as krypton or xenon do. It ionizes most organic compounds found in bacteria, with minimal fragmentation, and produces an average spectrum up to twice the mass range typically observed for such samples by pyrolysis EI MS. Some of the other noble gases give interference peaks by interacting with residual air molecules or from small amounts of intruding ions (e.g., stable Kr^+ isotopes interfere at m/z 78, 80, 82, 83, and 86 and can also interact with residual air in the ion source to produce spectral artifacts of higher mass). In contrast, argon MAB spectra of bacterial cells can be acquired without significant background interference from as low as m/z 63. Only two air interaction products—$(NOAr)^+$, at m/z 70 and $(O_2Ar)^+$, at m/z 72—and a dimer cation—$(Ar_2)^+$, at m/z 80—need to be deleted from argon MAB spectra.

5.3.3 Time-of-Flight MS

The Dephy MAB/TOF MS uses reflectron ion analyzer geometry to produce unit mass resolution up to m/z 2000. The mass spectrometer has a relatively small footprint (24″ × 28″) consistent with field portability inside a small vehicle. It has an almost 100% duty cycle, collecting 16,000 separate time of flight full range spectra per second. This feature alone increases its sensitivity relative to scanning quadrupole instruments by a factor of several hundred for the pyrolysis application. No matter how rapidly different pyrolysate constituents appear in the ion source, ions formed are all collected with equal efficiency and contribute to the average bacterial spectrum in proportion to their abundance in the sample. This increases fingerprint reproducibility. Figure 5.4 shows a typical total ion chromatogram obtained by pyrolysis of a *Vibrio parahaemolyticus* isolate in the Dephy MAB/TOF MS. The horizontal scale shows summed "scans", each accumulated in a buffer memory chip adjacent to the MS ion counter or electron multipliers, from over a thousand time-of-flight

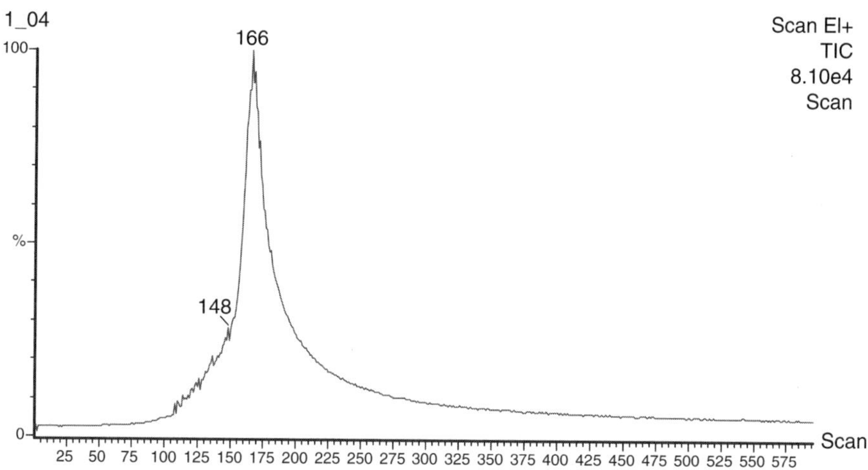

Figure 5.4 Typical total ion chromatogram by pyrolysis MAB/Tof MS for deposition of 0.5 µl of suspension (about 50,000 cells). The cells were a thermostable direct hemolysin producing *V. parahaemolyticus* serotype O4:K12. Peak width at half maximum intensity is 20 scans (~4 seconds).

scan events. To minimize acquisition data file size, only the accumulated result is imported from the buffer and saved. Each "summed" scan represents 0.2 second, so the figure shows total ion signal over a 2 minute acquisition.

A significant practical limitation of this reflectron TOF mass analyzer is the number of independent, interactive variables that must be adjusted to tune and calibrate it. For experienced operators and ordinary MS applications, tuning and calibration are not too difficult. But for a fingerprinting application, instrumental tuning criteria should include the added constraint that the spectra obtained from the tuned instrument closely match the relative intensity dimensions in corresponding library standards. One admits a calibration gas such as perfluorotributylamine (PFTBA) into the ion source, and tunes four major (and several minor) interactive variables so that spectral peak width, shape, and intensity are acceptable across the entire mass range over which bacterial pyrolysis ions are observed. Dephy is still developing an autotune program capable of producing the relative-intensity-matched result. The programming challenge is not trivial, nowhere near as straightforward as it would be for a quadrupole ion filter. One significant consolation is that once a reasonable tuning/calibration is defined (even if a perfect match in relative intensities to library conditions is not obtained), the pyrolysis MAB spectral patterns are stable. A method has been developed for compensating spectral drift, whether caused by the passage of time or use of an instrument different from the one that created the spectral library. This compensation technique will be described below.

With all the features that increase analytical sensitivity, it is possible to obtain very good quality pyrolysis Ar* MAB mass spectra from 50,000 cells

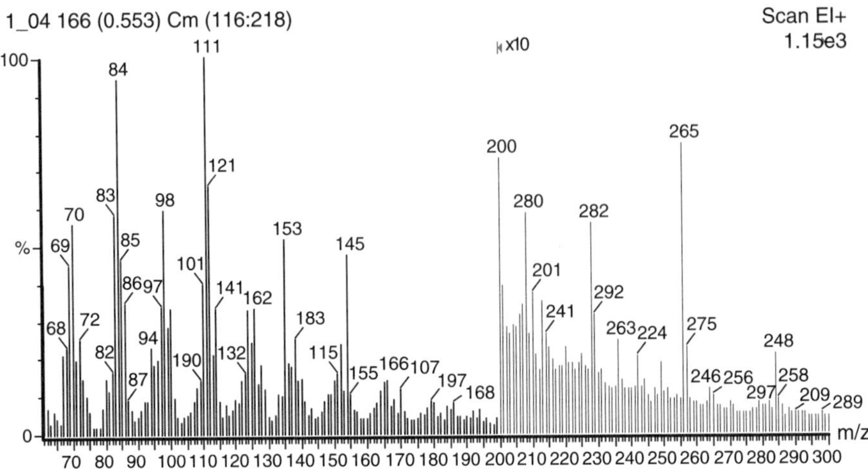

Figure 5.5 Typical average pyrolysis Ar MAB/Tof mass spectrum from 0.5 μl, about 50,000 cells, of a *tdh*+ strain of *V. parahaemolyticus* serotype O4:K12.

in 0.5 μl suspension. Figure 5.5 shows a typical average Ar MAB/TOF mass spectrum from pyrolysis of a strain of *V. parahaemolyticus* serotype O4:K12 responsible for a 1997 Pacific Northwest disease outbreak. This strain produces the thermostable direct haemolysin (*tdh*) toxin, and it is a challenge to distinguish such bacteria from the similar, avirulent (nonpathogenic) strains present in the environment that lack the *tdh* gene.

The average spectrum is defined, without background subtraction, by integrating across the pyrogram peak from the time it first rises above baseline until it returns to within 10% of the baseline. This covers a period of around 45 seconds during a typical acquisition, from scans 75 to 300 in Figure 5.4.

The Dephy data-processing package offers several proprietary noise filtering options. We have not yet determined unequivocally whether any of these data treatments increase reproducibility or concentrate the taxonomic information content of the averaged spectra, although first results suggest that use of filtered data can improve spectral reproducibility and thus assist the ruggedness of sample classification.

5.4 SPECTRAL LIBRARY ASSEMBLY

5.4.1 Causes of Spectral Drift

Pyrolysis spectra become distorted with respect to their diagnostic features for two major sets of reasons. The first is variations in instrument operation (e.g., heat transfer efficiency from wire to sample, ion source temperature, MAB gas identity, analyzer calibration, tuning, and ion transmission discrimination attributable to contaminated optics). Most of these factors can be controlled to achieve reproducibility. As discussed already, analyzer tuning is the least controllable of the instrumental factors in the case of the Dephy reflectron TOF mass analyzer.

The second set of reasons has to do with the samples to be analyzed. The spectra are regarded as "fingerprints," yet compared to the dietary independence of human fingerprints the metaphor fails because the diagnostic pattern for bacteria varies significantly depending on cell culture conditions. If one concentrates and purifies cells of a single strain from a nutrient-poor environment, they do not yield the same PyMS fingerprint as would be obtained for well-nourished cells of the same strain. (One experiment indicated that the pattern was similar in many ways, so a preliminary gross classification might be possible without culture standardization.) For precise classification the rapid culture and automated cell preparation steps previously discussed are important.

Over a long time period it may well not be possible to duplicate library cell culture conditions. What happens when the lot of media used in the final culture step prior to pyrolysis has been consumed? Can culture media suppliers assure nutritional *identity* between batches? Media types for growth of fastidious strains invariably include natural products such as brewer's yeast, tryptic soy, serum, egg, chocolate, and/or sheep blood. Trace components in natural products cannot be controlled to assure an infinite, invariable supply. The microtiter plate wells used here do not hold much media. Even so, the day will come when all media supplies are consumed and a change in batch is unavoidable. When that happens, if there were no effective way to compensate spectra for the resulting distortions, it would be necessary to re-culture and re-analyze replicates for every strain in the reference library. Until recently the potential for obsolescence was a major disincentive for developing PyMS spectral libraries of bacteria. Why this is no longer an insurmountable problem is discussed in the next section.

5.4.2 Correlation of Pyrolysis Patterns

Goodacre and Kell published several methods for correcting pyrolysis MS patterns attributable to instrumental drift.[39] However, they did not demonstrate the ability to achieve drift correction relative to a species different from the one used to characterize the drift, a necessary prerequisite to identify unknown bacteria by reference to library spectra. Using one of Goodacre and Kell's algorithms, we attempted to correct *Staphylococcus aureus* or *Psuedomonas mendocina* spectral variations by tracking changes in an *E. coli* sp. For the same species the relative ion intensities in a library spectrum and another spectrum acquired almost a month later were compared by ion-for-ion division, the library spectrum divided by the later spectrum. This gave a series of ratios or compensation factors for each ion. If there had been no change, the ion's compensation factor would calculate as the number 1.000. If the relative intensity for an ion had doubled during the month, its compensation factor was one-half, and so on. To compensate a spectrum acquired in the same session, one needs only to multiply each observed relative ion intensity by its corresponding correction factor. Simple.

Yet the compensation strategy failed in most cases. In score plots the direction of the spectral compensation vectors was correct but their lengths were not. As shown in Figure 5.6, we could successfully compensate for *Aeromonas hydrophila* spectral variations using *E. coli* variations to track the changes. Clearly, the arithmetic in the correction algorithm was working, but the quality of results was inconsistent. It became important to identify ways of implementing the algorithm that would consistently produce an acceptable drift compensation even when we could not anticipate the identity of the bacterium whose spectrum required the correction. (Was the unknown a strain of *P. mendocina* or some other that could not be corrected using *E. coli*? Or was it an *Aeromonas* sp. that could?)

The major causes of spectral variation were (1) instrumental drift, as Goodacre and Kell realized, but also (2) sample history, as discussed above. In particular, variations in the supplier or even the batch of tryptic soy agar (TSA) used for cell culturing led to spectral variations that differed in degree among disparate species. This phenomenon was attributed to the differential metabolic capabilities of the species with respect to the changed nutrients.

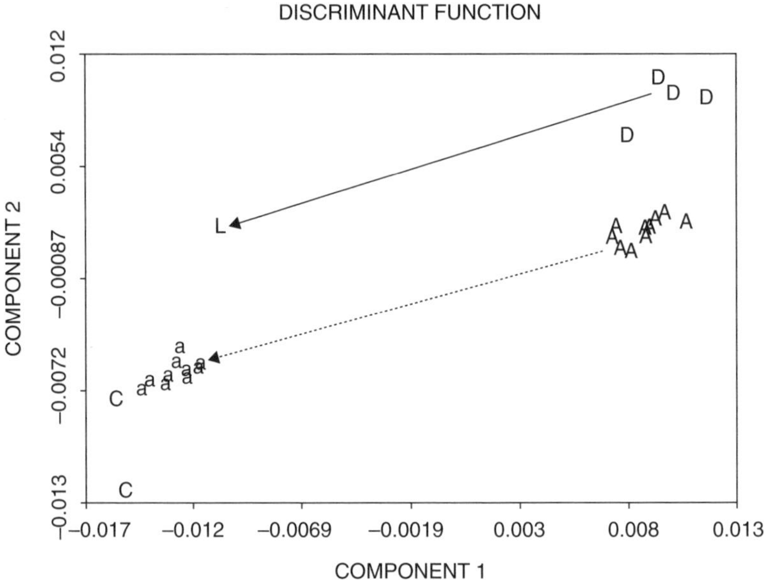

Figure 5.6 Successful transformation of *Aeromonas hydrophila* raw spectra "A" acquired on day 27 to new locations "a" (relationship indicated with a dotted arrow) near an *A. hydrophila* day 1 library spectrum "C" using another day 27 bacterium, *E. coli* 1090 "D" as reference compared to its own day 1 *E. coli* 1090 Library spectrum "L" (relationship indicated with a solid arrow).

This dilemma has now been resolved.[40,41] These two documents disclose a strategy for identifying a bacterial reference strain suitable for tracking the spectral drift and drift-compensating acquired spectra of an *unknown* strain. The simplest version of the strategy will now be summarized.

Whenever unknown samples are cultured, at the same time potential reference strains—ones already in the library and known to exhibit spectral variety spanning the probable range of the unknown samples—are cultured using the same growth conditions as the unknown. Spectra for the unknowns and that day's potential references are obtained and compared. The known spectrum most similar to a particular unknown is identified as the latter's most appropriate reference. A set of mass/charge ratio correction factors are calculated using the chosen reference spectrum divided into its library spectrum. The unknown's spectrum is then adjusted using those correction factors. Finally the unknown's adjusted (drift-compensated) spectrum is compared to all those in the library for classification. This sequence of comparison/compensation/identification is executed for each unknown. Since it does not involve iterative or complex calculations, the algorithmic classification process takes longer to read as described above than to execute.

A more advanced version of the process enables automated selection of the most appropriate reference spectrum for each unknown. That version performs all the spectral adjustments in batch fashion and can give two types of output. The first employs the same pattern recognition algorithm used in the initial comparisons, assesses the compensated spectra with respect to the entire library, and outputs a set of classification scores for each unknown. Alternatively, the output could be a set of compensated spectra ready to be processed (using the same or any other pattern recognition technique) by comparison to a subset containing only those library samples similar to the identified reference strain. This later process would be used if one chose to switch from noniterative multilinear statistical methods to ones based on artificial intelligence. The latter typically involve iterative calculations, require much greater computational overhead, and benefit greatly if there are only a limited number of output categories to which an unknown might belong. The nonlinear artificial intelligence methods have been found to give far superior results for classification of very similar strains: an artificial neural network (ANN) gave 92% correct assignments under leave-one-out (LOO) cross-validation compared to only 49% for the same spectra cross-validated by LOO using multivariate statistics.[41]

5.5 PATTERN RECOGNITION METHODS FOR OBJECTIVELY CLASSIFYING BACTERIA

When describing mathematical modeling in general (not just for classification of bacteria), it is important to point out the mathematical meaning of pattern recognition: the mapping of an n-dimensional function to describe a set of data. This mapping can be achieved in any number of ways, as will be illustrated.

Before pattern recognition analysis is started on a data set, an important question must first be answered: Is the model going to be used to predict unknown samples? If the answer is no, then "unsupervised" pattern recognition can be used; if yes, "supervised" techniques must be used. For unsupervised techniques, model development occurs without known outcomes, whereas supervised techniques are developed by establishing relationships to known outcomes and then using those relationships to predict unknowns. There are examples of both approaches among classical (statistically based) pattern recognition techniques and those based on artificial intelligence.

5.5.1 Classical Methods

Two examples of unsupervised classical pattern recognition methods are hierarchical cluster analysis (HCA) and principal components analysis (PCA). Unsupervised methods attempt to discover natural clusters within data sets. Both HCA and PCA cluster data.

In HCA the Euclidean distances between sample data points are calculated and used to form a two-dimensional plot that represents clusters and connectivity (according to similarity) in the data set. This is accomplished by grouping the samples that are closest together in multidimensional space, then continuing to link groups together in sequence by proximity, until all groups are finally merged into one tree-like structure known as a dendrogram. A little recognized limitation of HCA is that the pattern of clusters differs depending on which sample is used to define the first group. Goodacre et al. used HCA and generated a dendrogram grouping bacteria based on PyMS.[42]

PCA also groups samples together, through an algorithm based not on assessing similarity as in HCA but on observing differences. Principal components are orthogonal vectors in n-dimensional sample space that separate data by greatest variation. The first principal component vector is defined to describe as much variation in a data set as possible. The second principal component describes as much as possible of the *remaining* variation. The third and succeeding components describe increasingly minor remainders, with their number theoretically limited to the inherent dimensionality of the data. Because of the ranked sequence in which these vectors are calculated, far fewer than n of them are typically required to encompass all significant (nonrandom) factors contributing to variation. Thus a complex multidimensional set of data can be described using a few principal component dimensions. Clusters in the data set can then be observed graphically by plotting the scores (also called loadings or coordinates) along two or three principal component axes. A good demonstration of the application of PCA to PyMS data can be found in another article by Goodacre.[43]

Classical supervised pattern recognition methods include K-nearest neighbor (KNN) and soft independent modeling of class analogies (SIMCA). Both methods use a known set of examples to develop a model (based on similarity), which is then used for classification of unknowns. KNN predicts an unknown by assigning it to the group of objects closest to it in Euclidean space. Sample classes, either replicates or groupings based on assumed similarity (genus, species, etc.), are defined at the outset of model development. Each unknown is assigned to the class in Euclidean space having the largest number of nearest neighbors. Assignment occurs by a vote, which class contains the most of K "neighbors" polled. The KNN process does not assess the spatial distribution characteristics of the class members.

In contrast, SIMCA uses principal components analysis to model object classes in the reduced number of dimensions. It calculates multidimensional boxes of varying size and shape to represent the class categories. Unknown samples are classified according to their Euclidean space proximity to the nearest multidimensional box. Kansiz et al. used both KNN and SIMCA for classification of cyanobacteria based on Fourier transform infrared spectroscopy (FTIR).[44]

5.5.2 ANNs

Early in the development of rapid, spectral-based methods for pathogen classification/identification it was recognized that the spectral data structures contain a certain degree of "nonlinearity." Doubling of a particular ion's relative intensity between bacterial species A and species B may have a biological meaning completely different from a similar observation comparing species C to species D. The doubling might be a very important spectral relationship in the first case and of much lesser significance in the second. Linear (or even multilinear) pattern recognition would tend to assign equal significance to variations of both types and would have trouble establishing a clear basis for differentiating all four classes if all four species were built into one model. A linear model develops average rather than multiple relationships between inputs and outputs. In the example, the best linear model might perform poorly for all four bacterial classes.

By design, ANNs are inherently flexible (can map nonlinear relationships). They produce models well suited for classification of diverse bacteria. Examples of pattern analysis using ANNs for biochemical analysis by PyMS can be traced back to the early 1990s.[45–47] In order to better demonstrate the power of neural network analysis for pathogen ID, a brief background of artificial neural network principles is provided. In particular, backpropagation artificial neural network (backprop ANN) principles are discussed, since that is the most commonly used type of ANN.

An ANN is an array of three or more interconnected "layers" of cells called nodes (much like columns of cells in a spreadsheet). Data are introduced to the ANN through the nodes of the "input layer." For instance, each input layer node can contain the relative intensity of one of the m/z values from a bacterial pyrolysis mass spectrum. The output layer nodes can be assigned to identify each class of strains regarded as equivalent. The input and output layers are interconnected through another layer known as the hidden layer. Hidden layers have no physical meaning in relation to any of the spectral or biological characteristics modeled. They are composed of nodes that are used to map a function that describes the correlation between the input nodes (spectral features) and the corresponding output node (bacterial classes). The number of nodes in the output layer would equal the number of bacterial categories.

To develop a predictive model using a backprop ANN, one typically uses an algorithm based on Rummelhart, Hinton, and Williams's generalized delta rule.[48] The process is referred to as "training" and involves supervised learning that feeds information forward through the ANN layers, compares the resulting outcome to a known value, and then propagates the error backwards through the network. During a sequence of forward and backward cycles, the error is minimized by adjusting weights that are applied to the interconnections between the input and hidden nodes, and the hidden and output nodes. Nonlinear functions (a sigmoid, an arctangent, etc.) are used in each hidden layer node to transfer values through to the output node of the network. The use of nonlinear transfer functions, along with the multiple pathways connecting input and output nodes, confer on the ANN its nonlinear modeling capacity.

During training for a bacterial application, spectra of known bacteria are placed into the input layer. These relative intensity values are multiplied times connection weights (w_i) between the input and hidden layers. Inside the hidden nodes the various products are summed. Each total is operated on by a nonlinear transfer function (e.g., sigmoid or arctangent). The transformed values are multiplied by more connection weights and the products are summed in the appropriate elements (nodes) of the output layer. Since the spectrum belongs to a known species, the desired sequence of output products can be defined for some pattern that will be associated with that species. Early in a training sequence the resulting products differ greatly from any predefined pattern. The total of these differences is designated as error, the magnitude of which can be readily calculated. Adjustments to the connection weights are made with a goal of decreasing error between the net-predicted output and the actual output (training example outcome). Results at each step are assessed using a gradient descent least squares technique. Once the error reaches a user-defined acceptable minimum (a threshold value), training is terminated.

Despite their inherent modeling power, backprop ANNs have practical drawbacks that cause researchers to use them only as a last resort, if at all, after trying every statistical method. The biggest issue is the fact that an ANN can provide good quality correlations or valid predictions only if its configuration is optimized. Optimization is an iterative process that examines the feature space of an ANN, a methodical search involving adjustment of all of its architectural and physical parameters. Typically this proceeds by tedious trial and error, one configuration at a time, on a single computer, for as many permutations and combinations of ANN parameters as the analyst can endure. It is a lengthy, inefficient, and boring process. More important, efforts are often abandoned after some minimally acceptable result is obtained, leaving significant portions of architecture space unexamined. For any model having less than 100% predictive accuracy, a question remains whether results could have been improved with more time spent on architecture optimization. Assessing the inherent ruggedness of the input/output correlation (here, the PyMS system's "chemotaxonomic power") is compromised.

5.5.3 Parallel Distribution of the Iterative ANN Architecture Optimization Process

To address this issue and to take advantage of the pattern recognition analytical powers of ANNs, a parallel-distributed neural network optimization scheme was developed. The goal was to create an efficient, powerful process for obtaining optimized networks. This involved writing code for an error back propagating, feed-forward, *parallel-distributed* artificial neural network (PD-ANN). The PD-ANN is simply an algorithm that allows for systematic training and validation of multiple neural network architectures. The PD-ANN takes advantage of an Ethernet-based parallel distribution scheme to perform the optimization task simultaneously on multiple computers. The result is a dramatic decrease in the time required for finding an optimal network configuration. The PD-ANN code was conceived and developed in the Chemistry Division at the FDA's National Center for Toxicological Research (NCTR) and was implemented using JGravity™, a Java-based parallel distribution platform.[49] We are participating in the beta test phase of JGravity™ development.

The PD-ANN algorithm itself resides on a PC designated as the master PC. It assigns different neural network configurations to participating slave PCs (nodes). The algorithm distributes different neural network configurations to each node PC. Each node PC is given the task of evaluating the assigned architectures by their results using a cross-validation analysis. To accomplish this for a particular architecture, a small number of examples are automatically removed by the PC node from the training set (the number excluded is user selected), then the test architecture is trained to the specified number of cycles on the somewhat depleted set using an assigned learning rate and momentum. When training completes, the removed data are propagated through the trained network, and predictions are obtained, whose quality is subject to validation by comparing the predicted and actual values from the omitted examples. Each node's cross-validated results are then returned to the master PC, which tabulates the impacts of various architectural and physical parameter combinations on the ANN's predictive ability. This process of removing data, training the network, and propagating the removed data is repeated for each architectural variant with new training set samples removed, until all of the training set samples have been used to test that network. Combined results for each training set are collated and reported back to the master PC as the overall result for that architecture. After a report is received, the master PC assigns a new configuration to that node PC. This assignment process is repeated on all the available PC nodes until the master PC finishes its task list. The task list is a user-specified list of all the desired test configurations, which can easily span all reasonable quadrants of an ANN's feature space. To provide the reader with an idea of the efficiency of this system, consider that a five-node Pentium PC cluster executing the PD-ANN can examine most of an ANN's feature space in a few hours, whereas a less-than-desirable feature space analysis performed by one user on one PC in serial fashion could take a week or more.

5.5.4 Advantage of Using Fully Optimized ANN Models: Assessment of Inherent Taxonomic Power of PyMS

As already mentioned, use of a less-than-optimal ANN leaves the researcher wondering whether any failures in classification are attributable to the non-optimized architecture rather than a consequence of limitations in spectral quality or information content. Complete architecture optimization by manual trial and error is possible but usually impractical. The PD-ANN's ability to develop a global optimum configuration/solution for distinguishing a set of bacterial strains allows the researcher to probe the limits of spectral information content with respect to questions of inherent taxonomic power.

To illustrate this rather abstract concept, we present in detail results mentioned above for a PD-ANN analysis of Py-MAB spectra.[41] Thirty *Salmonella enterica* strains were analyzed. The data set consisted of 100 spectra, composed of three replicates for each strain, with more than three replicate spectra for several of the strains. The PD-ANN was used to ask three questions of increasing difficulty with respect to taxonomy. Could PyMS spectra be used to distinguish *Salmonella* serotypes? Could they distinguish PFGE profiles? Could they distinguish antibiotic resistance profiles? There were four *Salmonella enterica* serotypes represented by the data set, 12 PFGE profiles, and 12 antibiotic resistance profiles. The PD-ANN was used to develop two ANN models, one to predict serotype and PFGE profiles, and another to predict antibiotic resistance profiles. More than 25 neural network architectures were investigated by the PD-ANN over several hours executed on three high-speed Pentium computers accomplished on two different days. In addition to the neural network architectures, other important ANN parameters such as learning rate, momentum, and number of learning cycles were also investigated.

For the PFGE profile analysis the ANN architecture had an input layer of 190 nodes, one for each m/z intensity value used, and an output layer consisting of 12 nodes, one for each *S. enterica* category distinguished by PFGE plus one more to represent an *S. enterica* serotype Typhimurium control. Optimal performance of the network for predicting PFGE pattern profile was obtained using a 190-63-12 architecture, and 60,000 training cycles (performance was almost as good for 40,000 cycles). This architecture with 40,000 iterations was used to rebuild a single model based on all 100 spectra and build 100 models of 99 spectra each for leave-one-out (LOO) cross-validation. ANN quantitative accuracy on the single 100-spectra model had a coefficient of regression, $r^2 = 0.97$. This represents 97 of the 100 spectra correctly classified by PFGE pattern. For values collated from the 100 LOO cross-validation experiments, an excellent average cross-validation covariance coefficient, $q^2 = 0.92$, was obtained. That is, 92% of the spectra were correctly classified even when each spectrum made no contribution to the model by which it was categorized. For comparison, the best corresponding discriminant analysis model achieved only 49% by LOO cross-validation. Using the 190-63-12 architecture and 60,000 iterations, the PD-ANN was also trained on 70 of the spectra and asked to predict 30 test spectra (in essence, a single leave-30-out cross-validation experiment). Twenty of the 30 spectra (67%) were correctly classified. However, when broken down by strain, the correct predictions were 2 of 3 for *S. enterica* serotype Anatum (67%), 3 of 3 for serotype Worthington (100%), 2 of 2 for serotype Muenster (100%), 1 of 1 for the serotype Typhimurium (100%), and 12 of 21 for the combined serotype Heidelbergs (57%). All of the missed Heidelberg classifications involved confusion among other Heidelberg strains. That is, the network was 97% accurate for classifying 30 previously unseen spectra by serovar alone.

Attempts by PD-ANN to model all 100 spectra when they were grouped into 12 antibiotic resistance profile categories (i.e., ignoring PFGE or serotype) gave very poor results from a one time leave 30 out experiment (q^2_{30} = 0.10, which is not significantly better than random guesses among 12 possible classes). This result suggested the possibility that PFGE and serotype characteristics were represented in PyMS patterns by more prominent biomarkers than antibiotic resistance, which agreed with immunological and genetic microbiological fundamentals. It did not necessarily prove that PyMS spectra were altogether missing resistance-related information.

To eliminate serotype and PFGE contributions to spectral variation, 30 replicates containing 10 PFGE-identical Heidelberg strains were used for training. The 10 strains included three (group A) with resistance to erythromycin, bacitracin, novobiocin, and rifampin. Four strains (group B) were resistant to these plus streptomycin, spectinomycin, gentamicin, and tetracycline. Two strains (group C) were resistant to all antibiotics in group B except tetracycline. One strain (group D) was resistant to the four antibiotics from group A plus tetracycline and a sulfamethoxazole:trimethoprim cocktail. An ANN with a 190-63-4 architecture was then trained for 20,000 cycles. It was able to sort spectra into the 4 different antibiotic resistance profiles with 40% accuracy under leave-four-out cross-validation (an improvement over random guesses, which would give ~25%). The pattern of misclassifications in this experiment suggested that among the 10 strains there remained some unidentified confounding variables, ones that contributed to the spectra as strongly as antibiotic resistance profile and rendered groups B and C indistinguishable. The two groups formally differed only with respect to tetracycline resistance.

In addition one member of group D was consistently modeled as a member of group C.

We built a fourth PD-ANN model with new groups that took into account some of the previously observed confusions. Group D was moved into group C; one of the consistently misclassified members of group A was moved to group B, and another strain was moved from C to B. There were now three groups designated A′, B′, and C′. Thus the ANN model used a 190-63-3 architecture with the strains assigned to one of the three output nodes, one each for Group A′, B′, and C′. The best LOO cross-validated performance achieved 58% accuracy using 15,000 training cycles.

In yet another approach, a PD-ANN model was developed with only two classes of antibiotic resistance. The 10 strains were assigned to only 2 groups: group A″ representing two strains resistant to the four antibiotics erythromycin, bacitracin, novobiocin, and rifampin, and group B″ representing the other strains having a variety of additional resistance capabilities. Applied to unknown characterization, a group A″ match would predict that streptomycin, spectinomycin, gentamicin, tetracycline, or sulfamethoxazole : trimethoprim cocktail could be used for effective therapy. Similarly a group B″ match would suggest the need to use some other antibiotic, (e.g., ciprofloxacin, norfloxacin, polymyxin B, ofloxacin, nalidixic acid, ampicillin, chloramphenicol, kanamycin, ceftriaxone, cephalothin, or cefoxitin) for which sensitivity was observed among all of the *Salmonella* isolates. Using a 190-63-2 architecture and 15,000 training cycles, this two-category PD-ANN model separated the ten strains with LOO cross-validation of 90%.

These examples illustrate the power of proper ANN feature space optimization. In all the examples discussed, the limits of the type of information that could be gleaned from the *Salmonella* PyMAB spectra were probed. The PD-ANN's automated optimization removed the issue of methodological uncertainty and enabled a focus on questions of Py-MAB-MS spectral information content and its potential use for rapid strain ID. Question: Does Py-MAB-MS data support Serovar classification? Answer: Yes. How about PFGE classification? Yes. How about antibiotic resistance profile? Answer: Perhaps, if one first eliminates stronger contributions to spectral variation and then, by design and grouping, limits the possibilities to only a few classes.

5.5.5 Other Pattern Recognition Approaches

Other pattern recognition strategies have been used for bacterial identification and data interpretation from mass spectra. Bright et al. have recently developed a software product called MUSE, capable of rapidly speciating bacteria based on matrix-assisted laser desorption ionization time-of-flight mass spectra.[13] MUSE constructs a spectral database of representative microbial samples by using single point vectors to consolidate spectra of similar (not identical) microbial strains. Sample unknowns are then compared to this database and MUSE determines the best matches for identification purposes. In a test run on 212 samples, MUSE correctly matched 79%, 84%, and 89% for strain, species, and genus, respectively.[13] In another study temperature constrained cascade correlation neural networks (TCCCNs) were used to classify pathogenic bacteria into five species classes using mass spectra obtained from an air buffered quadrupole ion trap mass spectrometer fitted with an infrared pyrolyzer.[24] The best TCCCN model achieved an accuracy of 96 ± 2% when classifying the bacteria into the five species categories.[24]

There are a multitude of algorithmic pattern recognition approaches that can be applied for rapid bacterial classification based on mass spectra. MUSE is the only pattern recognition scheme commercially developed for use with a very large library of bacterial spectra, but it is only available with purchase of the MicrobeLynx MALDI MS platform. Implementation of computationally intensive, sophisticated pattern recognition methods on a very large database presents a number of practical challenges. The PD-ANN addresses the question of model development but is not a complete solution to this challenge. It is possible to use noniterative classical methods such as KNN as a first stage of unknown classification and, once the sample has been roughly categorized, change over to calculation-intensive approaches designed to refine the classification, distinguishing among similar strains. An alternative possibility is to implement sophisticated algorithms designed to minimize calculation overhead in model development. To address this alternative, we are presently investigating two novel algorithms developed by Dr. Robert Shelton at NASA Johnson Space Center (NASA-JSC). One is named Fuzmac and the other is a Bayesian classifier algorithm.[50]

Fuzmac is a hybrid pattern recognition tool that combines a decision tree approach with principal component analysis and backpropagation neural network algorithms. It is used in a way analogous to a backpropagation ANN. However, unlike simple backprop ANNs, Fuzmac uses a principal component vector to split the initial data into two branches according to variance. These branches are then separately analyzed for correlation to the output using backprop ANN algorithms. If a good correlation between branch and output is not found at this point, the branch is again split according to variance by a principal component vector; and the resulting branches are then analyzed (by the backprop ANN code) for correlation to the output. Fuzmac uses the decision tree approach to reduce drastically the amount of time required compared to optimizing a global neural network. Through such a hybrid method it should be possible to take advantage of the demonstrated strengths of neural network modeling when applied to very large, diverse training sets such as would be found in a comprehensive library of bacterial MS spectra.

The Bayesian classifier works by building approximate probability distributions for a set of n features using examples of each class. To illustrate, if there are three classes, each described by 10 features (for the purposes of this discussion, a feature is just a real number) then the classifier will try to model three probability distributions in 10-dimensional space. These distributions can be thought of as spheres or clusters in feature space. The process of approximating these distributions from the example data is the training phase.

Once the distributions are constructed, an unknown sample is classified by evaluating each distribution on the features of the unknown sample. Again, using the three-class problem as an example, this evaluation step would yield three real numbers q_1, q_2, and q_3, which are the values of each distribution on the features of the unknown. These raw values can be converted to probabilities by simply normalizing this set of equations:

$$p_1 = \frac{q_1}{d}, \quad p_2 = \frac{q_2}{d}, \quad p_3 = \frac{q_3}{d}, \quad \text{where } d = q_1 + q_2 + q_3$$

Assuming the distribution models are accurate and that they model all the possible behaviors in the data set, Bayes's theorem says that p_1, p_2, and p_3 are the probabilities that the unknown sample is a member of class 1, 2, or 3, respectively. The distributions are modeled using multivariate Gaussian functions in a method known as "expectation maximization."

The Bayesian algorithm is a nonclassical statistical method rather than an artificial intelligence approach. Compared to classical statistical methods, it has proved to be very rugged for classification projects not involving bacterial spectra. Therefore an attempt to use it for this application is warranted.

5.6 CONCLUSIONS

We have proposed a multi-subsystem approach to the challenge of rapid bacterial identification. The subsystems involve the following:

1. Instrumental cell isolation and sample handling in liquid culture.
2. Automated pyrolysis MS analysis of lysed vegetative cells in a disinfecting suspension.
3. Spectral drift compensation for initial and ongoing assembly as well as consultation of a PyMS spectral library.
4. Systematic optimization of artificial neural network and other advanced computational models for grouping strains and for classifying unknown samples as members of the most appropriate group.

Development of the system's four components is proceeding in parallel and evidence supporting the capabilities or potential of each has been presented.

Each of the subsystems can, apart from the others, make a significant diagnostic contribution. For example, the instrumental cell isolation and sample handling component could be used with DNA-based or other non-MS systems for detection and/or identification. As another example, the principles underlying pattern drift compensation can apply to MALDI MS and even non-MS detection systems such as capillary GC of fatty acid methyl esters.

The complete four-component system is necessary when the diagnostic requirement is rapid, low unit-cost analysis for both the strain-level characterization of pathogenic agents and identification of hoax bio-terror materials. Using the complete system, we are proposing to validate MS-based microbial taxonomy and to transfer the technology from an analytical research to a clinical or public health production-diagnosis environment.

ACKNOWLEDGMENT

This work was supported by the National Center for Toxicological Research, United States Food and Drug Administration. The views presented in this article do not necessarily reflect those of the Food and Drug Administration.

REFERENCES

1. CDC; Lay, D. J. (Ed.). Private communication.
2. Cartmill, T. D.; Orr, K.; Freeman, R.; Sisson, P. R.; Lightfoot, N. F. Nosocomial infection with *Clostridium difficile* investigated by pyrolysis mass spectrometry. *J. Med. Microbiol.* 1992, **37**, 352–356.
3. Freeman, R.; Gould, F. K.; Wilkinson, R.; Ward, A. C.; Lightfoot, N. F.; Sisson, P. R. Rapid inter-strain comparison by pyrolysis mass spectrometry of coagulase-negative staphylococci from persistent CAPD peritonitits. *Epidemiol. Infect.* 1991, **106**, 239–246.
4. Gould, F. K.; Freeman, R.; Sisson, P. R.; Cookson, B. D.; Lightfoot, N. F. Inter-strain comparison by pyrolysis mass spectrometry in the investigation of *Staphylococcus aureus* nosocomial infection. *J. Hosp. Infect.* 1991, **19**, 41–48.
5. Low, J. C.; Chalmers, R. M.; Donachie, W.; Freeman, R.; McLaughlin, J.; Sisson, P. R. Pyrolysis mass spectrometry of *Listeria monocytogenes* isolates from sheep. *Res. Vet. Sci.* 1992, **53**, 64–67.
6. Orr, K.; Gould, F. K.; Sisson, P. R.; Lightfoot, N. F.; Freeman, R.; Burdees, D. Rapid inter-strain comparison by pyrolysis mass spectrometry in nonsocomial infection with *Xanthomas maltophilia. J. Hosp. Infect.* 1991, **17**, 187–195.
7. Sisson, P. R.; Freeman, R.; Gould, F. K.; Lightfoot, N. F. Strain differentiation of nosocomial isolates of *Pseudomonas aeruginosa* by pyrolysis mass spectrometry. *J. Hosp. Infect.* 1991, **19**, 137–140.
8. Sisson, P. R.; Freeman, R.; Lightfoot, N. F.; Richardson, I. R. Incrimination of an environmental source of a Legionnaires' disease by pyrolysis mass spectrometry. *Epidemiol. Infect.* 1991, **107**, 127–132.
9. Sisson, P. R.; Freeman, R.; Magee, J. G.; Lightfoot, N. F. Differentiation between mycobacteria of the *Mycobacterium tuberculosis* by pyrolysis mass spectrometry. *Tubercle* 1991, **72**, 206–209.
10. Sisson, P. R.; Freeman, R.; Magee, J. G.; Lightfoot, N. F. Rapid differentiation of *Mycobacterium xenopi* from mycobacteria of the *Mycobacterium avium*-intracellulare complex by pyrolysis mass spectrometry. *J. Clin. Path.* 1992, **45**, 355–357.

11. Sisson, P. R.; Kramer, J. M.; Brett, M. M.; Freeman, R.; Gilbert, R. J.; Lightfoot, N. F. Application of pyrolysis mass spectrometry to the investigation of outbreaks of food poisoning and non-gastrointestinal infection associated with *Bacillus* species and *Clostridium perfringens*. *Int. J. Food Microbiol.* 1992, **17**, 57–66.

12. Beverly, M. B.; Basile, F.; Voorhees, K. J. Fatty acid analysis of beer spoiling microorganisms using pyrolysis mass spectrometry. *J. Am. Soc. Brewing Chemists* 1997, **55**, 79–82.

13. Bright, J. J.; Claydon, M. A.; Soufian, M.; Gordon, D. B. Rapid typing of bacteria using matrix-assisted laser desorption ionization time of flight mass spectrometry and pattern recognition software. *J. Microbiol. Meth.* 2002, **48**, 127–138.

14. Cain, T. C.; Lubman, D. M.; Weber, W. J. Differentiation of bacteria using protein profiles from matrix-assisted laser desorption/ionization time-of-flight mass spectrometry. *Rapid Comm. Mass Spectrom.* 1994, **8**, 1026–1030.

15. Cotter, R. J. Laser mass spectrometry: An overview of techniques, instruments and applications. *Anal. Chim. Acta* 1987, **195**, 45–59.

16. Fenselau, C.; Cotter, R. J. Chemical aspects of fast atom bombardment. *Chem. Rev.* 1987, **87**, 501–512.

17. Holland, R. D.; Wilkes, J. G.; Rafi, F.; Sutherland, J. B.; Persons, C. C.; Voorhees, K. J.; Lay, J. O. Rapid identification of intact whole bacteria based on spectral patterns using matrix assisted laser desorption ionization with time of flight mass spectrometry. *Rapid Comm. Mass Spectrom.* 1996, **10**, 1227–1232.

18. Holland, R. D.; Duffy, C. R.; Rafi, F.; Sutherland, J. B.; Heinze, T. M.; Holder, C. L.; Voorhees, K. J.; Lay, J. O. Identification of bacterial proteins observed in MALDI-TOF mass spectra from whole cells. *Anal. Chem.* 1999, **71**, 3226–3230.

19. Krishnamurthy, T.; Davis, M. T.; Stahl, D. C.; Lee, T. D. Liquid chromatography/microspray mass spectrometry for bacterial investigations. *Rapid Comm. Mass Spectrom.* 1999, **13**, 39–49.

20. Posthumus, M. A.; Kistemaker, P. G.; Meuzelaar, H. L. C. Laser desorption-mass spectrometry of polar nonvolatile bio-organic molecules. *Anal. Chem.* 1978, **50**, 985–991.

21. Arnold, R. J.; Reilly, J. P. Fingerprint matching of *E. coli* strains with matrix-assisted laser desorption/ionization time-of-flight mass spectrometry of whole cells using a modified correlation approach. *Rapid Comm. Mass Spectrom.* 1998, **12**, 630–636.

22. Goodcare, R. Characterization and quantification of microbial systems using pyrolysis mass spectrometry: Introducing neural networks to analytical pyrolysis. *Microbiol. Europe* 1994, **2**, 16–22.

23. Harrington, P. B.; Voorhees, K. J. MuRES: A multivariate rule-building expert system. *Anal. Chem.* 1990, **62**, 729–734.

24. Harrington, P. B.; Voorhees, K. J.; Franco, B.; Hendricker, A. D. Validation using sensitivity and target transform factor analyses of neural network models for classifying bacteria from mass spectra. *J. Am. Soc. Mass Spectrom.* 2002, **13**, 10–21.

25. Dare, D.; Wilkes, J. (Ed.). Private communication. 2004.

26. Jarman, K. H.; Cebula, S. T.; Saenz, A. J.; Peterson, C. E.; Valentive, N. B.; Kingsley, M. T.; Wahl, K. L. An algorithm for automated bacterial identification using matrix-assisted laser desorption/ionization mass spectrometry. *Anal. Chem.* 2000, **72**, 1217–1223.

27. Perfetto, S. P.; Ambrozak, D. R.; Koup, R. A.; Roederer, M. Measuring containment of viable infectious cell sorting in high-velocity cell sorters. *Cytometry* (Pt. A) 2003, **52A**, 122–130.

28. Varnam, A. H.; Evans, M. G. *Foodborne Pathogens: An Illustrated Text.* Chicago: Mosby Year Book, 1991.

29. Braga, P. C.; Bovio, C.; Culici, M.; Dal Sasso, M. Flow cytometric assessment of susceptibilities of *Streptococcus pyogens* to erythromycin and rokitamycin. *Antimicrob. Agents Chemother.* 2003, **47**, 408–412.

30. Darvey, H. M.; Kell, D. B. Flow cytometry and cell sorting of heterogenous microbial populations: the importance of single cell analysis. *Microbiol. Rev.* 1997, **60**, 641–696.

31. Gunasekera, T. S.; Veal, D. A.; Attfield, P. V. Potential for broad applications of flow cytometry and fluorescence techniques in microbiological and somatic cell analyses of milk. *Int. J. Food Microbiol.* 2003, **85**, 269–279.

32. Holm, C.; Jespersen, L. A flow-cytometric gram-staining technique for milk-associated bacteria. *Appl. Environ. Microbiol.* 2003, **69**, 2857–2863.

33. McLaughlin, R. W.; Vali, H.; Lau, P. C.; Palfree, R. G.; De Ciccio, A.; Sirois, M.; Ahmad, D.; Villemur, R.; Desrosiers, M.; Chan, E. C. Are there naturally occurring pleomorphic bacteria in the blood of healthy humans? *J. Clin. Microbiol.* 2002, **40**, 4771–4775.

34. Porter, J.; Deere, D.; Hardman, M.; Edwards, C.; Pickup, R. Go with the flow—Use of flow cytometry in environmental microbiology. *FEMS Microbiolog. Ecol.* 1997, **24**, 93–101.

35. Sorensen, S. J.; Sorensen, A. H.; Hansen, L. H.; Oregaard, G.; Veal, D. A. Direct detection and quantification of horizontal gene transfer by using flow cytometry and gfp as reporter gene. *Curr. Microbiol.* 2003, **47**, 129–133.

36. Tay, S. T.; Ivanov, V.; Yi, S.; Zhuang, W. Q.; Tay, J. H. Presence of anaerobic bacteroides in aerobically grown microbial granules. *Microbial Ecol.* 2002, **44**, 278–285.

37. Wahlberg, M.; Gaustad, P.; Steen, H. B. Rapid discrimination of bacterial species with different ampicillin susceptibility levels by means of flow cytometry. *Cytometry* 1997, **29**, 267–272.

38. Faubert, D.; Paul, G. J. C.; Giroux, J.; Bertrand, M. J. Selective fragmentation and ionization of organic compounds using an energy-tunable rare-gas metastable beam source. *Int. J. Mass Spectrom. Ion Proc.* 1993, **124**, 69–77.

39. Goodacre, R.; Kell, D. B. Correction of mass spectral drift using artifical neural networks. *Anal. Chem.* 1996, **68**, 271–280.

40. Wilkes, J. G.; Glover, K. L.; Holcomb, M.; Raffi, F.; Cao, X.; Sutherland, J. B.; McCarthy, S. A.; Letarte, S.; Bertrand, M. J. Defining and using microbial spectral databases. *J. Am. Soc. Mass Spectrom.* 2002, **13**, 875–887.

41. Wilkes, J. G.; Rushing, L.; Nayak, R.; Buzatu, D. A.; Sutherland, J. B. Rapid phenotypic characterization of *Salmonella enterica* strains by pyrolysis metastable atom bombardment mass spectrometry with multivariate statistical and artificial neural network pattern recognition. *J. Microbiol. Meth.* submitted for publication.

42. Goodacre, R.; Timmins, E. M.; Burton, R.; Kaderbhai, N.; Woodwards, A. M.; Kell, D. B.; Rooney, P. J. Rapid identification of urinary tract infection bacteria using

hyper spectral whole-organism fingerprinting and artificial neural networks. *Microbiology* 1998, **144**, 1157–1170.

43. Goodacre, R.; Trew, S.; Wrigley-Jones, C.; Neal, M. J.; Maddock, J.; Ottley, T. W.; Porter, N.; Kell, D. B. Rapid screening for metabolite overproduction in fermentor broths using pyrolysis mass spectrometry with multivariate calibration and artificial neural networks. *Biotechnol. Bioengin.* 1994, **44**, 1205–1216.
44. Kansiz, M.; Heraud, P.; Wood, B.; Burden, F.; Beardall, J.; McNaughton, D. Fourier transform infrared microspectroscopy and chemometrics as a tool for the discrimination of cyanobacterial strains. *Phytochemistry* 1999, **52**, 407–417.
45. Goodacre, R.; Edmonds, A. N.; Kell, D. B. Quantitative analysis of the pyrolysis–mass spectra of complex mixtures using artificial neural networks: Application to amino acids in glycogen. *J. Anal. Appl. Pyrolysis* 1993, **26**, 93–114.
46. Goodacre, R.; Karim, A.; Kaderbhai, M. A.; Kell, D. B. Rapid and quantitative analysis of recombinant protein expression using pyrolysis mass spectrometry and artificial neural networks: Application to mammalian cytochrome b_5 in *Escherichia coli*. *J. Biotechnol.* 1994, **34**, 185–193.
47. Goodacre, R.; Kell, D. B. Quantitative analysis of the pyrolysis–mass spectra of complex mixtures using artificial neural networks: Application to amino acids in glycogen. *Anal. Chim. Acta* 1993, **279**, 17–26.
48. Rumelhart, D. E.; Hinton, G.; Williams, R. *Parallel Distributed Processing*. Cambridge: MIT Press, 1986.
49. JGravity v1.0 beta-1 build-10. Titan Systems Corporation, Lin Com Division, Los Angeles, CA.
50. Information Systems Directorate, N. J. S. C., Houston, TX 77058. Fuzmac and the Bayesian classifier are NASA copyrighted and are used with permission from the Information Systems Directorate, NASA Johnson Space Center, Houston, TX 77058.

6

MALDI-TOF MASS SPECTROMETRY OF INTACT BACTERIA

Jackson O. Lay, Jr. and Rohana Liyanage
University of Arkansas, Fayetteville, AR 72701

6.1 INTRODUCTION

The focus of this chapter is the development of a technique often called whole-cell matrix-assisted laser desorption ionization (MALDI) time-of-flight (TOF) mass spectrometry (MS) or whole-cell MALDI-TOF MS. Some groups prefer to use terms such as "intact" or "unprocessed" rather than "whole," but the intended meaning is the same regardless of which word is used. As noted in the first chapter of this book, there are many different methods for the analysis of bacteria. However, for the analysis of intact or unprocessed bacteria, whole-cell MALDI-TOF MS is the most commonly used approach. This method is very rapid. MALDI-TOF MS analysis of whole cells takes only minutes because the samples can be analyzed directly after collection from a bacterial culture suspension. Direct MALDI MS analysis of fungi or viruses is similar in approach[1,2] but is not covered in this chapter. MALDI-TOF MS of "whole cells" was developed with very rapid identification or differentiation of bacteria in mind. The name (whole cell) should not be taken to imply that the cells are literally intact or whole. Rather, it should be taken to mean that the cells that have not been treated or processed in any way specifically for the removal or isolation of any cellular components from any others. In "whole-cell" analysis the cells have been manipulated only as necessary to

Identification of Microorganisms by Mass Spectrometry, Edited by Charles L. Wilkins and Jackson O. Lay, Jr.

transfer them into the mass spectrometer for analysis. In later publications investigators have looked at which of the whole-cell approaches also killed the cells being studied, but even cell killing was "incidental" rather than "deliberate" in the first studies. Thus in whole-cell MALDI experiments this means cells (in suspension) are applied to the target along with a matrix compound (to facilitate laser desorption/ionization) and solvents in which the matrix (a solid) has been dissolved. A side effect of this process is intimate contact between bacteria, MALDI matrix, and solvents. Under these conditions osmotic pressure results in stress on the cellular membranes. Cells are usually deposited onto the MALDI target in a liquid suspension. The transfer step has been used to kill cells for safety reasons. In order to minimize potential exposure to the organisms in the mass spectrometry laboratory, often the cells that are recovered from a liquid culture medium or a standard culture plate in the microbiology laboratory are suspended in solvents known to kill bacteria prior to transfer to the mass spectrometry laboratory for analysis. This step may also result in cellular disruption. Because none of these steps are intended to disrupt cells, and no steps are added to isolate proteins or other analytes, the technique has been called "whole-cell" or "intact cell" analysis to differentiate it from procedures where additional steps are included in the procedure to deliberately disrupt cellular membranes or separate/recover analytes from the cellular material.

One of the original ideas behind whole cell MALDI-TOF MS was the rapid identification of bacteria. Taxonomic classification (identification) of bacteria was to be based only on a subset of detectable cellular components observed from the much larger pool of compounds that are simply not represented in the mass spectra. This requires that this subset of the "marker" signals (ions) provide some level of taxonomic specificity. We now know that such specificity is the rule rather than the exception, and it extends to the genus, species and even strain level[3] as is discussed in some detail in Chapter 10 of this book. Initially it was anticipated that, for the method to have any taxonomic capability, the spectral profiles or patterns would have to be very reproducible. Spectra would be obtained under identical or near-identical conditions. This has been difficult to accomplish experimentally for both instrumental and biological reasons, but has been done, as detailed below. An alternative approach has developed that involves a comparison of specific taxonomy-characteristic marker signals with the larger set of signals predicted from the genome or proteome rather than from reference spectra. This approach, also described below, places more emphasis on mass assignments and less on ion intensity values. The subset of ions detected can change radically from one sample of a bacterium to the next (and often does so for biological reasons) so long as the "markers" are uniquely predicted by a specific genome or proteome. Because whole bacteria represent a complex chemical mixture, assignment of ions (masses) to the identities of specific proteomic biomarkers has been uncertain. Properly done, such assignments should rely, at a minimum, on accurate mass values. Mass accuracy has been a problem with MALDI-TOF MS of whole cells because of unresolved contributions from the many components having similar masses and also because the isotope profiles for proteins extend over several Daltons. Multiple components and isotope profiles often result in overlapping peaks in the mass spectra. As the extent of overlap changes with changes in the mixture composition (or contaminants) the observed, composite peaks appear to change in mass, leading to an apparent uncertainty in the mass assignment for diagnostic marker ions. Thus spectra from the same organism may show large taxonomically important signals whose mass values seem to change by 1 to 2 Da in replicate analyses. This often reflects differences in the extent of overlap, in a group of unresolved signals, rather than instrumental problems. Enhanced resolution (and mass accuracy) can largely eliminate such overlap, and this is the basis for recent interest in whole-cell MALDI Fourier transform mass spectrometry (FTMS) as an alternative (or complement) to MALDI-TOF MS. MALDI FTMS of whole bacteria is discussed in Chapter 13.

Usually bacteria are grown via culture methods prior to analysis. While this might be considered a disadvantage, it should be noted that competing molecular biology (and other) methods also typically require preliminary culture steps prior to their application. Because rapid analysis of samples directly from liquid culture media is possible, this method is attractive for the analysis of recombinant proteins directly from intact cells, especially when applied with sufficient mass resolution to detect differences in recombinant products not easily observed using 2D gels. Direct MS analysis of cells also provides unique insights into bacterial biology and chemistry based on the detection of specific chemical species as well as changes in response to environmental, temporal or other external influences.

6.2 MALDI MS OF CELLULAR EXTRACTS

As noted above, whole-cell MALDI-TOF MS was intended for rapid taxonomic identification of bacteria. Neither the analysis of specific targeted bacterial proteins, nor the discovery of new proteins, was envisioned as a routine application for which whole cells would be used. An unknown or target protein might not have the abundance or proton affinity to facilitate its detection from such a complex mixture containing literally thousands of other proteins. Thus, for many applications, the analysis of proteins from chromatographically separated fractions remains a more productive approach. From a historical perspective, whole-cell MALDI is a logical extension of MALDI analysis of isolated cellular proteins. After all, purified proteins can be obtained from bacteria after different levels of purification. Differences in method often reflect how much "purification" is done prior to analysis. With whole-cell MALDI the answer is literally none. Some methods attempt to combine the benefits of the rapid whole cell approach with a minimal level of sample preparation, often based on the analysis of crude fractions rather than extensively separated components. The analysis of cellular extracts by MALDI-TOF MS is somewhat beyond the scope of this chapter, but a few limited examples provide a perspective, a basis for comparing these two approaches, and reflect the complementary nature of the two approaches.

Most often proteins are the bacterial biopolymers studied using MALDI MS either from fractions or whole cells. They are not the only isolated cellular biopolymers studied by MALDI, nor the first. Very soon after the introduction of MALDI there were a few reports of the analysis of bacterial RNA or DNA from bacterial fractions. One of the first applications of MALDI to bacteria fractions involved analysis of RNA isolated from *E. coli.*[4] Other studies included analysis of PCR-amplified DNA,[5,6] DNA related to repair mechanisms[7] and posttranscriptional modification of bacterial RNA.[8] While most MALDI studies involve the use of UV lasers, IR MALDI has been reported for the analysis of double stranded DNA from restriction enzyme digested DNA plasmids, also isolated from *E. coli.*[9]

Large numbers of proteins from bacteria can also be characterized in bacterial fractions by MALDI. Typically this is best accomplished using peptide mapping and proteomics approaches. For example, consider a study of the organism associated with gastritis, ulcer and stomach carcinoma, *Helicobacter pylori.*[10] The known genome was used in conjunction with MALDI-TOF MS in a study of the functional proteome. From 1800 protein types, peptide mass fingerprinting helped identify 152 proteins, including 9 known virulence factors and 28 antigens. The results of this study illustrate why the whole cell MALDI method works for taxonomic classification. In this study of three strains of *H. pylori*, relatively few proteins were found in common. *This rarity of common proteins in closely related strains provides a basis for speciation at the strain level even when only a few proteins can be detected.*

One option is to attempt a compromise between the rapidity of analysis (whole cells) and exhaustive separation (isolated proteins). The number of proteins detectable from bacterial fractions is certainly larger than the number that could be detected from whole cells. As a rule, the number of proteins that can be detected by MALDI increases with the number of fractions collected. The rapidity of MALDI analysis on mixtures can be combined with a minimal level of separation based on the analysis of crude fractions. However, while this increases the number of peaks detected, and aids taxonomic classification, it is not as rapid and probably usually not necessary. Moreover this approach is sometimes applied when a secondary objective is the characterization of specific proteins. However, the additional separation is not usually sufficient to provide a reliability basis for the identification or proteins in such crude fractions using low-resolution TOF MS. Consider the following example: When high-performance liquid chromatography (HPLC) was used to separate components from a suspension of *E. coli* into crude fractions prior to rapid offline MALDI TOF analysis 300 peaks were detected in the 2 to 19 kDa mass range.[11] This represented a near order of magnitude increase in the number of components observable compared to whole cells.[11] This certainly provided additional ions for taxonomic classification. In the same study several of the crude fractions were subjected to proteolytic digestion. MALDI-TOF MS analysis of the digested fractions led to the classification of three cellular components as specific proteins normally expected to be present in *E. coli*. Based on a presumed protonated molecule at *m/z* 7273 and the masses of the several tryptic fragments, one protein component was identified as the *E. coli* protein CSP-C (a cold shock protein). However, subsequent exact mass whole-cell MALDI FTMS analysis of *E. coli* has not detected this protein. Instead, MALDI FTMS detected only a ribosomal protein at a similar mass (to close for differentiation by TOF) and no evidence for the proposed CSP-C protein.[12] It is not unusual for the same mass (ion) in MALDI-TOF mass spectra from whole cells (or extracts) from a given organism to be assigned different identities by various research groups. While it can be argued that low-resolution mass spectrometry provides evidence regarding the identity of proteins from whole cells or extracts, it does not provide proof of identity.

6.3 TAXONOMY: FROM ISOLATES TO WHOLE-CELL MALDI

Lubman's group demonstrated that MALDI-TOF mass spectrometry could be used to obtain characteristic and taxonomically specific mass spectral profiles from the proteins isolated from disrupted cells. They showed that MALDI-TOF MS of protein containing isolates was suitable for identification of bacteria.[13,14] They simply harvested cells by centrifugation, washed them with buffer, resuspended the cells and disrupted them by sonication. Finally a protein-containing fraction was collected from a methanol precipitate for MALDI-TOF MS analysis. Using a similar approach, proteins were additionally separated by capillary liquid chromatography and collected for offline analysis by MALDI MS.[14] This second approach required only pmol levels of protein to differentiate bacteria to the species level based on their overall protein profiles. Using a similar strategy, Krishnamurthy and Ross analyzed bacterial extracts from pathogenic and nonpathogenic organisms by MADLI-TOF MS and came to the same conclusions regarding bacterial taxonomy using bacterial extracts.[15]

The use of magnetic beads and avidin-biotin technology has also been proposed as a means of recovering bacterial proteins from cells for rapid analysis, in this case it was biotinlated proteins.[16] They mixed beads with the MALDI matrix solution and then by removing the beads with a magnet to transfer peptides from the beads to the MALDI probe. Next the matrix and any biotinylated peptides are analyzed by MALDI. Using either chromatographic fractionation or an affinity basis and beads, they showed that proteins can be isolated and pooled prior to MALDI analysis to give a characteristic fingerprint.

Holland et al. proposed a simpler and faster approach using the protein profile obtained *directly* from whole cells rather than from cellular extracts or affinity beads.[17] This was the first report of the whole-cell MALDI-TOF MS in a scientific journal. To demonstrate the methodology, they identified blind-coded samples of bacteria by comparison of the spectra from coded bacteria to spectra previously obtained using reference cultures. While the spectra obtained from reference and blind coded strains were not identical, the spectra were sufficiently alike that all the blind coded bacteria in this study were correctly identified. Figure 6.1 shows a reproduction of part of one set of the original spectra for the reference and blind coded spectra obtained from *Enterobacter cloacae* in this study. The spectra are not very impressive by todays standards, having been obtained using a very low-resolution liner TOF/MS. Nevertheless the ions at *m/z* 7763 and 9580 in the figure were observed with this organism and not any of the others in this study. Although a number of less intense ions were only seen in one of the two spectra (Figure 6.1*b*), these ions seemed much less significant at the time, and the correct identification was made based only on the two largest ions. While this may seem subjective, it should be kept in mind that the samples were blind coded. The spectral differences were nevertheless troubling, and a subsequent confirmation step was also developed, involving a direct comparison of the identified organism's reference spectrum with the blind-coded sample, with both spectra obtained from material run at the same time. This step provided an additional level of confirmation but did not change the results. Later it was learned that whole-cell spectral changes reflect both instrumentally and environmentally related differences.[3] There were some differences in bacterial growth and storage conditions that were then not deemed significant.

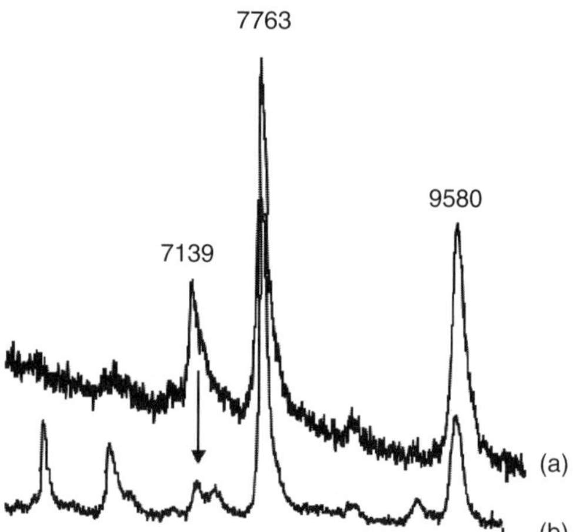

Figure 6.1 Sections (*m/z* 5000–10,000) from unpublished spectra—standard (*a*) and blind-coded (*b*)—obtained from *Enterobacter cloaca* from work reported in Holland et al.[17] showing the very low mass resolution, additional peaks, and three consistent signals used to identify this organism in the first reported whole-cell MALDI experiments in 1996.

Fortunately these unforeseen variables did not affect the most abundant peaks, nor the correct identification of the blind-coded bacteria. While differences associated with matrix and instrumental parameters were expected, the authors conducting this experiment did not anticipate the extent and speed with which bacteria could respond, through their protein profiles, to subtle environmental changes.

The direct whole-cell method of Holland et al. was extremely rapid, even in comparison to Lubman's MALDI analysis of fractions collected after bacterial sonnication. With the whole-cell approach bacteria were simply sampled from colonies on an agar plate, mixed with the matrix, air-dried, and introduced in batches into the mass spectrometer for analysis. In all of the spectra obtained in these and later experiments, each bacterial strain showed a few characteristic high-mass ions that were attributed to bacterial proteins. Studies demonstrating the whole cell methodology for strain-level differentiation were reported independently by Claydon et al. at almost the same time.[18] Shortly thereafter a third study on whole-cell MALDI included bacteria from pathogenic and nonpathogenic strains appeared.[19]

6.4 WHOLE-CELL MALDI MS

Spectra from "whole-cell" experiments typically show a small number (10–50) of peaks between m/z 3000 and 10,000. For example, whole-cell MALDI mass spectra obtained from two different species of *Vibrio* bacteria are shown in Figure 6.2. They were obtained using experimental conditions recommended by Williams et al.[20] These spectra are more representative of current whole-cell MALDI-TOF MS spectra than what is shown in the previous figure. The spectra are quite distinct, even though both bacteria are from genus *Vibrio*. The spectra show differences in ions over the 3 to 12 kDa mass range, and differentiation can be based on ion masses (presence or absence) without careful consideration of minor changes in abundance. For these related organisms there is very little overlap in mass values for prominent ions. *In other words, these spectra represent differences typical at the genus level where spectra are not at all similar*. Spectra in other studies sometimes contain higher mass ions. These ions are usually not needed for taxonomic differentiation. Usually ions from about 3 to 10 kDa are adequate. Generally, all MALDI spectra from whole bacteria show characteristic ions in this region, and thus it is now referred to as the "fingerprint" region for whole-cell MALDI mass spectrometry.

Because much smaller differences are anticipated at the strain level, strain-level differentiation and spectral reproducibility have received much attention. A 1998 work from Riley's group[21] is representative of a common approach to this issue. In Riley's studies *E. coli* strains shared many peaks in common but also showed strain-unique ions as well. Many of these ions fell in the fingerprint region near 3630, 3850, 4170, 4780, 5100, 5380, 7280, 8320,

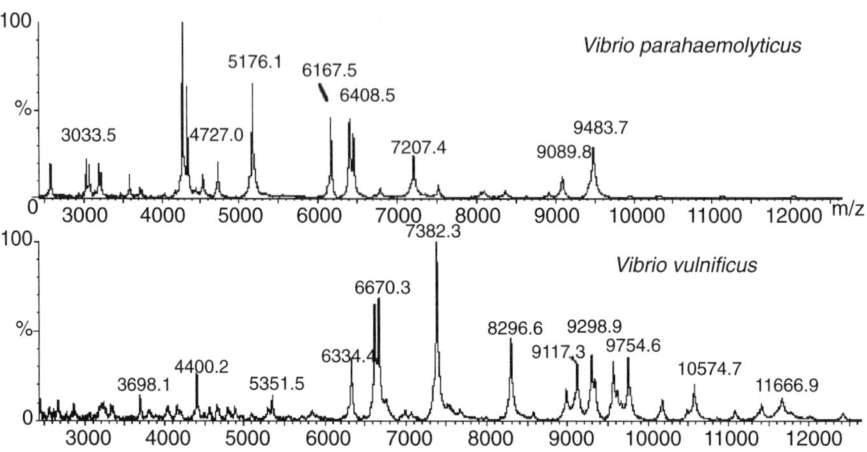

Figure 6.2 Whole-cell MALDI-TOF spectra of two *Vibrio* species taken using a modern TOF mass spectrometer. Even in the linear mode *V. parahaemolyticus* can be easily differentiated from *V. vulnificus*.

9070, 9530, and 9740 Da and seemed to differentiate one or more strains of *E. coli* from the others. This study was complicated by the observation that spectra from replicate samples of the same strain showed differences. The question was then determining which spectral changes were taxonomically significant and which were not. A mathematical treatment of the data was developed to compare MALDI-TOF mass spectra from whole bacteria using replicate spectra. This mathematical fingerprint-matching technique eliminated the need for subjectivity inherent in visually comparing spectra to determine whether or not they were the same or different. Using this mathematical approach, they readily distinguished 25 different strains of *E. coli*. Cells were grown in culture, samples were prepared, and MALDI-TOF mass spectra were recorded in a manner comparable with the previously reported studies. Pairs of spectra were compared based on a modified cross-correlation procedure. This approach increased the sensitivity of important small spectral differences that often allow strains to be distinguished.

Wang et al. also addressed the mass spectral reproducibility. They conducted a carefully controlled interlaboratory experiment where the effects of a number of parameters were systematically investigated.[22] They demonstrated that nearly identical spectra could be obtained in carefully controlled experiments. Minor variations in the sample/matrix preparation procedures for MALDI and in the experimental conditions used for bacterial protein extraction or analysis were shown to result in changes in the resulting spectra. They also noted that a subset of peaks was less sensitive to experimental variables. These ions appeared to be "conserved" in spectra obtained even under different experimental conditions so long as they were obtained using genetically identical bacteria. The existence of these conserved peaks helped explain the successful application of MALDI MS to bacterial characterization in many laboratories even without a general acceptance of a single set of standardized experimental conditions.

There is as yet not a consensus on a single best method for MALDI-TOF MS of intact bacteria.[3] One noteworthy study reports the development of a relatively rugged set of conditions that will provide high-quality spectra using simple and rapid approaches.[20] This study is noteworthy because the authors specifically addressed issues relating to spectral quality and the safe handling of bacteria by providing information on the viability of bacteria handled using the approaches they evaluated. Most whole-cell MALDI-TOF MS experiments start with a bacterial suspension. Because their method involves an initial suspension into a bactericidal solvent, it has been used for the analysis of killed samples of very harmful bacteria in a standard mass spectrometry laboratory. Samples are handled in the fume hood as a precaution, but the bactericidal suspension solvent essentially kills most (or all) of the bacteria.

The overall efficacy of whole-cell MALDI-TOF MS method for taxonomic classification can be judged based on comparisons with accepted methods. Recently such a comparison was reported for the analysis of 16 isolates of putative *B. pumilus* from spacecraft assembly facilities, the *Mars Odyssey* spacecraft, and the International Space Station.[23] This group reported that MALDI-TOF MS protein profiling was more accurate than metabolic profiling, more discriminating than 16S rDNA sequence analysis and complemented the results of DNA–DNA hybridization. They reported that the MALDI-TOF MS generated protein profiles correlated strongly with DNA–DNA hybridization data. Successful taxonomic differentiation was attributed in part to the observation of specific biomarker ions. Overall, in this comparison of several methods the authors reported that MALDI-TOF MS protein profiling was a rapid, simple, and useful taxonomic tool.

Another interesting comparison has recently been made between MALDI-TOF MS analysis of whole cells, and MALDI FTMS of the same organisms. This work is reported in greater detail in a dedicated chapter later in this book. It should be noted here that it appears to be much more difficult to obtain spectra from intact bacteria by MALDI FTMS than it is by MALDI-TOF MS. Thus far only a single research group has reported protein-like ions desorbed directly from intact cells by MALDI FTMS.

6.5 BIOLOGY-BASED CHANGES IN WHOLE-CELL MALDI SPECTRA

Biologically (environmental) based changes in the protein composition of cells might have been anticipated. Bacteria respond rapidly to environmental changes. Changes in cellular processes occur very soon after changes when cells are stored, handled or cultured over different time periods prior to analysis. These changes result in rapid changes in the protein profile. Arnold et al.

reported the time evolution of MALDI spectra taken from a growing bacterial culture by periodically removing and analyzing whole cells.[24] Their mass spectra contained tens of peaks in the 3 to 11 kDa mass range. When cultures of *E. coli* strain K-12 were grown under otherwise identical conditions and sampled periodically from 6 to 84 hous after inoculation, the spectra varied considerably in both the numbers and intensities of peaks. In several experiments the relative intensities of several of the stronger peaks changed dramatically and consistently over time. They noted that such temporal characteristics must be taken into account when MALDI produced spectral patters are applied to bacterial identification. Bacterial incubation time is one of the variables that most needed to be carefully controlled. While illustrating a potential problem for taxonomic characterization, there results demonstrated how MALDI MS of whole cells might be used to monitor biological changes. Welham et al. also examined time-related changes in bacterial spectra, but over a much longer time frame.[25] Even with three months between experiments spectra contained a number of relatively reproducible ions.

6.6 ANALYSIS OF MIXTURES

One of the most common questions about the analysis of bacteria by mass spectrometry deals with the analysis of mixtures of organisms. This is an important question. Mass spectrometry, like most methods for the identification of bacteria, normally includes a preliminary culture step from which it is expected that only a single organism will be grown. This step is used to increase sample size, and also because most methods give unreliable results when a mixture is analyzed rather than a pure culture. A mixture of bacteria could be sampled multiple times and cultured onto many different media. Different plates or liquid cultures might then grow different organisms, but the analysis of organisms from each plate or liquid medium still represents the analysis of a single species or strain. Nevertheless, there are some applications where the analysis of a mixture of organisms might be desirable. One possible approach involves chromatographic separation. The application of a chromatographic step to the analysis of protein isolates from a single organism was mentioned in a previous section. An alternative application of chromatographic principles is the resolution of organisms in a mixture into discrete fractions with separate identification of each using MALDI mass spectrometry. Field-flow fractionation mass spectrometry was used to study differences in an organism associated with growth stages and other parameters presumed to produce bacteria having different sizes.[26] Later another group proposed to use a form of field-flow fractionation called hollow-fiber flow field-flow fractionation [HF FlFFF] to fractionate bacteria, in a mixture of organisms, for taxonomic identification. Separated fractions were identified by MALDI-TOF MS analysis. They reported that the most characteristic ion signals of each species was detected with little or no signal from the other when a mixture of bacterial species was separated and analyzed.[27] The low volumes (minimal dilution), disposable nature of the HF FlFFF column, and rapid fractionation make this method an attractive approach for mixtures when relatively large quantities of bacteria are available and bacterial culture steps are not preformed prior to mass spectral identification.

A more direct approach to the analysis of mixtures is simultaneous analysis of mixture components by whole-cell MALDI-TOF MS. One group has now reported the analysis of microbial mixtures in two double-blind studies.[28,29] In the most recent one[29] they reported that nine different organisms were used to produce 50 different simulated mixed bacterial cultures, and that while an automated data analysis approach using a spectral fingerprint library showed promise, there were some difficulties encountered for mixtures that included closely related organisms or unrelated organisms having somewhat similar spectra. While it is perhaps somewhat unexpected that bacteria from different Genera (*Serretia*, *Escherichia*, and *Yersinia*) would be as difficult to differentiate in mixtures as closely related strains, the approach is nevertheless an important one. Because of its rapidity, the direct analysis of bacterial samples that include mixtures has shown sufficient efficacy to suggest its use for screening purposes.

Rather than comparing the spectrum from a bacterial mixture to the spectra of individual bacteria, spectra from mixtures (or pure cultures) can be examined for the presence of specific biomarker compounds. While the detection of a smaller number of maker ions provides less taxonomic specificity than a complete spectrum, taxonomic specificity is not the only basis for considering the components in a mixture of bacteria.

Specific biomarker proteins can be associated with specific genes and hence can be suggestive of bacterial properties. While it is certainly true that mass assignments alone do not provide proof of the identity of a protein marker, they can be used as evidence and for screening purposes. The detection of a an ion (mass) for a specific protein in a mixture might signal the possible presence of a target ranging from a specific organism to any of a group of organisms having a specific protein or gene. This approach was used to demonstrate a procedure for screening samples of cotton cloth, lettuce, and water for the presence of organisms having a specific acid-resistance gene.[30] Biomarker ions previously identified for this virulence-related gene were detected in rinses recovered from samples heavily contaminated with either of two organisms contaminating the gene. Organisms recovered from the sample washes were directly analyzed by whole cell MALDI-TOF MS without purification or amplification by pre-MS culture steps. Samples contaminated with acid-resistant bacteria showed marker ions that indicated the presence of acid-resistant organisms, whereas the controls did not. While this study did not demonstrate taxonomic specificity, it did demonstrate the detection of organisms that could survive in the human gastrointestinal tract. These organisms were detected in a rinse sample containing the adulterant and all of the other organisms washed from the samples at the same time.

6.7 EXPERIMENTAL APPROACHES

Different experimental approaches have been proposed, in part to increase specificity. Increasing the mass range over which ions are detected from whole cells is one approach. Welham's study[25] was noteworthy because of the detection of some ions with masses approaching 40 kDa. These ions were much higher than the normal "fingerprint" region. A number of high mass ions were also noted in a study by Winkler et al.,[31] who reported that differentiation of *Campylobacter jejuni*, *Campylobacter fetus*, *Campylobacter coli*, *Helicobacter pylori*, and *Helicobacter mustelae* was facilitated by the presence of ions ranging up to 62,000 Da. While this is not a common observation, in this study these high-mass ions were as important as the ions in the fingerprint region. Recently ions over 158,000 Da have been reported in a study using a dried-droplet method with a solvent having very high aqueous content. This was attributed, in part, to very slow matrix crystal formation on the MALDI target.[32]

The use of a ferulic acid matrix has also been associated with high-mass ions in spectra from whole cells. This matrix has been systematically studied for production of high-mass ions. For example, Voorhees's group has evaluated methods to enhance the detection of high-mass ions using this compound in their MALDI matrix.[33] Their matrix consisted of 12.5 mg of ferulic acid dissolved in 1 ml of a solution with a composition of 17% formic acid, 33% acetonitrile, and 50% water. Their procedure involves depositing growing bacteria colonies from culture dishes directly onto the MALDI probe, treatment with 40% ethanol, and then the ferulic acid matrix solution described above. Using this approach, high-mass ions were observed along with a substantial number of intense signals in the more "conventional" fingerprint region extending to 20 kDa. Like Winkler,[31] they suggested that high-mass ions might be particularly important to the differentiation of some bacteria based on the rarity of high-mass signals and the minimal background present in the higher mass region of bacterial MALDI spectra. They also illustrated the utility of using high-mass ions for bacterial differentiation by using a sample containing both *E. coli* and *Proteus mirabilis*. This mixture showed six characteristic ions for each organism between 15 and 70 kDa and also demonstrated that at least for this mixed experiment, the presence of one organism did not significantly suppress characteristic ions from the other. The ability to detect ions from coexisting bacteria in mixtures will play an important role as MALDI-TOF MS is extended to many potential "real-world" applications, and this observation may be more important than the extension of the mass region for the "bacterial fingerprint."

Despite the large number "whole-cell" MALDI protocols that have been tested a single approach has not yet been widely adopted. There remain different and sometimes conflicting reports in the literature regarding optimum methodologies.[3] In addition to the ferulic acid matrix mentioned above sinapinic acid, α-yano-4-hydroxycinnamic acid, 2,4-hydroxyphenyl benzoic acid, and 2,5-dihydroxybenzoic acid have all been used as matrix compounds. Both Nd/YAG and nitrogen lasers have produced useful data. Studies have addressed other solvent or matrix effects[34] the use of internal controls to correct for the concentration dependence on bacterial spectra[35] and the use of a two-layer or dried droplet sample preparation procedure for MALDI sample preparation.[32,36] Concentrations that are either to high or to low can result in spectra with little or no indication of the presence of bacteria in the sample.[35] Generally a cloudy suspension gives good results. Bacteria can be concentrated by centrifugation and the suspension solution decanted sufficiently to give such a suspension. It is likely that the debate regarding the best conditions for whole-cell MALDI continues not because of the difficulty of obtaining good spectra but rather because of the ease with which spectra can be obtained using many similar methods.

Spore-forming bacteria have also been studied by whole-cell MALDI-TOF MS analysis. Like other bacteria, unique ion patterns differentiate *Bacillus* species members, one from another giving the typical MALDI whole-cell fingerprint region below 10kDa.[37] Lipophilic biomarkers, attributed to compounds present on the outside of the spore, also characterize *Bacillus* spores. This has been show with bacterai grown using different media and after storage for more than 30 years. Specific characteristic ions were detected in replicate analyses under a variety of different conditions. Because of the protecting protein capsule, modifications to the direct MALDI analysis method have been proposed for spore-forming bacteria. Sonication of bacteria is one approach and a brief exposure to a corona plasma discharge (CPD) prior to MALDI MS is another. Either approach should disrupt the spore capsule and enhance MALDI MS characterization. The number of detectable biomarker ions in the MALDI spectra of CPD-treated spores generally increases compared to MALDI spectra from untreated cells. This simplifies differentiation and gives results very similar to those obtained by sonication of the spores prior to MALDI analysis.[38] Cotter's group has demonstrated rapid screening for *Bacillus* spore species by the analysis of in situ tryptic digestion products formed from a selected group of proteins isolated from the spores.[39] While these were not the rapid whole-cell analyses, it is noteworthy that these analyses were accomplished using a miniaturized MALDI-TOF MS. The development of miniature mass spectrometers will increase the possible contributions of whole-cell MALDI-TOF MS and similar approaches to the problems ranging from detecting bioterrorism to clinical applications.

An interesting variation on the whole-cell MALDI approach was recently reported in a study aimed more at FTMS than TOF MS, but the results are nevertheless interesting and important to users of both methods for analysis of bacteria 40. Wilkins's group showed both MALDI-TOF and MALDI-FTMS spectra of whole bacteria grown on isotopic media depleted in C^{13} and N^{14}. Because most bacterial identification protocols involve a culture step prior to analysis, it is possible to manipulate the sample based on control of the growth media. For mass spectral analysis manipulation of the isotope profile is an important option. The measured mass resolution from TOF mass spectra of bacteria generally seems much lower than expected based on the intrinsic resolution of the instrument. This is because the observed resolution in the spectrum reflects both the instrument performance and the nature of the sample being analyzed. The presence of overlapping signals can cause a broad peak to appear because of the presence of "chemical noise" from overlapping components. These components can be other substances or even isotopes. In some sense the isotope peaks can be viewed as "noise." In TOF mass spectra they broaden the width of peaks for proteins and all other species having large numbers of carbon atoms. For a typical small protein the isotope profile may cover 10 Da. Reduction of the mass width to only a few Daltons allows many more proteins to be detected in whole cell MALDI-TOF mass spectra.[40]

6.7.1 Data Interpretation

As noted above, the initial applications of whole-cell MALDI to bacterial identification were based on spectral comparisons. Comparisons based on library spectra rather than co-analysis minimize the need for reference cultures but also introduce some problems because the spectra of identical cultures can very somewhat from analysis to analysis. Jarman et al. developed a method for constructing and extracting reproducible MALDI spectral from whole cells. Their method is based on statistics, automated data extraction, and a relatively high level of spectral reproducibility.[28,41] The method can be used to extract a fingerprint (specific biomarkers) characteristic of a targeted bacterium from a complex spectrum containing multiple analytes or other background signals. This approach has been demonstrated in studies involving mixtures of dissimilar[41] and closely related[28] organisms using blind-coded samples.

In response to observations that spectra from readily change in replicate experiments an alternative method was developed to minimize the impact of such changes.[42] It exploits the information contained in publicly available genomic and proteomic databases. This approach uses mass values extracted from database entries rather than reference spectra in comparisons used for identification. Spectral libraries cannot easily include spectra from (a) a large number of bacteria, (b) grown under wide a variety of different conditions, and (c) analyzed on a number of different instruments ($a \times b \times c$ entries). In theory, the genomic/proteomic approach described above can be scaled to a large number of organisms (and conditions) if the proteome or genome is known. For each organism only a single entry into the database is needed, and often the databases are freely accessible over the Internet. In this second approach the masses of a set of protein ions above a threshold are extracted from an experimental spectrum. Each ion mass in the spectrum is assumed to be associated with a protein (and mass) from the database. Masses from the spectrum are compared to database masses for a variety of target organisms. Convoluting the lists for all ions and ranking the organisms corresponding to the number of matched ions is the basis for identification. For small numbers of organisms the method has been tested successfully using mixtures of microorganisms, using spectra from organisms at different growth stages, and using spectra originating in multiple laboratories. While spectra might change in concert with environmental or experimental changes, it is presumed that proteins from an organism can be associated uniquely with the correct set of predicted proteins, and identification should be possible regardless of environmental or experimental conditions. A variation on this approach is to look at tryptic fragments from whole cells, rather than the masses of the directly desorbed proteins.[43] Either way, this approach has some disadvantages. Only a limited number of bacteria exist for which the genome is known. Perhaps more important, higher accuracy in the mass assignment step is needed to minimize the number of possible protein assignments for each measured mass. Even for a limited number of organisms, there are many possible assignments for any given nominal mass value. Many of the masses in the mass spectrum are from proteins represented in the database by different mass values. A complex variety of biological processes can cause the protein masses in the database to differ from masses predicted from the genome. This difference has resulted in some misidentifications of proteins and bacteria. In summary, there are problems with the use of either spectral libraries and or mass comparisons with database-derived masses. Using low-mass resolution TOF MS, comparison of whole-cell MALDI spectra with libraries is viable for organisms grown and analyzed using similar approaches. The use of mass values in comparisons with proteomic or genomic databases can bridge experimental or environmental differences. Nevertheless, mass values should be used carefully, primarily for screening because of the limited specificity of mass values derived from low-resolution spectra, especially with complex mixtures such as whole cells. Otherwise, complex strategies using cells, protease, chemically treated cells, and isotope manipulation and other experimental approaches to increase taxonomic specificity need to be developed.

6.7.2 Sensitivity

The most common criticism of whole-cell MALDI is that the method requires a relatively large number of cells, usually obtained directly from culture media. In principle, an analysis of even a few unknown bacteria (a colony-forming unit) is possible after a culture step. More important is the number of bacteria needed in a sample or on the sample probe for successful analysis. Detection of a very small number of bacteria could eliminate the need for a preliminary culture step. This would be a considerable asset for environmental analysis (unless to many bacteria were detected) and for the detection of a bioterrorism-related release of bacteria.

A few particularly novel approaches have been developed to increase the sensitivity (and specificity) involved in the detection of bacteria by MALDI. The use of avadin-biotin technology to recover biotinylated proteins from bacteria has already been mentioned.[16] Perhaps more noteworthy in the context of recovering whole cells is the Fenselau group's used of immobilized lectins incorporated into "affinity surfaces" to isolate several kinds of samples, including bacteria, for mass spectrometry analysis.[12,44] The recovery of bacterial species having a carbohydrate-binding motif was demonstrated using a Concanavalin A probe. The probe was immobilized on a gold foil via a self-assembled monolayer. Then urine spiked with *E. coli* was placed into contact with the probe surface and allowed to interact with it briefly. Just prior to MALDI-TOF MS analysis the surface was washed to remove salts and other unbound components. This lectin-derivatized surface allowed bacteria to be concentrated and readily characterized at levels corresponding to about 5000 cells applied to the capture surface. However, the target biomarkers in the study using lectin-derivatized surfaces were not proteins but rather phosphatidylethanolamines. The use of small molecules (lipids, etc.) directly desorbed from cells for bacterial identification has also been demonstrated by the recent report of exact mass data for phospholipids, obtained using the same whole-cell MALDI protocol described for desorption of proteins.[12]

Voorhees's group has proposed a clever way of amplifying components in a bacterial sample that can produce taxonomic biomarkers without the use of lengthy culture steps to grow additional bacteria.[45] One part of their strategy is straightforward, namely immunomagnetic separation (IMS). With IMS magnetic beads with an immuno-tag are used to concentrate the bacteria from a samples onto a small number of easily recovered magnetic beads. This step will concentrate bacteria. For smaller numbers of organisms the total number of bacteria recovered may be undetectable by MALDI MS. In a second step they infect the captured bacteria with a low concentration of bacteriophage specific for the target organism. If the target bacterium is present, the phage will infect cells and produce very large quantities of phage-related proteins using the host cell. This process can be completed in less than an hour. MALDI MS is then used to analyze the samples for proteins associated with the phage. If these proteins are present at levels higher than was additionally added to infect the cells (an undetectable level) it indirectly indicates that the target bacterium is present. The method was used to reduce the detection limit for *E. coli* to less than 5.0×10^4 cells/ml. (The cells/ml value is fundamentally different that the number of cells reported present *on the MALDI probe* in other experiments.) In their study immunomagnetic beads were coated with antibodies specific for *E. coli*. The antibody-coated beads were used to isolate *E. coli* bacteria from a solution. The *E. coli* containing beads were suspended in a solution containing a small quantity of MS2, an *E. coli*-specific bacteriophage. After a brief (40-minute) incubation period an aliquot of the solution was analyzed by direct MALDI/TOF MS with detection of the 13kDa MS2 capsid protein. This represents a one- to twofold increase in sensitivity compared to the traditional approach if you consider only the numbers of bacteria needed to produce a signal. However, in the traditional experiments bacteria are typically sampled in a 1 microliter suspension from a much larger sample volume, whereas with the use of immunomagnetic beads this number may represent a large fraction of the total number of bacteria present in the entire sample. Hence this approach, and the other affinity-based approaches may really represent an improvement of three or more orders of magnitude in sensitivity compared to the simple whole-cell MALD-TOF MS method.

6.8 IDENTIFICATION OF PROTEIN MARKERS

While many studies have used nominal mass values and matches with predicted protein masses to assign (infer) protein identities, only a few studies have actually shown data sufficient to properly identify any of the bacterial proteins giving rise to the "protein" signals in the MALDI-TOF mass spectra from whole cells. More exact identification of proteins is important for several reasons. First, there is clear evidence that the simple matching procedure described above has led to the same ion being identified as more than one protein. Second, the data produced in mass spectrometry studies is being used to support protein identity in protein databases, sometimes based only on nominal mass matches with no other supporting evidence. Finally, the identification of specific biomarkers might enhance the value of these ions for taxonomic identification, especially if they are from a conserved set of biomarkers or if they are used to develop non–mass spectrometry based tests used to detect specific proteins by less expensive means. For proteins that are associated with specific toxicological or medical properties in several organisms, it might also be possible to use these protein markers to detect bacteria with specific attributes or genes. Holland et al.[46] reported the identification of several proteins, observed both from cells and in filtered cellular suspensions, which were isolated by HPLC and identified on the basis of their mass spectra and their partial amino acid sequence, determined using the Edman method. For one pair of bacteria, *E. coli* and *Shigella flexneri*, whole-cell MALDI-TOF mass spectra showed ions near m/z 9060 and m/z 9735. This is noteworthy because both organisms are considered "acid resistant", meaning they are not readily killed in the GI tract. It is now known that these two ions are protonated molecules for the biologically processed forms of the acid-resistance proteins HdeA and HdeB, observed in whole cells and cellular extracts from both *E. coli* 1090 and *Shigella flexneri* PHS-1059. These two proteins were also detected in *E. coli* in other studies but were not so identified because of the problem mentioned above. Both masses can be associated with other proteins in the database for *E. coli* (using TOF MS data) based on nominal mass values. However, they cannot be easily associated with the parent proteins in the database because the forms of the proteins present in the cells, which is detected by whole-cell MALDI (and in extracts), results from biological processing of the parent protein within the cell. Even when data from enzymatic digestion has been used to supplement the "whole-cell" data, incorrect identifications have been reported for this protein. Because of the partial amino acid sequence obtained by Edman analysis it is clear that these experiments actually demonstrate the detection of two virulence-related biomarker proteins in different organisms as well as the need to take into account cellular biological processing steps when assigning the identity to proteins based on genomic data. A few other proteins were also confirmed based on Edman analysis in this same study.[46] A cold-shock protein, CspA, was associated with an ion near *m/z* 7643 from *Pseudomonas aeruginosa*, and a cold-acclimation protein, CapB, was identified as the source of the ion near *m/z* 7684 in *P. putida*. This last protein was homologous with a CapB protein to be present in *P. fragi*. Thus a common protein seems to be detectable in two species of *Pseudomonas* bacteria as well.

6.9 DETECTION OF TARGET PROTEINS

In general, whole-cell MALDI MS normally detects only a small subset of the cellular proteins. This is a very small group compared to what can be detected by the analysis of cellular extracts or fractions. It includes only the most basic and abundant proteins within the cells, ribosomal proteins, for example. For taxonomic classification of bacteria this is not a complication. So long as the subset of proteins is taxonomy-characteristic, and the ribosomal proteins are, you simply accept whatever set of proteins are observed as a "fingerprint." Thus the detection of a specific target protein is not usually attempted outside this small group. However, because the recombinant proteins are typically so greatly overexpressed in cells, they do represent reasonable targets for analysis directly from whole cells. Their great abundance can compensate to some extent for a somewhat lower ability to attract protons in the initial MALDI ionization step. In other words, recombinant proteins are a special case. *Recombinant proteins have been detected in MALDI experiments using cells with very minimal or no cleanup*.[47–50] MALDI-TOF MS coupled with a rapid sample preparation steps allows very rapid analysis of recombinant proteins in cells directly from the growth media with a total time for analysis that was reported to be less than 10 minutes in favorable cases. The resolution of the chemical nature of posttranslational modifications and other deviations from the expected chemical structure of the recombinant proteins is possible by mass spectrometry, whereas, when using gels, it was not always possible to separate similar overexpressed proteins.

Notwithstanding the aforementioned difficulty in detecting specific target proteins other than the types normally observed in the "taxonomic fingerprints" from whole bacteria MALDI spectra (i.e., ribosomal proteins), some other target proteins and protein-like materials have been studied directly from whole cells. For example, Lantibiotics, antimicrobial peptides secreted by Gram-positive bacteria have been detected directly from whole bacteria by MALDI-TOF MS.[51] The lantibiotics nisin and lacticin 481 were detected from whole cells and crude supernatants. Surprisingly, better results were reported from whole cells than the supernatants. In this study the presence of variants of lacticin 481 resulting from mutations of the structural genome was confirmed. The decline in efficacy of traditional antibiotics may lead to increased interest in studying lantibiotics in bacteria, and whole-cell MALDI-TOF MS may provide a very rapid method for their study based on rapid analysis directly from the host cells.

6.10 ANALYSIS OF CLINICAL ISOLATES

There have been a limited number of reports suggesting that rapid whole-cell MALDI MS might be useful for differentiation of strains of pathogenic organisms. One example involves speciation of *Haemophilus*. *Haemophili* are important etiological agents causing pneumonia, meningitis, conjunctivitis, and other diseases. However, the identification and speciation of *Haemophilus* in clinical settings has been slow and expensive. A paper by Haag and co-workers reported the use of MALDI-TOF MS for the rapid identification and speciation of pathogenic *Haemophilus* strains.[52] In addition to differentiation of *H. ducreyi* from other genus and species, they also reported strain differences from different isolates of the same organism. They reported that mass spectral "fingerprints" permitted the rapid speciation of pathogenic forms of *Haemophilus* and other organisms.

MALDI-TOF MS has been applied to bacteria causing food-borne illness. The health effects of human food-borne illness in the United States remain significant despite many advances in food preparation techniques and sanitation. A report published by CDC suggested that there were over 76 million food-borne illnesses in the United States in 1997.[53] In addition 325,000 hospitalizations and 5000 deaths were reported as a direct result of microbial contamination of food. Interestingly 64% of the deaths were from unknown organisms. This clearly points to a need for additional methods for identifying bacteria causing such illness. In one blind-coded study an unknown clinical isolate was classified by rapid whole cell MALDI-TOF MS based on a comparison with reference cultures.[54] Figure 6.3 shows some of the original data from this study. The very low resolution of the linear TOF MS used is evident from the spectra. Nevertheless the authors noted a subtle difference in spectra obtained from Gulf Coast and Pacific Northwest strains of *Vibrio parahaemolyticus.* While these original spectra are poorly labed on the mass axis, a horzontal line on the figure denotes the region near m/z 9500. Differentiation was based on whether or not the region near m/z 9500 showed a resolved doublet or only a single peak. The CDC review also noted that this second peak present in samples from the Texas Gulf Coast was also not observed in strains missing the TDH gene or in another *Vibrio* species that did not even have a TDH-like gene. Some of these isolates were associated with a recent (1997) outbreak of *V. parahaemolyticus* in the Pacific Northwest region of the United States. For this reason the rapid whole-cell MALDI MS results were compared to a review of the results reported for the microbiology studies

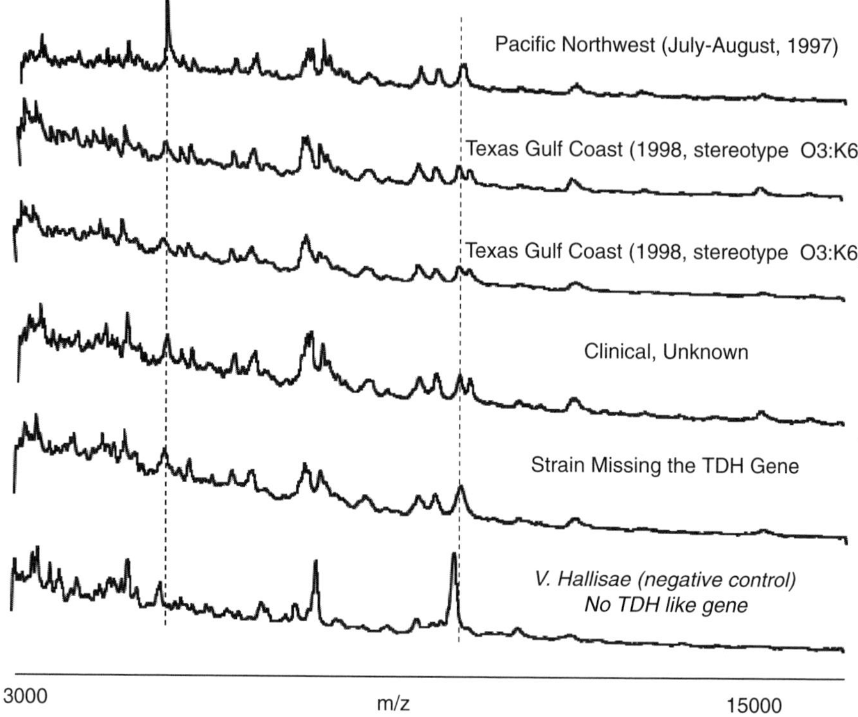

Figure 6.3 Differentiation of *Vibrio parahaemolyticus* strains by region of clinical outbreak, showing the correlation of a mass spectral feature (near dotted line), the region of clinical outbreak, and the presence or absence of the TDH gene in control organisms. (From a linear TOF, reported at the American Society for Mass Spectrometry in 2000.)

conducted on these strains during this outbreak.[55] The CDC reported that typical microbiological procedures resulted in detection of between three and six types, and sometimes several subtypes, from this single outbreak. It noted that genetic changes occurring even during the outbreak complicated the classification of strain types and subtypes. On the other hand, the MALDI-TOF analysis of intact cells seemed to readily differentiate strains on a more general basis. Moreover the results suggested that this might have something to do with the TDH gene. Thus Wilkes et al. reported that whole-cell MALDI MS could be used to rapidly differentiate *Vibrio* bacteria in general and to classify outbreaks of this specific organism by region. Figure 6.4 shows the distinctive mass region for these two strains obtained later with higher mass resolution. With these data it is clear that the initial strain differentiation was based on a reproducible change in the mass of a characteristic marker ion that produced a resolved doublet in the low-resolution spectra for one strain and a single peak in the other. While protein ions near *m*/*z* 8911, 9088, 9419, and

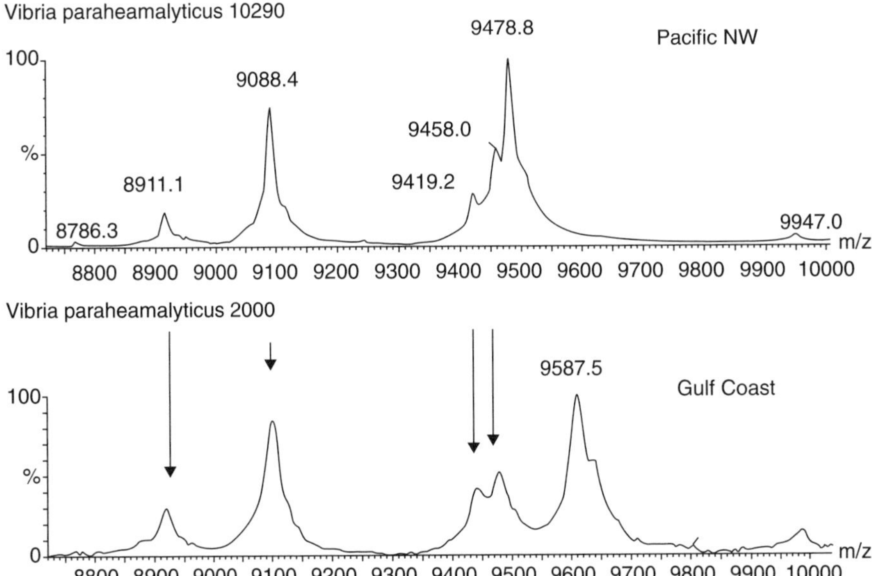

Figure 6.4 Expanded portion of the MALDI-TOF mass spectra of two different clinical isolates (by region) of *Vibrio parahaemolyticus* obtained on a reflectron TOF/MS, showing a single change in mass of a protein from *m/z* 9479 to *m/z* 9587 giving rise to the strain and geography-related differences observed in Figure 6.3 and reported in Wilkes et al.[54]

9458 seem to be common among these strains, a signal at *m/z* 9479 seems to differentiate the Pacific NW strain from the Texas Gulf Coast strains, which consistently show a peak at a higher mass near *m/z* 9588. The reproducibility of the spectra obtained from these strains is demonstrated by the spectra in Figure 6.5, over the same mass range. It shows data from two isolates of the same strain, from the Pacific Northwest. While the assigned masses and peak abundances change somewhat from sample to sample of this strain, the basic pattern is very reproducible.

An increasingly important set of human-health-related differences among bacteria results from their emerging resistance to antibiotics, bactericides, and sterilization procedures. Progress has been reported on the differentiation of antibiotic-resistant and nonresistant strains of one important human pathogen, *Staphylococcus aureus*.[56] The rapid and accurate discrimination between methicillin-sensitive and methicillin-resistant strains of this organism (and others) could lead to major improvements in the treatment strategy for infected patients. This strain-specific differentiation by MALDI-TOF mass spectrometry was demonstrated using intact cells from 20 *staphylococcal* isolates. The spectra showed characteristic peaks (masses ranging from 0.5–10kDa) with species-specific and strain-specific markers as well as

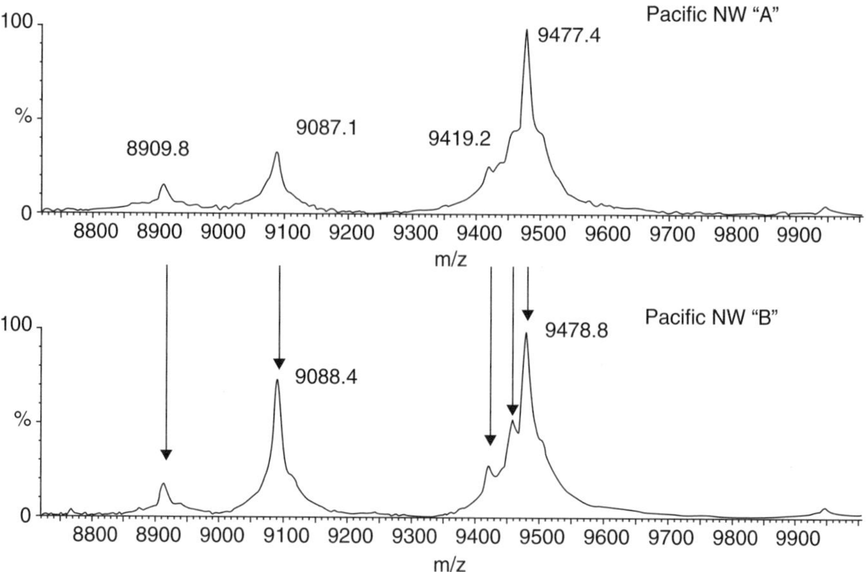

Figure 6.5 Expanded portions of the MALDI-TOF mass spectra of two clinical strains of *Vibrio parahaemolyticus* from the same outbreak (and region) obtained on a reflectron TOF/MS, showing consistent spectral features observed in all strains from this region/outbreak.

characteristic markers for the methicillin susceptibility status of the strain being detected. The results of this study have been replicated in another study of the same organism[57] and in a study that also included two other anaerobic bacteria of clinical importance.[58] While reports of such successes are very promising, it has been suggested that Gram-negative bacteria, such as *Staphylococcus* and *Streptococcus*, give even better results when a small quantity of lysozyme is added about 30 min. prior to MALDI-TOF MS analysis.[59,60] This approach is designed to disruption the thick peptidoglycan layer of the cell walls characteristic of Gram-negative bacteria. By this approach reproducible strain-specific ions were detected in larger numbers compared to analysis without lysozyme for reference strains and clinical isolates.

The use of MALDI-TOF MS as a rapid tool for the identification of common clinical bacterial isolates was also evaluated in a study of 25 carefully selected isolates of pathogenic *E. coli* (14 isolates) and *Enterobacteriaceae* (9 isolates) bacteria.[60] Organisms were prepared in accordance with standard clinical microbiological protocols and analyzed with minimal additional processing. Spectra were reproducible from preparation to preparation, generally showing 40 to 100 peaks in each spectrum with masses up to 25 kDa. By the spectral comparison approach, *E. coli* isolates closely related to each other were readily distinguishable from one another and from *Enterobacteriaceae*, including *Salmonella* and *Shigella*. In this experiment, which processed the samples as clinical samples, the method allowed the analyisis of 40 unknown isolates per hour. This group reported that MALDI-TOF MS offered a rapid and reliable approach for identification of unknown bacterial isolates.

6.11 CONCLUSIONS

It is now possible to evaluate the reproducibility and taxonomic utility of the MALDI-TOF MS method using spectra generated over a period of about eight years.[3,61,62] Comparisons between MALDI-TOF MS and other standard methods have now been reported,[23] and the method clearly can be used for rapid screening, if not yet for positive identification.

Current data also support claims regarding the existence of sets of bacteria-specific conserved biomarker ions. For example, Lynn et al. reported MALDI-TOF MS spectra for several bacteria that had previously been analyzed, and their spectra reproduced a number of the same specific marker ions claimed as species or strain-specific for these organisms in prior studies.[63] The biological origin and general nature of the ions observed in MALDI spectra from bacteria are now fairly well established. Fenselau's group has shown, using *E. coli*, that the MALDI-produced ions are predominantly associated with strongly basic proteins, moderately hydrophilic in nature, likely coming from the cytoplasm and organelles such as the ribosome.[64] In addition to these readily observed protein markers, recombinant proteins and a few other proteins of biological or biomedical interest can also be studied based on direct analysis from cells. Although the number of cells needed for characterization by MALDI-TOF MS may be at high at about 10^6 (on probe), several groups have reported analysis of smaller numbers of cells and demonstrated methods to enhance detection of smaller numbers of bacteria based on the concentration of biological amplification steps. These include methods for isolation and concentration from larger samples and even one approach for production of surrogate proteins from bacterial parasites with specificity for specific target host cells.[45] It is easy to anticipate widespread application of this method in microbiology laboratories for specific research applications in the near term. Indeed, a number of publications of whole-cell MALDI-TOF MS applications have already appeared in the microbiology literature.[23,37,38,51,56,58–60]

REFERENCES

1. Welham, K. J.; Domin, M. A.; Johnson, K.; Jones, L.; Ashton, D. S. Characterization of fungal spores by laser desorption/ionization time-of-flight mass spectrometry. *Rapid Comm. Mass Spectrom.* 2000, **14**, 307–310.
2. Tan, S. W.; Wong, S. M.; Kini, R. M. Rapid simultaneous detection of two orchid viruses using LC- and/or MALDI-mass spectrometry. *J. Virol. Meth.* 2000, **85** (1–2), 93–99.

3. Lay, J. O., Jr. MALDI-TOF mass spectrometry of bacteria. *Mass Spectrom. Rev.* 2002, **20**, 172–194.
4. Nordhoff, E.; Ingendoh, A.; Cramer, R.; Overberg, A.; Stahl, B.; Karas, M.; Hillenkamp, F.; Crain, P. F. Matrix-assisted laser desorption/ionization mass spectrometry of nucleic acids with wavelengths in the UV and IR. *Rapid Comm. Mass Spectrom.* 1992, **6**, 771–776.
5. Hurst, G. B.; Doktycz, M. J.; Vass, A. A.; Buchanan, M. V. Detection of bacterial DNA polymerase chain reaction products by matrix-assisted laser desorption/ionization mass spectrometry. *Rapid Comm. Mass Spectrom.* 1996, **10**, 377–382.
6. Hurst, G. B.; Weaver, K.; Doktycz, M. J.; Buchanan, M. V.; Costello, A. M.; Lidstrom, M. E. MALDI-TOF analysis of polymerase chain reaction products from methanotrophic bacteria. *Anal. Chem.* 1998, **70**, 2693–2698.
7. D'Ham, C.; Romieu, A.; Jaquinod, M.; Gasparutto, D.; Cadet, J. Excision of 5,6-dihydroxy-5,6-dihydrothymine, 5,6-dihydrothymine, and 5- hydroxycytosine from defined sequence oligonucleotides by *Escherichia* coli endonuclease III and Fpg proteins: Kinetic and mechanistic aspects. *Biochemistry* 1999, **38**, 3335–3344.
8. Kirpekar, F.; Douthwaite, S.; Roepstorff, P. Mapping posttranscriptional modifications in 5S ribosomal RNA by MALDI mass spectrometry. *RNA* 2000, **6**, 296–306.
9. Kirpekar, F.; Berkenkamp, S.; Hillenkamp, F. Detection of double-stranded DNA by IR- and UV-MALDI mass spectrometry. *Anal. Chem.* 1999, **71**, 2334–2339.
10. Jungblut, P. R.; Bumann, D.; Haas, G.; Zimny-Arndt, U.; Holland, P.; Lamer, S.; Siejak, F.; Aebischer, A.; Meyer, T. F. Comparative proteome analysis of *Helicobacter pylori. Mol. Microbiol.* 2000, **36**, 710–725.
11. Dai, Y.; Li, L.; Roser, D. C.; Long, S. R. Detection and identification of low-mass peptides and proteins from solvent suspensions of *Escherichia coli* by high performance liquid chromatography fractionation and matrix-assisted laser desorption/ionization time-of-flight mass spectrometry. *Rapid Comm. Mass Spectrom.* 1999, **13**, 73–78.
12. Jones, J. J.; Stump, M. J.; Fleming, R. C.; Lay, J. O., Jr.; Wilkins, C. L. Investigation of MALDI-TOF and FT-MS techniques for analysis of *Escherichia coli* whole cells. *Anal. Chem.* 2003, **75**, 1340–1347.
13. Cain, T. C.; Lubman, D. M.; Weber, W. J., Jr. Differentiation of bacteria using protein profiles from matrix-assisted laser desorption/ionization time-of-flight mass spectrometry. *Rapid Comm. Mass Spectrom.* 1994, **8**, 1026–1030.
14. Liang, X.; Zheng, K.; Qian, M. G.; Lubman, D. M. Determination of bacterial protein profiles by matrix-assisted laser desorption/ionization mass spectrometry with high-performance liquid chromatography. *Rapid Comm. Mass Spectrom.* 1996, **10**, 1219–1226.
15. Krishnamurthy, T.; Ross, P. L.; Rajamani, U. Detection of pathogenic and non-pathogenic bacteria by matrix-assisted laser desorption/ionization time-of-flight mass spectrometry. *Rapid Comm. Mass Spectrom.* 1996, **10**, 883–888.
16. Girault, S.; Chassaing, G.; Blais, J. C.; Brunot, A.; Bolbach, G. Coupling of MALDI-TOF mass analysis to the separation of biotinylated peptides by magnetic streptavidin beads. *Anal. Chem.* 1996, **68**, 2122–2126.
17. Holland, R. D.; Wilkes, J. G.; Rafii, F.; Sutherland, J. B.; Persons, C. C.; Voorhess, K. J.; Lay, J. O. Rapid identification of intact whole bacteria based on spectral

patterns using matrix-assisted laser desorption/ionization with time-of-flight mass spectrometry. *Rapid Comm. Mass Spectrom.* 1996, **10**, 1227–1232.

18. Claydon, M. A.; Davey, S. N.; Edwards-Jones, V.; Gordon, D. B. The rapid identification of intact microorganisms using mass spectrometry. *Nature Biotechnol.* 1996, **10**, 1992–1996.
19. Krishnamurthy, T.; Ross, P. L. Rapid identification of bacteria by direct matrix-assisted laser desorption/ionization mass spectrometric analysis of whole cells. *Rapid Comm. Mass Spectrom.* 1996, **10**, 1992–1996.
20. Williams, T. L.; Andrzejewski, D.; Lay, J. O., Jr.; Musser, S. M. Experimental factors affecting the quality and reproducibility of MALDI TOF mass spectra obtained from whole bacteria cells. *J. Am. Soc. Mass Spectrom.* 2003, **14**, 342–351.
21. Arnold, R. J.; Reilly, J. P. Fingerprint matching of *E. coli* strains with matrix-assisted laser desorption/ionization time-of-flight mass spectrometry of whole cells using a modified correlation approach. *Rapid Comm. Mass Spectrom.* 1998, **12**, 630–636.
22. Wang, Z.; Russon, L.; Li, L.; Roser, D. C.; Long, S. R. Investigation of spectral reproducibility in direct analysis of bacteria proteins by matrix-assisted laser desorption/ionization time-of-flight mass spectrometry. *Rapid Comm. Mass Spectrom.* 1998, **12**, 456–464.
23. Dickinson, D. N.; La Duc, M. T.; Satomi, M.; Winefordner, J. D.; Powell, D. H.; Venkateswaran, K. MALDI-TOF MS compared with other polyphasic taxonomy approaches for the identification and classification of *Bacillus pumilus* spores. *J. Microbiol. Meth.* 2004, **58**, 1–12.
24. Arnold, R. J.; Karty, J. A.; Ellington, A. D.; Reilly, J. P. Monitoring the growth of a bacteria culture by MALDI-MS of whole cells. *Anal. Chem.* 1999, **71**, 1990–1996.
25. Welham, K. J.; Domin, M. A.; Scannell, D. E.; Cohen, E.; Ashton, D. S. The characterization of micro-organisms by matrix-assisted laser desorption/ionization time-of-flight mass spectrometry. *Rapid Comm. Mass Spectrom.* 1998, **12**, 176–180.
26. Lee, H.; Williams, S. K. R.; Wahl, K. L.; Valentine, N. B. Analysis of whole bacterial cells by flow field-flow fractionation and matrix-assisted laser desorption/ionization time-of-flight mass spectrometry. *Anal. Chem.* 2003, **75**, 2746–2752.
27. Reschiglian, P.; Zattoni, A.; Cinque, L.; Roda, B.; Dal Piaz, F.; Roda, A.; Moon, M. H.; Min, B. R. Hollow-fiber flow field-flow fractionation for whole bacteria analysis by matrix-assisted laser desorption/ionization time-of-flight mass spectrometry. *Anal. Chem.* 2004, **76**, 2103–2111.
28. Jarman, K. H.; Cebula, S. T.; Saenz, A. J.; Peterson, C. E.; Valentine, N. B.; Kingsley, M. T.; Wahl, K. L. An algorithm for automated bacterial identification using matrix-assisted laser desorption/ionization time-of-flight mass spectrometry. *Anal. Chem.* 2000, **72**, 1217–1223.
29. Wahl, K. L.; Wunschel, S. C.; Jarman, K. H.; Valentine, N. B.; Petersen, C. E.; Kingsley, M. T.; Zartolas, K. A.; Saenz, A. J. Analysis of microbial mixtures by matrix-assisted laser desorption/ionization time-of-flight mass spectrometry. *Anal. Chem.* 2002, **74**, 6191–6199.
30. Holland, R. D.; Rafii, F.; Heinze, T. M.; Sutherland, J. B.; Voorhees, K. J.; Lay, J. O., Jr. Matrix-assisted laser desorption/ionization time-of-flight mass spectrometric detection of bacterial biomarker proteins isolated from contaminated water, lettuce and cotton cloth. *Rapid Comm. Mass Spectrom.* 2000, **14**, 911–917.

31. Winkler, M. A.; Uher, J.; Cepa, S. Direct analysis and identification of *Heliobacter* and *Camphlobacter* species by MALDI-TOF mass spectrometry. *Anal. Chem.* 1999, **71**, 3416–3419.

32. Vaidyanathan, S.; Winder, C. L.; Wade, S. C.; Kell, D. B.; Goodacre, R. Sample preparation in matrix-assisted laser desorption/ionization mass spectrometry of whole bacterial cells and the detection of high mass (>20 kDa) proteins. *Rapid Comm. Mass Spectrom.* 2002, **16**, 1276–1286.

33. Madonna, A. J.; Basile, F.; Imma, I.; Meetani, A. M.; Rees, J. C.; Voorhees, K. J. On-probe sample pretreatment for the detection of proteins above 15 KDa from whole cell bacteria by matrix-assisted laser desorption/ionization time-of-flight mass spectrometry. *Rapid Comm. Mass Spectrom.* 2000, **14**, 2220–2229.

34. Domin, M. A.; Welham, K. J.; Ashton, D. S. The effect of solvent and matrix combinations on the analysis of bacteria by matrix-assisted laser desorption/ionization time-of-flight mass spectrometry. *Rapid Comm. Mass Spectrom.* 1999, **13**, 222–226.

35. Gantt, S. L.; Valentine, N. B.; Saenz, A. J.; Kingsley, M. T.; Wahl, K. L. Use of an internal control for matrix-assisted laser desorption/ionization time-of-flight mass spectrometry analysis of bacteria. *J. Am. Soc. Mass Spectrom.* 1999, **10**, 1131–1137.

36. Dai, Y.; Whittal, R. M.; Li, L. Two-layer sample preparation: a method for MALDI-MS analysis of complex peptide and protein mixtures. *Anal. Chem.* 1999, **71**, 1087–1091.

37. Hathout, Y.; Demirev, P. A.; Ho, Y.; Bundy, J. L.; Ryzhov, V.; Sapp, L.; Stutler, J.; Jackman, J.; Fenselau, C. Identification of *Bacillus* spores by matrix-assisted laser desorption ionization-mass spectrometry. *Appl. Environ. Microbiol.* 1999, **65**, 4313–4319.

38. Ryzhov, V.; Hathout, Y.; Fenselau, C. Rapid characterization of spores of *Bacillus cereus* group bacteria by matrix-assisted laser desorption-ionization time-of-flight mass spectrometry. *Appl. Environ. Microbiol.* 2000, **66**, 3828–3834.

39. English, R. D.; Warscheid, B.; Fenselau, C.; Cotter, R. J. *Bacillus* spore identification via proteolytic peptide mapping with a miniaturized MALDI TOF mass spectrometer *Anal. Chem.* 2003, **75**, 6886–6893.

40. Stump, M. J.; Jones, J. J.; Fleming, R. C.; Lay, J. O., Jr.; Wilkins, C. L. Use of double-depleted 13C and 15N culture media for analysis of whole cell bacteria by MALDI time-of-flight and Fourier transform mass spectrometry. *J. Am. Soc. Mass Spectrom.* 2003, **14**, 1306–1314.

41. Jarman, K. H.; Daly, D. S.; Peterson, C. E.; Saenz, A. J.; Valentine, N. B.; Wahl, K. L. Extracting and visualizing matrix-assisted laser desorption/ionization time-of-flight mass spectral fingerprints. *Rapid Comm. Mass Spectrom.* 1999, **13**, 1586–1594.

42. Pineda, F. J.; Lin, J. S.; Fenselau, C.; Dimerov, P. A. Testing the significance of microorganism identification by mass spectrometry and proteome database search. *Anal. Chem.* 2000, **72**, 3739–3744.

43. Harris, W. A.; Reilly, J. P. On-probe digestion of bacterial proteins for MALDI-MS. *Anal. Chem.* 2002, **74**, 4410–4416.

44. Bundy, J.; Fenselau, C. Lectin-based affinity capture for MALDI-MS analysis of bacteria. *Anal. Chem.* 1999, **71**, 1460–1463.

45. Madonna, A. J.; Van Cuyk, S.; Voorhees, K. J. Detection of *Escherichia coli* using immunomagnetic separation and bacteriophage amplification coupled with matrix-assisted laser desorption/ionization time-of-flight mass spectrometry. *Rapid Comm. Mass Spectrom.* 2003, **17**, 257–263.

46. Holland, R. D.; Duffy, C. R.; Rafii, F.; Sutherland, J. B.; Heinze, T. M.; Holder, C. L.; Voorhees, K. J.; Lay, J. O., Jr. Identification of bacterial proteins observed in MALDI TOF mass spectra from whole cells. *Anal. Chem.* 1999, **71**, 3226–3230.

47. Parker, C. E.; Papac, D. I.; Tomer, K. B. Monitoring cleavage of fusion protein by matrix-assisted laser desorption ionization/mass spectrometry: Recombinant HIV-1IIIB p26. *Anal. Biochem.* 1996, **239**, 25–34.

48. Easterling, M. L.; Colangelo, C. M.; Scott, R. A.; Amster, I. J. Monitoring protein expression in whole bacterial cells with MALDI time-of-flight mass spectrometry. *Anal. Chem.* 1998, **70**, 2704–2709.

49. Jebanathirajah, J. A.; Andersen, S.; Blagoev, B.; Roepstorff, P. A rapid screening method to monitor expression of recombinant proteins from various prokaryotic and eukaryotic expression systems using matrix-assisted laser desorption ionization-time-of-flight mass spectrometry. *Anal. Biochem.* 2002, **305**, 242–250.

50. Winkler, M. A.; Hickman, R. K.; Golden, A.; Aboleneen, H. Analysis of recombinant protein expression by MALDI-TOF mass spectrometry of bacterial colonies. *BioTechniques* 2000, **28**, 890, 892, 894–895.

51. Hindre, T.; Didelot, S.; Le Pennec, J.-P.; Haras, D.; Dufour, A.; Vallee-Rehel, K. Bacteriocin detection from whole bacteria by matrix-assisted laser desorption ionization-time of flight mass spectrometry. *Appl. Environ. Microbiol.* 2003, **69**, 1051–1058.

52. Haag, A. M.; Taylor, S. N.; Johnston, K. H.; Cole, R. B. Rapid identification and speciation of *Haemophilus* bacteria by matrix-assisted laser desorption/ionization time-of-flight mass spectrometry. *J. Mass Spectrom.* 1998, **33**, 750–756.

53. Mead, P. S.; Slutsker, L.; Dietz, V.; McCaig, L. F.; Bresee, J. S.; Shapiro, C.; Griffin, P. M.; Tauxe, R. V. Food-related illness and death in the United States. *Emerg. Infect. Dis.* 1999, **5**, 607–625.

54. Wilkes, J. G.; Holland, R.; Holcomb, M.; Lay, J. O., Jr.; McCarthy, S., Comparison of PY-MS for rapid classification of *Vibrio parahaemolyticus* outbreak strains. *Proc. Am. Soc.of Mass Spectrom. Conf. on Mass Spectrometry and Allied Topics.* ASMS, Long Beach, CA 2000.

55. Marshall, S.; Clark, C. G.; Wang, G.; Mulvey, M.; Kelly, M. T.; Johnson, W. M. Comparison of molecular methods for typing *Vibrio parahaemolyticus*. *J. Clin. Microbiol.* 1999, **37**, 2473–2478.

56. Edwards-Jones, V.; Claydon, M. A.; Evason, D. J.; Walker, J.; Fox, A. J.; Gordon, D. B. Rapid discrimination between methicillin-sensitive and methicillin-resistant *Staphylococcus aureus* by intact cell mass spectrometry. *J. Med. Microbiol.* 2000, **49**, 295–300.

57. Bernardo, K.; Pakulat, N.; Macht, M.; Krut, O.; Seifert, H.; Fleer, S.; Hunger, F.; Kronke, M. Identification and discrimination of *Staphylococcus aureus* strains using matrix-assisted laser desorption/ionization-time of flight mass spectrometry. *Proteomics* 2002, **2**, 747–753.

58. Claydon, M. A. MALDI-ToF-MS, a new and novel technique for studies of intact cells. *Anaerobe* 2000, **6**, 133–134.
59. Smole, S. C.; King, L. A.; Leopold, P. E.; Arbeit, R. D. Sample preparation of Gram-positive bacteria for identification by matrix assisted laser desorption/ionization time-of-flight. *J. Microbiol. Meth.* 2002, **48**, 107–115.
60. Conway, G. C.; Smole, S. C.; Sarracino, D. A.; Arbeit, R. D.; Leopold, P. E. Phyloproteomics: Species identification of *Enterobacteriaceae* using matrix-assisted laser desorption/ionization time-of-flight mass spectrometry. *J. Mol. Microbiol. Biotechnol.* 2001, **3**, 103–112.
61. Lay, J. O., Jr. MALDI-TOF mass spectrometry and bacterial taxonomy TrAC. *Trends Anal. Chem.* 2000, **19**, 507–516.
62. Fenselau, C.; Demirev, P. A. Characterization of intact microorganisms by MALDI mass spectrometry. *Mass Spectrom. Rev.* 2002, **20**, 157–171.
63. Lynn, E. C.; Chung, M.; Tsai, W.; Han, C. Identification of *Enterobacteriaceae* bacteria by direct matrix-assisted laser desorption/ionization time-of-flight mass spectrometry analysis of whole cells. *Rapid Comm. Mass Spectrom.* 1999, **13**, 2022–2027.
64. Ryzhov, V.; Fenselau, C. Characterization of the protein subset desorbed by MALDI from whole bacterial cells. *Anal. Chem.* 2001, **73**, 746–750.

7

DEVELOPMENT OF SPECTRAL PATTERN-MATCHING APPROACHES TO MATRIX-ASSISTED LASER DESORPTION/IONIZATION TIME-OF-FLIGHT MASS SPECTROMETRY FOR BACTERIAL IDENTIFICATION

Kristin H. Jarman and Karen L. Wahl

Pacific Northwest National Laboratory, Richland, WA 99352

7.1 INTRODUCTION

The concept of rapid microorganism identification using matrix-assisted laser desorption/ionization mass spectrometry (MALDI MS) dates back to the mid-1990s. Prior to 1998 researchers relied on visual inspection in an effort to demonstrate feasibility of MALDI MS for bacterial identification.[1–3] In general, researchers in these early studies visually compared the biomarker intensity profiles between different organisms and between replicates of the same organism to show that MALDI signatures are unique and reproducible. Manual tabulation and comparison of potential biomarker mass values observed for different organisms was used by numerous researchers to qualitatively characterize microorganisms using MALDI mass spectra.[4–7]

Identification of Microorganisms by Mass Spectrometry, Edited by Charles L. Wilkins and Jackson O. Lay, Jr.

Some of the challenges with manual MALDI MS data analysis include the variability in relative intensity and peak appearance of MALDI MS ions typically observed for direct microorganism analysis. This spectral variability creates a challenge for manual data analysis where visual intensity differences are more noticeable. Replicate MALDI spectra of the same sample visually show significant differences that are hard to manually deal with effectively.

Beginning in the late 1990s, researchers began developing numerical methods for microorganism identification using MALDI MS. Many of these methods have been used in empirical studies to validate the idea that MALDI MS can effectively enable microorganism identification. To date, an array of techniques have been developed that draw from a broad range of statistical and pattern recognition methods. Methods also vary in the degree of automation—some methods require user interaction at several points in the analysis while others perform in a completely automated fashion.

All methods for microorganism identification using MALDI MS rely on some sort of database to which an unknown sample can be compared. The current methods can be characterized by the type of database used to identify an unknown: (1) a library of MALDI MS signatures constructed from spectra of known samples or (2) a library of proteins generated from one of the publicly available genomic and/or proteomic databases. With the first type of database, users are required to generate signatures of all microorganisms of interest—no standardized, widely available database of MALDI MS signatures is yet available. With the second type of database, users can exploit the extensive amount of information available from the genomic databases, but these databases may or may not contain the organisms of interest.

In this chapter we describe algorithms that only use the first type of database—a library of MALDI MS signatures constructed from MALDI analyses of target organisms. The genomic database search methods are discussed in more detail in Chapter 1 . Here we summarize different numerical approaches for automated microorganism identification using MALDI MS. We discuss advantages and disadvantages of the basic approaches and describe future directions in algorithm development.

7.2 MALDI MS SIGNATURE LIBRARY CONSTRUCTION AND IDENTIFICATION

In this section algorithms are presented that use a reference library constructed from MALDI MS signatures. An unknown sample spectrum is compared to signatures in the reference library, and a score for each comparison is generated. The sample contents are identified based on this score. These methods fall into two categories: full-spectrum methods and peak table based methods. With a full-spectrum comparison, the entire (usually smoothed or compressed) spectrum is compared to reference signatures without a peak table containing biomarker locations and intensities first being generated.

With a peak table based method, the biomarkers are extracted from an unknown spectrum, and those biomarkers are then used to compare the unknown to a reference signature.

7.3 FULL SPECTRUM PATTERN METHODS

A typical MALDI spectrum of a bacterial sample has a number of peaks that vary greatly in intensity superimposed on a relatively noisy baseline. This can be problematic for many peak detection routines. Therefore methods that eliminate the need for peak detection also eliminate problems associated with poor peak detection performance. Full-spectrum identification algorithms use the (usually smoothed) spectral data without first performing peak detection.

7.3.1 Correlation and Cross-correlation

The most straightforward full-spectrum comparison approach uses a correlation or cross-correlation coefficient to construct a score. Given a reference signature spectrum y_i and an unknown sample spectrum x_i, for channels $i = 1, 2, \ldots,$ the linear cross-correlation can be expressed as

$$r = \frac{\sum_i (x_i - \bar{x})(y_i - \bar{y})}{\sqrt{\sum_i (x_i - \bar{x})^2 \sum_i (y_i - \bar{y})^2}}.$$

The correlation coefficient provides an indication of the similarity between two spectra. It is invariant to linear shifts and therefore is unaffected by baseline changes and differences in overall intensity. However, the correlation coefficient can be greatly affected by noise, and researchers who have implemented this approach generally smooth the spectrum before scoring. A commonly used smoothing algorithm is the Savitzky-Golay filter.[8–11] Arnold and Reilly[8,9] first proposed this statistic for comparing an unknown sample MALDI spectrum to a reference signature. They divide a spectrum into a specified number of intervals and compute a cross-correlation between an unknown spectrum and a MALDI signature. A score is constructed from the product of cross-correlations throughout the entire spectrum. More recently Dickinson et al.[10,12] have used this approach to differentiate species of *Bacillus* spores. In this study the overall cross-correlation is used, without dividing the spectrum into intervals.

7.3.2 Neural Networks

Because of their ability to classify complex data types that have no explicit mathematical model, neural networks have become a powerful and widely used approach to pattern recognition problems in general. A neural network is a series of mathematical operations performed on input data that ultimately produces a score. In the present application this score is used to associate an unknown spectrum with one or more reference signature spectra from a database. A set of training data is used to construct the parameters of a neural network, and in this way serves as a database against which an unknown sample will be compared. Neural networks vary greatly in complexity and transparency of results to users, and the reader is referred to Duda et al.[13] for a detailed discussion of and references related to artificial neural networks.

Each set of mathematical operations in a neural network is called a layer, and the mathematical operations in each layer are called neurons. A simple layer neural network might take an unknown spectrum and pass it through a two-layer network where the first layer, called a hidden layer, computes a basis function from the distances of the unknown to each reference signature spectrum, and the second layer, called an output layer, that combines the basis functions into a final score for the unknown sample.

Zhang et al.[14] develop a neural network approach to bacterial classification using MALDI MS. The developed neural network is used to classify bacteria and to classify culturing time for each bacterium. To avoid the problem of overfitting a neural network to the large number of channels present in a raw MALDI spectrum, the authors first normalize and then reduce the dimensionality of the spectra by performing a wavelet transformation.

Bright et al.[15] used a hybrid neural network package specifically written for intact cell MALDI MS (ICM MS) data to build and search databases for microbial analysis. The developed Manchester University Search Engine (MUSE™) constructs a reference library from replicate spectra that have previously been checked for quality. A nonlinear neural network based transformation of each spectrum is performed to produce a single value representing the data. Replicate data are used to construct uncertainty intervals for each reference signature, and the developed nonlinear transformation along with the uncertainty estimates are used for microorganism identification. Specific details of the algorithm are not provided, however, for the results presented in Bright's paper.[15]

The disadvantage of a full-spectrum matching method is that all peaks are used in the analysis even if they are not due to the microorganism. Hence more rigorous sample preparation and analyte extraction may be necessary to remove any extraneous peaks from other microbes or from environmental or clinical sample background. Alternatively, peak extraction prior to analysis may allow for postprocessing removal or ignorance of extraneous or background peaks.

7.4 PEAK TABLE BASED METHODS

Peak table based methods first extract biomarker peaks from a spectrum and then use those peaks for organism identification. These methods have the ability to ignore peaks that are due to background or extraneous factors.

However, they do generally require an effective peak detection routine to ensure that both large and small key peaks are used, and that noise spikes are not detected as peaks. The reader is referred to Jarman et al.[16] for a discussion and comparison of some different peak detection routines applied to MALDI data of bacterial samples.

7.4.1 Peak Probability Based Algorithms

Statistical studies of MALDI MS applied to bacterial samples show that some biomarker peaks are highly reproducible and appear very consistently, while others appear much less reliably.[17–19] In Jarman et al.[20] and Wahl et al.[21] a probability model for MALDI signatures is proposed that takes into account the variability in appearance of biomarker peaks. This method constructs MALDI reference signatures from the set of peak locations for reproducible biomarker peaks, along with a measure of the reproducibility of each peak.

If, when a given organism A is present in the sample, biomarker peak i appears with probability p_i. Let $x_i = 0$ if biomarker peak i is not observed in the unknown sample, and $x_i = 1$ if biomarker peak i is observed in the unknown sample. Then the likelihood of the observed peak table is given by

$$l_A(x) = \prod_i p_i^{x_i}(1-p_i)^{1-x_i}.$$

If we let H_0 represent the hypothesis that organism A IS NOT present in the sample, and H_A represent the hypothesis that organism A IS in the sample, then the likelihood ratio for H_0 versus H_A is given by the probability of the observed peak table under H_A divided by the probability of observing the outcome under H_0. Specifically,

$$LR(x) = \frac{\prod_i p_i^{x_i}(1-p_i)^{1-x_i}}{\prod_i q_i^{x_i}(1-q_i)^{1-x_i}},$$

where q_i is the probability that biomarker peak i appears in an unknown spectrum, given organism A IS NOT present in the sample. If the likelihood ration LR exceeds some threshold, then it is determined that organism A IS in the sample. Otherwise, organism A is declared to be absent from the sample. In Jarman et al.[20] and Wahl et al.[21] this approach is proposed and tested in two blind studies.

7.4.2 Random Forests

In developing algorithms for automated identification of bacterial samples using MALDI MS, it is important to consider only those reproducible peaks that are very likely to appear in an unknown. However, such reproducible peaks may or may not be unique to a single organism. The authors in Woolfitt et al.[22] extend the idea of using reproducible biomarker peaks to consider only those peaks that are unique or distinguishing for a given organism. Their algorithm, called random forests, identifies discriminating peaks in a signature database by tabulating which peaks appear in one or more organisms contained in the library. This algorithm then uses only those peaks that are unique to a given organism for identification. Initial studies indicate that using only the most differentiating peaks increases ability to differentiate near neighbors.

7.5 FUTURE DIRECTIONS

Through the development of a comprehensive, widely available database, MALDI MS could become a commonly used method for routine microorganism identification. The numerous research groups that have demonstrated the feasibility of this technique for bacterial identification with different instruments and very different spectral analysis methods suggests that the underlying methodology is robust. However, a number of issues currently present challenges to existing data interpretation methods. In particular, spectral variability has been observed when organisms are subjected to different growth conditions[9,23,24] or other experimental variables.[24] Variability in the MALDI spectra across different culture media has also been observed.[16,25] In addition differences between spectra from the same organism collected on different instruments can be seen.[26] Truly robust and accurate microorganism identification algorithms will take these different sources of variability into account when constructing reference signatures and performing identifications. To date, some research has addressed these issues[16,25,26] by constructing signatures from a variety of laboratory and/or growth conditions.

Automated identification methods could potentially benefit through the combination of MALDI MS signature database identification methods with proteomics-based methods that exploit the publicly available genomic databases. In particular, one of the criticisms of the pattern recognition methods presented here is that a MALDI MS signature database must be constructed, which can be a time consuming and expensive process. On the other hand, proteomics-based methods exploit readily available, rapidly growing public databases. Conversely, one of the criticisms of the proteomics-based methods is that the genomic databases include all proteins for the resident organisms, not just the proteins that are expressed in MALDI spectra, whereas the signature database approach better characterizes biomarkers that are observed in MALDI spectra. Identification algorithms that combine the ability to quickly screen large genomic databases for possible target organisms with the more statistically rigorous confirmatory ability of MALDI MS signature comparison methods could provide researchers with the best of both worlds.

REFERENCES

1. Holland, R. D.; Wilkes, J. G.; Rafii, F.; Sutherland, J. B.; Persons, C. C.; Voorhees, K. J.; Lay, J. O., Jr. Rapid identification of intact whole bacteria based on spectral patterns using matrix-assisted laser desorption/ionization with time-of-flight mass spectrometry. *Rapid Comm. Mass Spectrom.* 1996, **10**, 1227–1232.
2. Krishnamurthy, T.; Ross, P. L. Rapid identification of bacteria by direct matrix-assisted laser desorption/ionization mass spectrometric analysis of whole cells. *Rapid Comm. Mass Spectrom.* 1996, **10**, 1992–1996.
3. Claydon, M. A.; Davey, S. N.; Edwards-Jones, V.; Gordon, D. B. The rapid identification of intact microorganisms using mass spectrometry. *Nature Biotechnol.* 1996, **10**, 1992–1996.
4. Lynn, E. C.; Chung, M.; Tsai, W.; Han, C. Identification of *Enterobacteriaceae* bacteria by direct matrix-assisted laser desorption/ionization time-of-flight mass spectrometry analysis of whole cells. *Rapid Comm. Mass Spectrom.* 1999, **13**, 2022–2027.
5. Nilsson, C. L. Fingerprinting of *Helicobacter pylori* strains by matrix-assisted laser desorption/ionization mass spectrometric analysis. *Rapid Comm. Mass Spectrom.* 1999, **13**, 1067–1071.
6. Ryzhov, V.; Hathout, Y.; Fenselau, C. Rapid characterization of spores of *Bacillus cereus* group bacteria by matrix-assisted laser desorption-ionization time-of-flight mass spectrometry. *Appl. Environ. Microbiol.* 2000, **66**, 3828–3834.
7. Winkler, M. A.; Uher, J.; Cepa, S. Direct analysis and identification of *Heliobacter* and *Campylobacter* species by MALDI-TOF mass spectrometry. *Anal. Chem.* 1999, **71**, 3416–3419.
8. Arnold, R. J.; Reilly, J. P. Fingerprint matching of *E. coli* strains with MALDI-TOF mass spectrometry of whole cells using a modified correlation approach. *Rapid Comm. Mass Spectrom.* 1998, **12**, 630–636.
9. Arnold, R. J.; Reilly, J. P., *Proc. 46th ASMS Conf. on Mass Spectrometry and Allied Topics*, Orlando, FL, 1998. p. 125.
10. Dickinson, D. N.; Duc, M. T. L.; Haskins, W. E.; Gornushkin, I.; Winefordner, J. D.; Powell, D. H.; Venkateswaran, K. Species differention of a diverse suite of *Bacillus* spores by mass spectrometry-based protein profiling. *Appl. Environ. Microbiol.* 2004, **70**, 475–482.
11. Savitsky, A.; Golay, M. Smoothing and differentiation of data by simplified least squares procedures. *Anal. Chem.* 1964, **36**, 1627–1639.
12. Dickinson, D. N.; La Duc, M. T.; Satomi, M.; Winefordner, J. D.; Powell, D. H.; Venkateswaran, K. MALDI-TOFMS compared with other polyphasic taxonomy approaches for the identification and classification of *Bacillus pumilus* spores. *J. Microbiol. Meth.* 2004, **58**, 1–12.
13. Duda, R. O.; Hart, P. E.; Stork, D. G. *Pattern Classification*. New York: Wiley, 2001, Chap. 6.
14. Zhang, Z.; Wang, D.; Harrington, P. d. B.; Voorhees, K. J.; Rees, J. Forward selection radial basis function networks applied to bacterial classification based on MALDI-TOF-MS. *Talanta* 2004, **63**, 527–532.

15. Bright, J. J.; Claydon, M. A.; Soufian, M.; Gordon, D. B. Rapid typing of bacteria using matrix-assisted laser desorption ionisation time-of-flight mass spectrometry and pattern recognition software. *J. Microbiol. Meth.* 2002, **48**, 127–138.
16. Jarman, K. H.; Daly, D. S.; Anderson, K. K.; Wahl, K. L. A new approach to automated peak detection. *Chemomet. Intell. Lab. Sys.* 2004, **69**, 61–76.
17. Jarman, K. H.; Daly, D. S.; Peterson, C. E.; Saenz, A. J.; Valentine, N. B.; Wahl, K. L. Extracting and visualizing matrix-assisted laser desorption/ionization time-of-flight mass spectral fingerprints. *Rapid Comm. Mass Spectrom.* 1999, **13**, 1586–1594.
18. Saenz, A. J.; Petersen, C. E.; Valentine, N. B.; Gantt, S. L.; Jarman, K. H.; Kingsley, M. T.; Wahl, K. L. Reproducibility of matrix-assisted laser desorption/ionization time-of-flight mass spectrometry for replicate bacterial culture analysis. *Rapid Comm. Mass Spectrom.* 1999, **13**, 1580–1585.
19. Smole, S. C.; King, L. A.; Leopold, P. E.; Arbeit, R. D. Sample preparation of Gram-positive bacteria for identification by matrix assisted laser desorption/ionization time-of-flight. *J. Microbiol. Meth.* 2002, **48**, 107–115.
20. Jarman, K. H.; Cebula, S. T.; Saenz, A. J.; Peterson, C. E.; Valentine, N. B.; Kingsley, M. T.; Wahl, K. L. An algorithm for automated bacterial identification using matrix-assisted laser desorption/ionization time-of-flight mass spectrometry. *Anal. Chem.* 2000, **72**, 1217–1223.
21. Wahl, K. L.; Wunschel, S. C.; Jarman, K. H.; Valentine, N. B.; Petersen, C. E.; Kingsley, M. T.; Zartolas, K. A.; Saenz, A. J. Analysis of microbial mixtures by matrix-assisted laser desorption/ionization time-of-flight mass spectrometry. *Anal. Chem.* 2002, **74**, 6191–6199.
22. Woolfitt, A.; Moura, H.; Barr, J.; De, B.; Popovic, T.; Satten, G.; Jarman, K. H.; Wahl, K. L., Differentiation of *Bacillus* spp. by MALDI-TOF mass spectrometry using a bacterial fingerprinting algorithm and a random forest classification algorithm. Presented at 5th ISIAM Meeting, Richland, WA 2004.
23. Arnold, R. J.; Karty, J. A.; Ellington, A. D.; Reilly, J. P. Monitoring the growth of a bacteria culture by MALDI MS of whole cells. *Anal. Chem.* 1999, **71**, 1990–1996.
24. Wang, Z.; Russon, L.; Li, L.; Roser, D. C.; Long, S. R. Investigation of spectral reproducibility in direct analysis of bacteria proteins by matrix-assisted laser desorption/ionization time-of-flight mass spectrometry. *Rapid Comm. Mass Spectrom.* 1998, **12**, 456–464.
25. Valentine, N. B.; Wunschel, S. C.; Petersen, C. E.; Wahl, K. L. The effect of culture conditions on microorganism identification by MALDI mass spectrometry. *Appl. Environ. Microbiol.*, 2005, **71**, 58–64.
26. Wunschel, S. C.; Schauki, D.; Nelson, C. P.; Jackman, J.; Jarman, K. H.; Petersen, C. E.; Valentine, N. B.; Wahl, K. L. Bacterial analysis by MALDI-TOF mass spectrometry: An inter-laboratory comparison. *J. Am. Soc. Mass Spectrom.* 2005, **16**, 456–462.

8

STUDIES OF MALARIA BY MASS SPECTROMETRY

PLAMEN A. DEMIREV

Applied Physics Laboratory, Johns Hopkins University, Laurel, MD 20723

8.1 INTRODUCTION

Malaria is one of the most devastating diseases that have afflicted humankind from the dawn of its recorded history. Even today, malaria is a leading cause of morbidity and mortality in the developing world. Each year more than 500 million people are infected with malaria and more than a million, mostly children, die from malaria and its complications.[1–5] More than 40% of the world's population currently lives in malaria-endemic regions, and so far there is no vaccine against malaria.[3,4] Despite intense efforts to eradicate it, for the last 20 years malaria has been on the rise due to a number of socioeconomical, climatic, and biophysiological factors. Among these are wars and poverty, global warming, unbridled deforestation, and the emergence of drug-resistant parasite strains and insecticide-resistant mosquitoes. Malaria has already returned to geographical regions where it had been formerly eradicated.

Rapid and high-throughput methods for detection of malaria are indispensable for its eradication, for screening new drug and candidate vaccine efficacy, and for monitoring the emergence of drug-resistant parasite strains. In the "Roll Back Malaria" initiative, early malaria diagnosis with effective prompt treatment and early detection and forecasting of epidemics are listed as key elements in strategies to fight malaria.[1] One of the major challenges in

Identification of Microorganisms by Mass Spectrometry, Edited by Charles L. Wilkins and Jackson O. Lay, Jr.

that fight is the lack of novel, accurate, and affordable technologies for definitive point-of-care malaria diagnosis in developing world settings. For more than a century, optical microscopy of Giemsa-stained blood smears has been the "gold standard" for malaria diagnosis with a working sensitivity of hundreds of parasites per microliter infected blood.[3,6,7] Optical microscopy is low-throughput, requires significant expertise, and an individual sample must be examined field-by-field for about 30 minutes in order to achieve the required sensitivity. Several novel molecular level technologies for malaria parasite detection are currently being developed. These include fluorescence microscopy, PCR-based assays, serological ("dipstick") antigen detection, flow cytometry (with or without laser light depolarization monitoring), and microfluidics.[8–13] Most such techniques are time-consuming, have low sensitivity or specificity, or are expensive for mass screening.

A novel physical method for malaria detection, developed recently, is reviewed here. The method—ultraviolet laser desorption (LD) mass spectrometry (MS)—is based on the detection of heme (iron protoporphyrin) in blood as a qualitative and quantitative malaria biomarker.[14–16] LDMS detects only heme from parasite-containing samples, and not the heme, bound to hemoglobin or other proteins in control blood samples. In contrast to matrix-assisted laser desorption/ionization (MALDI), external photo-absorbing matrix does not need to be added to the sample. The test requires only a single drop of blood. Heme detection by LDMS is extremely sensitive and highly specific, reflecting both the structural and the photochemical properties of the heme molecule. The ion signal intensities are correlated with sample parasite densities from 5×10^2 to 1.5×10^4 parasites per microliter of blood. The LDMS method is further illustrated with recent data from in vivo studies involving a mouse model[15] as well as ongoing clinical human sample studies.[16] Implementations of the method on different instrumental setups, such as a miniaturized LD time-of-flight (TOF) multi-array analyzer,[17,18] and an atmospheric pressure (AP) LD ion source,[19,20] coupled to an ion trap analyzer, are presented. These instruments may provide the capabilities to design field-portable systems for parallel and high-throughput malaria screening in large populations. Optimization of experimental parameters, including different sample preparation protocols and data acquisition optimization, that further improve the detection sensitivity and specificity is discussed.

Detection by LDMS and structural elucidation of other secondary metabolite products, generated in the host during the onset of the parasite disease, is discussed. These molecules may serve as additional biomarkers for rapid malaria diagnosis by LDMS. For instance, choline phosphate (CP) is identified as the source of several low-mass ions observed in parasite-infected blood samples in addition to heme biomarker ions. The CP levels track the sample parasitemia levels. This biomarker can provide additional specificity and sensitivity when compared to malaria detection based on heme ion signals alone. Furthermore the observed elevated CP levels are discussed in the context of *Plasmodium* metabolism during its intra-erythrocytic life cycle. These data can provide further insight into the unique biosynthetic pathways of parasite-specific membrane lipids and parasite–host interactions in general.

Finally, mass spectrometry related studies of the malaria parasite biology and immunology are briefly reviewed. The recently completed sequencing of the genomes of two *Plasmodium* species[21,22] has opened a new chapter in the search for new molecular drug targets and vaccines to combat malaria. In particular, comprehensive malaria proteome studies[23–26] may be utilized in screening strategies for selective and potent inhibitors of enzymes essential for the parasite life cycle.

8.2 *PLASMODIUM* IN RED BLOOD CELLS

In humans, malaria is caused by four different single-celled protozoa species of the genus *Plasmodium*. These are *P. falciparum*, *P. vivax*, *P. ovale*, and *P. malariae*, with *P. falciparum* being the most lethal. The vector, responsible for the *Plasmodium* parasite transmission in humans, is the female *Anopheles* mosquito (with more than 50 known species). The life cycle of the malaria parasite in the vector and in the vertebrate host is quite complex, proceeding through a number of asexual and sexual developmental stages.[7] *Plasmodium* sporozoites are injected in the blood of the human host together with mosquito saliva. After initial proliferation in the liver, parasites in the merozoite stage are released back into the blood stream. In the asexual stage of infection, a single merozoite invades a red blood cell (RBC) and matures by forming a ring-shaped cell. In about 24 hours the matured parasite enters the trophozoite stage, during which most of the RBC cytoplasm is catabolized.[27–29] Through the final (schizont) stage in the RBC, the parasite undergoes several divisions to produce up to 32 new merozoites that burst the host RBC, enter the blood stream, and invade new erythrocytes. The major clinical symptoms of malaria (e.g., periodic bouts of fever) appear during the intra-erythrocytic developmental stages of the parasite. While residing in the RBC, the *Plasmodium* parasite relies on human hemoglobin (comprising 95% of the RBC cytosolic proteins) as a food source to provide amino acids needed for the de novo synthesis of malarial proteins. Intra-erythrocytic hemoglobin is catabolyzed by a series of proteases that are plausible targets for antimalarial drug development.[28] It has been estimated that more than 100 g of hemoglobin is digested by the parasite per 24-hour period during a moderate malarial infection in an adult![29]

Hemoglobin, the major oxygen transporter in human blood, is a tetrameric protein with molecular weight approximately 64 kDa, consisting of two α- and two β-chains (with lengths of 141 or 146 amino acids, respectively). Heme (iron protoporphyrin IX; Figure 8.1) is the hemoglobin prosthetic group with one heme molecule noncovalently attached to each chain. Under physiological conditions (in uninfected RBCs) heme is extremely tightly bound to hemoglobin with an equilibrium constant $>10^{16} M^{-1}$.[30] In infected RBCs, heme is

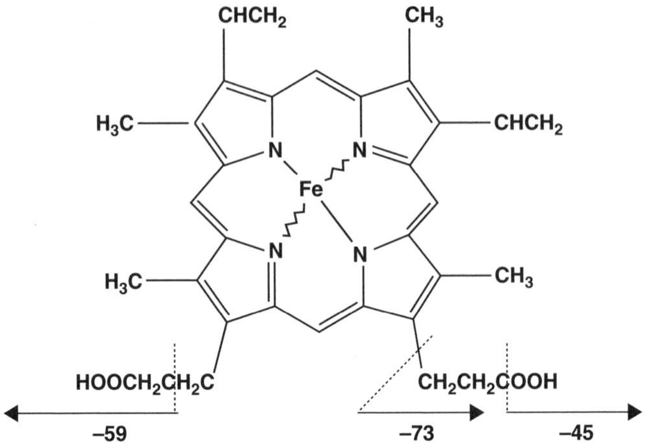

Figure 8.1 Heme (iron protoporphyrin IX) structure. The most frequently observed cleavages in LDMS of the intact species are denoted. Heme elemental composition is $C_{34}H_{32}N_4O_4Fe$; monoisotopic molecular mass is 616.176 Da; average molecular mass is 616.487 Da.

released in the "digestive vacuole"—a specialized acidic organelle that is the site of hemoglobin degradation during parasite growth. Liberated free heme is sequestered inside the vacuole as an inert birefringent molecular crystal (hemozoin or malaria pigment—with a characteristic brown color). Hemozoin formation alleviates the considerable oxidative stress from the extremely toxic free heme that would otherwise kill the parasite. Disruption of this detoxification mechanism by inhibiting hemozoin crystal formation is the main operation mode for many effective anti-malarial drugs. It was the observation in 1897 by Ross of pigment granules in the digestive tract of mosquitoes that established the link between the *Plasmodium* parasite as the agent of malaria (discovered several years earlier by Laveran) and the mechanism of its transmission.[31] Immediately after the discovery, in a letter to his former student Ross, Manson wrote: "and you have to thank the plasmodium that it is fool enough to carry its pigment along with it into the mosquito's tissues. Otherwise I suppose you would have never spotted him. . . ."[32]

8.3 EXPERIMENTAL PROTOCOLS FOR LDMS DETECTION OF MALARIA

8.3.1 Sample Preparation

Almost all applications of mass spectrometry in biochemistry and biomedicine require preparation of the samples prior to MS analysis. Sample preparation typically includes a number of steps to purify and concentrate target biomarker analytes. However, heme crystal formation, a unique evolutionary feature of *Plasmodium*, allows the parasite itself to concentrate and purify heme—the main biomarker molecule for malaria detection by LDMS. The hemozoin crystal is needle-like with characteristic dimensions $1.0 \times 0.6 \times 0.2\,\mu m^3$, and contains almost 1 pg heme per RBC. It presents a volume of high concentration of purified biomarker molecules, uniquely suited for their sensitive and specific detection by LDMS. In malaria, the microorganism itself performs the often complex and time-consuming sample processing tasks, encountered in the MS analysis of biological material!

Two sample preparation protocols (Figure 8.2), described in more detail previously, have been used in LDMS malaria detection thus far—a "concentration" method (*C*)[14], and a rapid "dilution" method (*D*).[15,16] In method *C*, established procedures for isolation of intact parasites from RBC are followed.[32] Samples (2 ml for parasite cultures, grown in vitro,[14] or 30 to 50 µl infected blood for the mouse[15] or human[16] studies) are initially centrifuged at 10^4 g and the supernatant is removed. RBC lysis is achieved by addition of either 1% acetic acid solution or 0.15% saponin solution. Samples are diluted with equal volume of PBS, and centrifuged again at 10^4 g. Pellets are washed twice with PBS and the intact parasites fractionated from the blood constituents by extensive washing of the pellet. Washed pellets are re-suspended in smaller volume of PBS buffer, compared to the original samples, thereby enriching the parasite density several-fold. In contrast, in method *D*, the original sample is being diluted 10-fold (v/v) in PBS. While both protocols can be performed in a high throughput set up, the highly abbreviated protocol *D* is much more rapid compared to *C* (the latter taking approximately 20 to 30 minutes). The protocol *D* is inexpensive, with estimated cost of consumables less than $0.01 per test.[15,16] For LDMS analysis, typically 0.5 µl of sample is

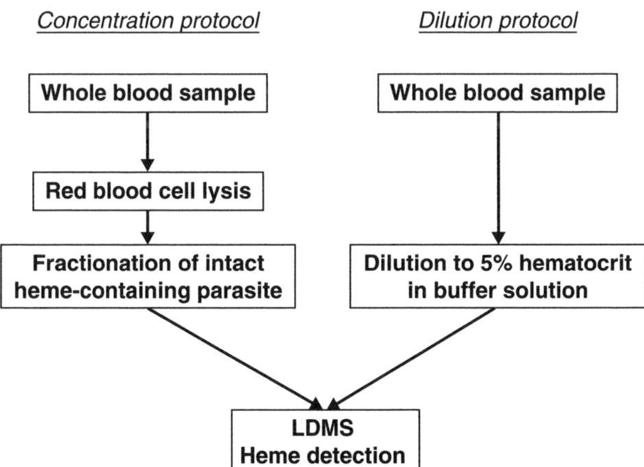

Figure 8.2 Sample preparation protocols, currently employed in malaria parasite detection by LDMS. Both protocols require less than 50 µl blood, obtained by either finger-stick or phlebotomy.

pipetted onto individual wells of the sample holder (total volume 10 μl = 0.5 μl/well × 20 sample holder wells) and allowed to air dry prior to insertion into the instrument. It is clear that the overall LDMS detection efficiency from the same sample will be much higher for protocol *C*, compared to protocol *D*, since the expected surface parasite density on the target holder will be orders of magnitude lower in the latter case. To offset for that reduced counting statistics, a larger surface area with *D*-protocol prepared samples has to be examined by LDMS (which can be performed rapidly in an automated fashion with the high rep rate pulsed UV lasers, currently available). *Caution*: Proper safety procedures ("biohazard level two") should be followed when handling human blood plasma.[33]

8.3.2 Mass Spectrometry

In most experiments, a "Kompact MALDI" commercial time-of-flight mass spectrometer (Kratos Analytical Instruments, Chestnut Ridge, NY) has been used to acquire mass spectra using the instrument settings already described.[14–16] Typically only positive ion LD mass spectra are acquired in the linear mode with a 20 kV total ion acceleration. Pulsed ion (delayed) extraction is optimized for ion focusing and transmission at m/z 1200. The N_2 laser ("VSL-337ND" Laser Science, Franklin, MA, provided the instrument) has an estimated fluence of 1 J/cm^2 before attenuation (150 μJ average energy/pulse at 337 nm laser wavelength, pulse duration 4 ns, 150 μm × 100 μm estimated beam spot dimensions). The laser beam is appropriately attenuated for simultaneous detection of both the heme molecular ion and heme-specific fragment ions. One hundred individual laser shot spectra are typically acquired while the laser beam is linearly rastered (20 μm step) across each sample well (2 mm length). This procedure generates two thousand individual mass spectra per sample (20 wells on each sample slide). The laser beam interrogates approximately 10% of the sample surface area. Individual spectra are initially processed by in-house written software, in which a heme spectral matching filter is applied.[16] The matching filter selects only spectra containing the molecular and the six most intense heme fragment ions (*vide infra*, Section 8.4.1). The integrals of the heme molecular and fragment ion intensities for each sample are calculated by summing the intensities in the individual laser shot spectra that "pass" through the filter. Ion signal intensity distributions on a laser shot-by-shot basis have been also subjected to covariance-mapping analysis, as already described.[34]

For comparison purposes, two other MS systems have been used with in vitro grown parasite cultures. In the first case, a home-built miniaturized linear TOF multi-array analyzer[17,18] was interfaced to the frequency-tripled output at 355 nm of a Q-switched Nd-YAG laser ("Polaris," New Wave Research, Fremont, CA). In this experiment the laser beam was not scanned, and the estimated laser fluence after attenuation was similar to that of the commercial LD TOF system. The second MS system consisted of a Mass-Tech (Columbia, MD) AP LD ion source, coupled to a Thermo-Finnigan (San Jose, CA) LCQ Deca XP ion trap mass spectrometer. Operating parameters were similar to the ones already published.[19] These included 2.7 kV potential between the target plate and LCQ inlet transport capillary, maintaining the transport capillary at 200°C, and manually setting the ion trapping time to 400 ms. Sample irradiation was with a 337 nm N_2 laser ("VSL-337ND," Laser Science, Franklin, MA) operating at 10 Hz with around 100 μJ energy/pulse. In both cases sample suspensions (1 μl, containing approximately 10^3 in vitro grown *P. falciparum* parasites, prepared according to protocol *C*) were deposited on either a support stage with a commercially available polymer coating (for the TOF system) and gold-plated metal target plate (for the AP LD LCQ system). Additional details, relevant for these experiments, are listed in parallel with presentation of the respective results.

8.4 MALARIA DETECTION BY LASER DESORPTION MASS SPECTROMETRY

8.4.1 LDMS of Heme and Hemozoin

LDMS is particularly well suited for the analysis of porphyrins.[35–39] The heme molecule—a 22 π-electron conjugated protoporphyrin system (Figure 8.1)—is an efficient photo-absorber in the visible and near UV (with an absorption maximum—the Soret band—near 400 nm). This feature, concurrently with its low ionization potential, warrants that direct LDMS will possess extremely low limits for heme detection. The uses of IR or UV LDMS for structural characterization of natural porphyrins and their metabolites, synthetic monomeric porphyrins (e.g., used in photodynamic therapy), porphyrin polymers, and multimeric arrays, have been well documented.[40–48] In addition fast atom bombardment MS has been used to characterize purified hemozoin, isolated from the spleens and livers of *Plasmodium yoelii* infected mice.[49]

The LD hemozoin mass spectra under all experimental conditions studied exhibit a radical molecular cation ($M^{+\bullet}$ at *m/z* 616) and several structurally characteristic heme fragment ions (Figures 8.3 and 8.4). A radical molecular anion has been observed in negative ion mode, and no doubly- or triply-charged ions have been detected from heme in either polarity.[18] Ions corresponding to heme molecule dimers $(2M)^{+\bullet}$ and dimeric fragments are observed in positive ion spectra as well, with signal intensities varying as a function of laser fluence, sample concentration, and so on. Investigations of factors influencing its formation from synthetic porphyrins suggest that the dimer occurs in the plume of laser-ablated material (unpublished data). Under otherwise identical LDMS instrumental conditions, the heme signal intensity in the positive ion mode is about an order of magnitude higher than in negative ions.[18]

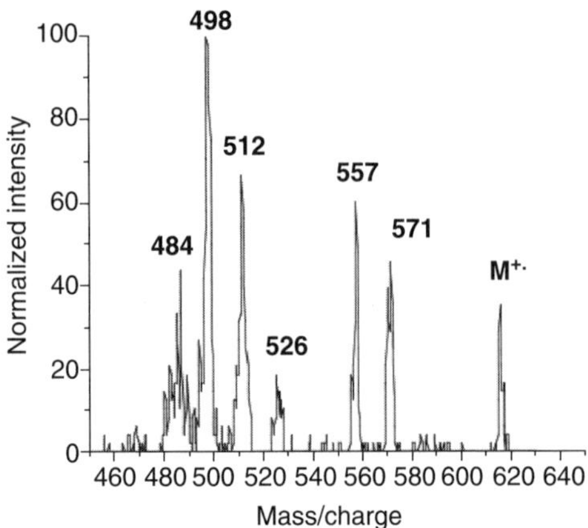

Figure 8.3 Positive ion LD TOF mass spectrum of blood from a *P. vivax* infected human patient (only asexual parasites have been observed by microscopy; estimated parasitemia approximately 72 parasites/µl). Protocol *C* is used for sample preparation; estimated number of parasites deposited per well is approximately 90. A commercial TOF system is used: laser wavelength 337 nm. All one hundred single laser shot spectra, obtained from linear scanning of an individual well, are averaged (no data smoothing). The characteristic "fingerprint" ions of detected heme are denoted.

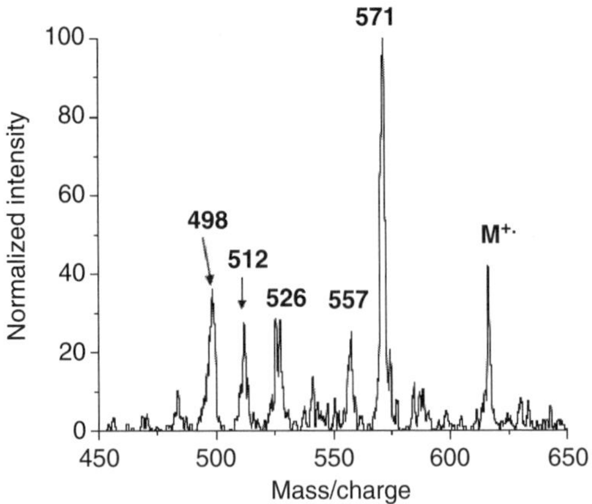

Figure 8.4 Positive ion direct atmospheric pressure LD mass spectrum of in vitro grown *P. falciparum* parasites. Protocol *C* is used for sample preparation; estimated number of parasites deposited is approximately 10^3. A commercially available AP LD quadrupole ion trap (LCQ) system is used.[19,20] Typical laser beam spot diameter is 0.5 mm; acquisition time is approximately 20 s; LCQ inlet capillary temperature –200°C.

The fragment ions at m/z 571, 557, 526, 512, 498, and 484 correspond to consecutive cleavages of 45 ($COOH^•$), 59 ($CH_2COOH^•$) or 73 ($CH_2CH_2COOH^•$) from $M^{+•}$ that are directly correlated with the heme structure. Hemin, synthetic β-hematin (heme crystals), and parasite-derived heme crystals yield mass spectra virtually identical to that of heme.[14,15] This characteristic heme "signature" has been used as a template in the signal-processing algorithms. Matching filter analysis of each individual single-shot spectrum, obtained by laser beam scanning across the surface, demonstrates that heme signals come in discrete, spatially contiguous "bursts." Burst frequency tracks surface parasite density, which is consistent with the presence of isolated local heme sources on the surface, that is, detection of individual parasites by LDMS (to be published).

While the desorbing laser wavelength (337 vs. 355 nm) does not influence the observed heme fragmentation pattern, the degree of fragmentation increases with increased laser fluence.[14] The existence of a laser fluence threshold (F_{th}) for heme detection has been also established. The F_{th} value is a function of the amount of parasite sample deposited—lower amounts are detected at higher F_{th}. Above threshold, the ion signal increases faster than linear as a function of fluence. Under the selected laser fluence conditions all observed fragments are due to losses from the two propionyl side chains, while the Fe atom-containing tetrapyrrole ring is preserved intact. The AP LD mass spectrum (Figure 8.4) exhibits a much lower extent of fragmentation of the heme molecular ion—the most intense peak being at m/z 571—when compared to the vacuum LD mass spectra with base peak at m/z 498 (Figure 8.3). Reduced fragmentation in AP LD is probably due to enhanced ion cooling (ion internal energy thermalization) through collisions during molecular ion transport through the atmospheric pressure region before the ions enter the capillary. This is in direct analogy with AP MALDI experiments. Comparison between conventional vacuum and AP MALDI of various peptides, nucleotides and sugars demonstrates the extra softness and tolerance to higher laser fluences of the latter technique.[19] The AP LD technique in conjunction with a quadrupole ion trap allows multistage tandem MS experiments to be easily performed, for example, MS^3 of the heme molecular ion: $M^{+\bullet} \rightarrow m/z\ 557 \rightarrow m/z$ 498 (spectra not shown). Such tandem MS capability provides an enhanced means for more sensitive detection and biomarker identity confirmation.

The LDMS signal from nonsynchronized *P. falciparum* cultures (prepared according to protocol *C*) varies linearly as a function of parasite density (number of parasites deposited) for samples containing from 500 to 1.55×10^4 parasites per microliter.[14] The heme biomarker is pan-malarial, since heme sequestration is a feature of all malaria parasite species. The in vivo studies of the time course of nonlethal *Plasmodium yoelii* infections in mice (done in parallel with light microscopy and a colorimetric hemozoin assay) have demonstrated that hemozoin determinations by the other techniques lags behind the LDMS detection of infections by 2 to 4 and 3 to 5 days, respectively.[15] Evolution of the LDMS heme signal during the various phases of infection is shown in Figure 8.5. Hemozoin has been detected by LDMS in 0.3 µl of mouse blood within 2 days of infection. No heme is detected in the blood collected from noninfected mice.[15] Typically, and independent of the sample preparation protocols, the LDMS heme signal intensity

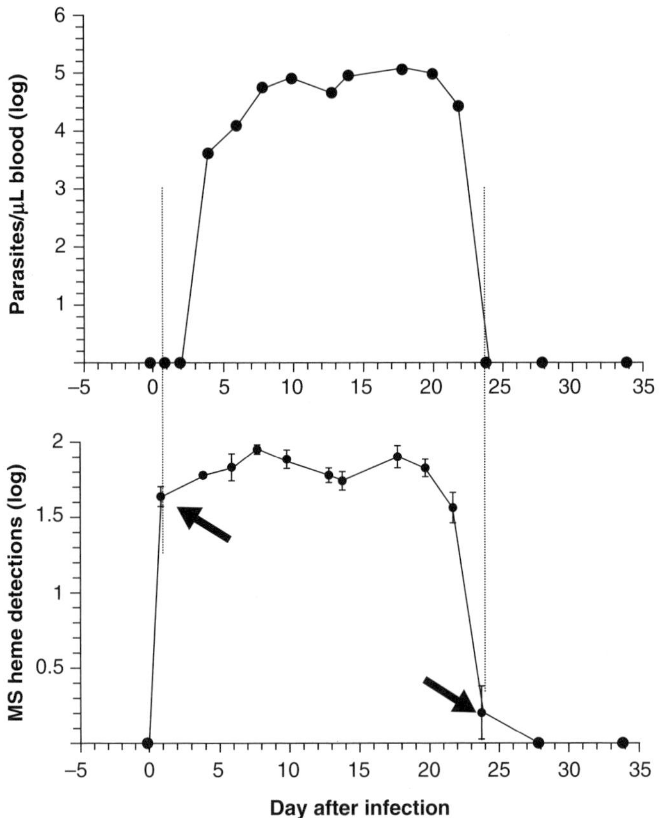

Figure 8.5 Monitoring the in vivo time course of *P. yoleii* malaria infection in mice inoculated with live parasites at day 0.[15] (*Upper trace*) Parasite count obtained by microscopy of blood smear, folded with anemia model from the literature: (parasites/vol) = (parasites/RBC) × (RBC/vol). (*Lower trace*) Integrated LDMS heme signal from 300 shots across three consecutive sample wells; each sample (30 μl) is processed following protocol *C*, and examined on a commercial LD TOF instrument. Infection is more easily and more rapidly discerned both at earlier and later times by LDMS, compared to the traditional optical microscopy examination.

increases through day 6, reaches a plateau on subsequent days and returns to nondetectable levels by day 28.

In the first LDMS-based detection of malaria in human subjects (unpublished), 100 μl *P. falciparum* or *P. vivax*-infected blood samples, grouped into three different parasitemia ranges—low (10–150 parasites/μl), mid (2×10^3 parasites/μl), and high (25×10^3–60×10^3 parasites/μl)—have been examined using both sample preparation protocols. Parasitemia levels in these samples were previously determined independently for each sample by optical microscopy examination of blood smears. The LDMS data clearly indicate that heme signal intensities increase with increasing parasitemia, and that heme signal intensities are higher for *P. vivax* than for *P. falciparum* at equivalent parasitemia levels. The consistently higher heme signal amplitude for *P. vivax* is presumably due to the different characteristics of the *P. vivax* and *P. falciparum* parasites circulating in the peripheral blood. While only ring-stage parasites with minimal hemozoin are typically observed during *P. falciparum* infections, for *P. vivax* infections later-stage parasites with higher hemozoin content are also present. In this study heme has been detected in 17 of the 19 malaria-positive human samples by LDMS employing the *C*-protocol for sample preparation. The LDMS method fails to detect heme only in two of the *P. falciparum* samples with the lowest parasitemias (13 and 35 parasites/μl) but detects heme in the other two (57 and 127 parasites/μl). All low parasitemia *P. vivax* samples (at 27, 48, and 73 parasites/μl) have been positively identified. No heme has been detected in any of the 10 true negative controls.

The tight binding of heme to hemoglobin and other proteins (e.g., hemopexin, albumin) can be exploited to eliminate the centrifugation steps and various reagents used for LDMS sample preparation. Heme complexation with proteins under physiological conditions renders it LDMS "silent" in non-infected blood samples (Figure 8.6). For that reason protocol *C*, initially used to prepare whole blood samples for malaria detection[14], can be replaced with protocol *D*, without incurring false positive detections, and thereby increasing

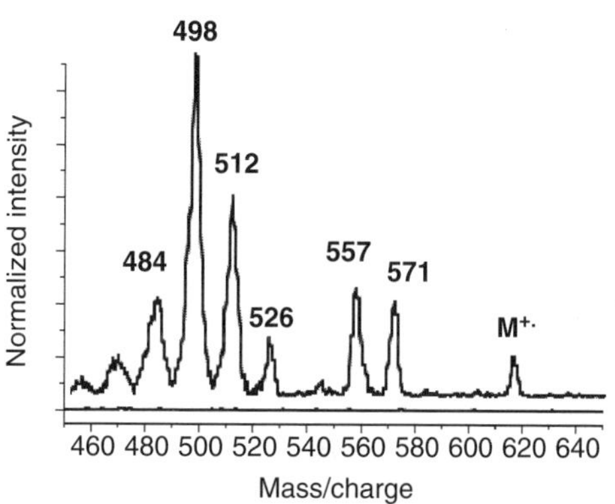

Figure 8.6 Positive ion LD TOF mass spectra of *P. falciparum* parasite sample (*upper trace*), and a control (uninfected blood) sample (*lower trace*). Protocol *D* is used for sample preparation. Both samples—in vitro cultured *P. falciparum* parasites in whole blood, and the whole blood control—are diluted to 5% hematocrit (10-fold) in PBS buffer. In the infected sample the estimated number of deposited parasites per sample well is approximately 100. A commercial LD TOF system is used, and both spectra are normalized to the same (40 mV) detector response value. Each trace represents the average of one hundred single laser shot spectra obtained from linear scanning of an individual well (no data smoothing). The characteristic "fingerprint" ions of detected heme in the upper trace are denoted.

assay throughput. On the other hand, heme is readily detected after addition of, for example, formic acid (0.5 μl, 12 M) to dry noninfected blood samples, possibly due to release of the noncovalently attached heme after protein denaturation.

8.4.2 Choline Phosphate—Another Unique LDMS Biomarker for Malaria Detection

Covariance mapping analysis[34] of individual LD mass spectra has been performed in order to identify other potential biomarkers that may co-desorb with heme ions. In the LD spectra of malaria-infected blood samples,[15,16] several signal bands have been found that are strongly positively correlated with heme fingerprint ions. Most prominent are the ion signals at *m*/*z* 184 (*a*) and 86 (*c*) (Figure 8.7). These ions are the two most characteristic for the presence of the choline phosphate (CP) moiety.[50] Ions observed at *m*/*z* 104 (*b*, choline) and 125 (*d*) can be also mapped to CP-specific fragmentation pathways (Figure 8.8). The strong correlation between the heme "fingerprint" ions and the CP-specific ions in the LDMS spectra of all *P. falciparum* and *P. vivax* infected human blood samples (at mid- and high-range parasitemias) is illustrated by the fact that integrated intensities of the CP-ion signal are at least 2 orders of magnitude higher in *Plasmodium*-infected samples, compared to

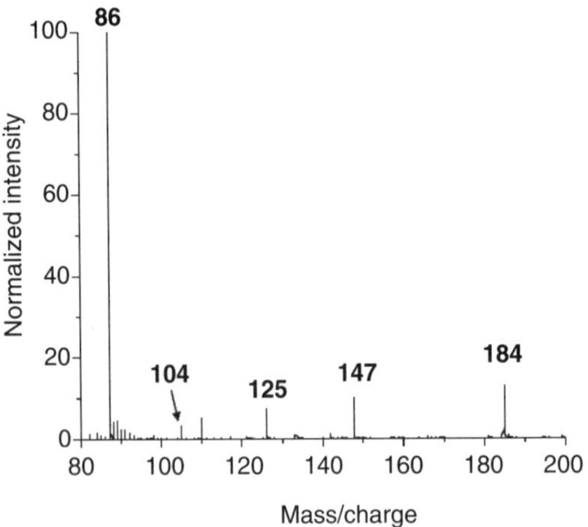

Figure 8.7 Low mass (choline phosphate) region of a LD mass spectrum of blood from a *P. vivax* infected human patient (estimated parasitemia approximately 32,000 parasites/μl). Protocol *C* is used for sample preparation. A commercial LD TOF system is used: laser wavelength 337 nm. All one hundred single laser shot spectra, obtained from linear scanning of an individual well, are averaged (no data smoothing). The characteristic ions of detected choline phosphate are denoted.

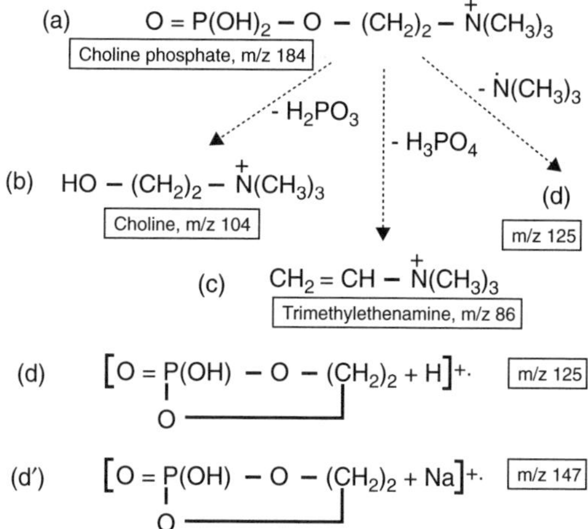

Figure 8.8 Tentative structures of choline phosphate-related biomarker ions observed by LDMS.

baseline signal levels, measured in uninfected blood samples (detailed report to be published elsewhere). Elevated CP levels in malaria-infected blood are most likely related to the specific requirements for lipid biosynthesis during the parasite's intra-erythrocytic growth.[51–57] During its intra-erythrocytic life cycle, *Plasmodium* initiates the formation of a complex dynamic network of various membranes. They include, for example, the parasitophorous vacuolar membrane (PVM) and the turbovesicular membrane linking together the PVM and the RBC host membrane.[52,54,55] The physiological role of these membranes in the parasite growth is multifold. The lipid membranes provide the structural integrity and the required intra-erythrocytic compartmentalization, and serve both as barriers and as transport conduits to and from the developing parasites for extracellular nutrients and metabolic waste disposal, for example. In order to form these membranes (including its own plasma membrane), the malaria parasite has developed a variety of pathways for the de novo lipid biosynthesis. Some pathways are unique for *Plasmodium* and present target opportunities for novel antimalarial drugs. In particular, sphingomyelin is required for membrane assembly, and CP has been identified as an essential starting material for the sphingomyelin biosynthesis.[51,53] Very recently CP formation has been implicated as essential in the de novo phosphatidylcholine biosynthesis from serine in *P. falciparum*.[56] A plant-like *S*-adenosyl-L-methionine-dependent three-step methylation reaction pathway has been elucidated. In it, ethanoleamine phosphate is converted into CP via an enzyme—phosphoethanolamine methyltransferase—with no human homologs.

8.5 MS-BASED PROTEOMICS OF THE *PLASMODIUM* PARASITE

The sequencing[21,22] of the *P. falciparum* (clone 3D7) and *P. yoelii* genomes provides the basis for qualitative and quantitative proteomics, with the ultimate aim to identify new potential drug and vaccine targets. It is estimated that the 23 Mb *P. falciparum* genome (distributed among 14 chromosomes) encodes for more than 5300 proteins. Approximately 65% of these proteins have been identified as hypothetical. High-throughput proteomics methodologies[58] would allow better understanding of the biology of this complex protozoan parasite, such as elucidation of its global protein expression profiles and of the posttranslational modifications and subcellular locations of the various gene products during different developmental stages. Furthermore the availability of the human and *Anopheles* genomes will facilitate the efficient mapping of host–parasite and transmission vector–parasite interactions central to disease pathogenesis and treatment. It has been pointed out[23,25] that the *Plasmodium* parasite may also serve as the model for the discovery and validation of effective algorithmic/computational approaches to translate genomic information into vaccines, by mapping multiple targets of protective T cell or antibody responses from the sequence data. The *Plasmodium* parasite—a complex multistage pathogen targeted by multiple immune responses—may require a new generation vaccine that is nearly as complex as the organism itself. It has been argued that proteomics should be the strategy of choice for *P. falciparum* vaccine development.[23–26] MALDI MS employing advances in genomics, proteomics, and bioinformatics has begun to emerge recently as an efficient tool in biomedicine for detection of intact microorganisms.[59,60] If rapid sample preparation techniques emerge for MALDI MS detection of malaria proteins in blood, the malaria proteome may be also utilized to practically diagnose infections at the *Plasmodium* species level.

Two complementary proteomics/mass spectrometry technologies have been implemented in the two studies, described in papers published in the special issue of *Nature* magazine, devoted to the parasite genomics. Florens et al.[24] have used a high-throughput multidimensional proteomics approach, combining tandem HPLC-tandem MS of the total digests and the SEQUEST algorithm to match peptide tandem mass spectra to ORF in the genome sequence of the *P. falciparum* parasite. More than 40% of all parasite proteins, expressed in four intra-erythrocytic lifecycle stages of the parasite—sporozoite, merozoite, trophozoite, and gametocyte—have been functionally profiled by this "shotgun" proteomics approach. Of these identified proteins roughly half have been annotated as hypothetical (i.e., *Plasmodium*-specific). Each of the identified proteins has grouped according to the functional classification established for genome databases.[24] More than 130 clusters of three co-expressed consecutive genes for each stage have been identified. The presence of such clusters demonstrates a widespread high order of chromosomal organization of the parasite. The proteomics studies can provide the means for deciphering and controlling gene expression regulation during the parasite life cycle, since the direct regulatory sequence identification in the *Plamsodium* genome is rather challenging. Furthermore 18% of the identified proteins are soluble proteins with a signal sequence (potentially secreted or located to organelles) or are predicted to have at least one transmembrane segment or a phosphatydil inositol anchor addition signal. Well over half of these proteins are annotated as hypothetical, meaning with no homology to known proteins. As such, they all represent potential targets for new drug and vaccine development.

In the second large-scale proteomics study, Lasonder et al.[26] employ electrophoresis of isolated parasite proteins, followed by in-gel digestion and by nano-LC MS/MS of the digest peptide mixtures. The use of high-accuracy (average deviation less than 0.02 Da at 1000 Da) quadrupole-TOF instrument for the tandem MS experiments has allowed the confirmation of approximately 1300 individual genome-derived gene products. Approximately 700 of these proteins are expressed only in the asexual stages of parasite development, approximately 200 are common to both the asexual and sexual blood stages, and approximately 600 are found only in gametes. Improved understanding of the biology of the parasite sexual stages with the underlying conserved membrane protein components, responsible for intercellular signaling and interactions, can help identify potential malaria vaccines. In addition the parasite genome is highly AT-rich and therefore difficult to be assembled and annotated. In that case direct searches into the genome sequence have been performed with peptide sequence tags derived by MS/MS. A set of more than 100 peptide tags has been matched to the parasite genome sequence with a high significance. However, no corresponding individual proteins have been identified by traditional bioinformatics/computational biology methods in these matched regions of the genome. Using these identified "orphan" peptides, re-assignments of exon–intron boundaries or assignments of additional exons in the parasite genome have been accomplished.

8.6 CONCLUSIONS

The transformation of mass spectrometry into a viable technology for biomedical diagnostics has been the long-cherished goal of many scientists.[60] The detection of malaria parasites by LDMS, both in vitro and in vivo, may provide such an example. Based on the unique features of the host–parasite metabolomics, sequestered heme as a pan-malarial *Plasmodium* biomarker allows less than 10 parasites per µl blood to be detected. Parasite "counting"—detection of individual parasites by LDMS under appropriate conditions—can be demonstrated. LDMS is not only extremely sensitive, but it is quantitative—heme signal intensity tracks the number of parasites per unit volume of blood. Choline phosphate—another distinctive malaria biomarker formed during the intra-erythrocytic parasite stages—can complement the set of heme fingerprint ions in the biomarker template used in the data processing algorithms for sensitive malaria detection by LDMS. A simplified sample protocol requiring minimal handling combined with miniaturized LD TOF instruments can permit the large-scale deployment of automated screening systems for rapid and affordable malaria diagnosis in large populations. Such systems could not only enable accurate point-of-care diagnoses in remote regions but could also make large-scale surveillance programs possible, creating a new paradigm for malaria control and eradication. A number of infectious agents, including *Plasmodium*, pose serious threats to blood transfusion recipients.[61] Preventing the transmission of these agents during blood transfusions is often problematic and novel approaches are needed. One such approach can be the efficient high-throughput screening of blood banks for malaria parasites and other infectious agents, implemented on miniaturized multi-array LD TOF MS instruments with advanced signal processing.

ACKNOWLEDGMENTS

I am very grateful to a large number of colleagues for fruitful collaborations during various phases of the research, described here: A. Feldman, T. Cornish, S. Ecelberger, J. Lin (Johns Hopkins University Applied Physics Laboratory, Laurel, MD); D. Haddad, D. Kongkasuriyachai, N. Kumar, J. Pisciotta, P. Scholl, L. Shi, D. Sullivan Jr. (Bloomberg School of Public Health, Johns Hopkins University, Baltimore, MD); V. Doroshenko, N. Taranenko (Mass Tech Inc., Columbia, MD); C. Wongsrichanalai (U.S. Naval Medical Research Unit, Jakarta, Indonesia); R. Gasser (Walter Reed Army Institute of Research, Silver Spring, MD). Financial support from the Johns Hopkins Malaria Research Institute (Baltimore, MD) and the Stuart S. Janney Sabbatical Fellowship Program at the Applied Physics Laboratory is acknowledged as well.

REFERENCES

1. "Roll Back Malaria"—a partnership between the World Health Organization, the UN Development Program, UNICEF and the World Bank. Available at http://mosquito.who.int.
2. Brundtland, G. H. Winning the war against AIDS, tuberculosis, and malaria. *Harvard Health Policy Rev.* 2002, **3**(2).
3. Guerin, P. J.; Olliaro, P.; Nosten, F.; Druilhe, P.; Laxminarayan, R.; Binka, F.; Kilama, W. L.; Ford, N.; White, N. J. Malaria: Current status of control, diagnosis, treatment and a proposal agenda for research and development. *Lancet Infec. Dis.* 2002, **2**, 564–573.
4. Phillips, R. S. Current status of malaria and potential for control. *Clin. Microbiol. Rev.* 2001, **14**, 208–226.
5. Weiss, U. Nature insight: Malaria. *Nature* 2002, **415**, 669–715.

6. Makler, M.; Palmer, C.; Ager, A. A review of practical techniques for the diagnosis of malaria. *Ann. Trop. Med.Parasitol.* 1998, **92**, 419–433.
7. Sherman, I. W. *Malaria: Parasite Biology, Pathogenesis, and Protection.* Washington, DC: ASM Press, 1988.
8. World Health Organization. Malaria rapid diagnosis, making it work. In RS/2003/GE/05 report. Geneva: WHO, 2003.
9. Farcas, G. A.; Zhong, K. J. Y.; Mazzulli, T.; Kain, K. C. Evaluation of the Real Art Malaria LC Real-Time PCR Assay for malaria diagnosis. *J. Clin. Microbiol.* 2004, **42**, 636–638.
10. Gascoyne, P.; Satayaviviad, J.; Ruchirawat, M. Microfluidic approaches to malaria detection. *Acta. Tropica* 2004, **89**, 357–369.
11. Hanschied, T. Diagnosis of malaria: A review of alternatives to conventional microscopy. *Clin. Lab. Haem.* 1999, **21**, 235–245.
12. Moody, A. Rapid dianostic tests for malaria parasites. *Clin. Microbiol. Rev.* 2002, **15**, 66–78.
13. Murray, C. K.; Bell, D.; Gasser, R. A.; Wongsrichanalai, C. Rapid diagnostic testing for malaria. *Trop. Med. Int. Health* 2003, **8**, 876–883.
14. Demirev, P. A.; Feldman, A. B.; Kongkasuriyachai, D.; Scholl, P.; Sullivan, D. J.; Kumar, N. Detection of malaria parasites by laser desorption mass spectrometry. *Anal. Chem.* 2002, **74**, 3262–3266.
15. Scholl, P. F.; Kongkasuriyachai, D.; Demirev, P. A.; Feldman, A. B.; Lin, J. S.; Sullivan, D. J.; Kumar, N. Rapid detection of malaria in vivo by laser desorption mass spectrometry. *J. Trop. Med. Hyg.* 2004, **71**, 546–551.
16. Nyunt M.; Pisciotta, J.; Feldman, A.; Thuma, Ph.; Scholl, P.; Demirev, P. A.; Lin, J.; Shi, L.; Kumar, N.; Sullivan, D. Detection of *plasmodium falciparum* in pregnancy by laser desorption mass spectrometry. *J. Trop. Med. Hyg.* 2005, **73**, in press.
17. Demirev, P.; Cornish, T.; Ecelberger, S.; Feldman, A.; Kongkasuriyachai, D.; Kumar, N.; Scholl, P.; Sullivan, D., Jr. *Proc. of the 50th AMS Conference*, Orlando, FL, 2002.
18. Cornish, T. J.; Antoine, M.; Ecelberger, S. A.; Demirev, P. A. Arrayed time-of-flight mass spectrometry for time-critical detection of hazardous agents. *Anal. Chem.* 2005, **77**, 3954–3959.
19. Doroshenko, V. M.; Laiko, V. V.; Taranenko, N. I.; Berkout, V. D.; Lee, H. S. Recent developments in atmospheric pressure MALDI mass spectrometry. *Int. J. Mass Spectrom.* 2002, **221**, 39–58.
20. Moyor, S. C.; Marzilli, L. A.; Woods, A. S.; Laiko, V. V.; Doroshenko, V. M.; Cotter, R. J. Atmospheric pressure matrix-assisted laser desorption/ionization (AP MALDI) on a quadrupole ion trap mass spectrometer. *Int. J. Mass Spectrom.* 2003, **226**, 133–150.
21. Carlton, J. M.; et al. Genome sequence and comparative analysis of the model rodent malaria parasite *Plasmodium yoelii*. *Nature* 2002, **419**, 512–519.
22. Gardner, M. J.; et al. Genome sequence of the human malaria parasite *Plasmodium falciparum*. *Nature* 2002, **419**, 498–511.
23. Doolan, D. L.; Aguiar, J. C.; Weiss, W. R.; Sette, A.; Felgner, P. L.; Regis, D. P.; Quinones-Casas, P.; Yates, J. R.; Blair, P. L.; Ritchie, T. L.; Hoffman, S. L.; Carucci, D. J. Utilization of genomic sequence information to develop malaria vaccines. *J. Experiment. Biol.* 2003, **206**, 3789–3802.

24. Florens, L.; Washburn, M. P.; Raine, J. D.; Anthony, R. M.; Grainger, M.; Haynes, J. D.; Moch, J. K.; Muster, N.; Sacci, J. B.; Tabb, D. L.; Witney, A. A.; Wolters, D.; Wu, Y.; Gardner, M. J.; Holder, A. A.; Sinden, R. E.; Yates, J. R.; Carucci, D. J. A proteomic view of the *Plasmodium falciparum* life cycle. *Nature* 2002, **419**, 520–526.

25. Johnson, J. R.; Florens, L.; Carucci, D. J.; Yates, J. R. Proteomics in malaria. *J. Proteome Res.* 2004, **3**, 296–306.

26. Lasonder, E.; Ishihama, Y.; Andersen, J. S.; Vermunt, A. M.; Pain, A.; Sauerwein, R. W.; Eling, W. M.; Hall, N.; Waters, A. P.; Stunnenberg, H. G.; Mann, M. Analysis of the *Plasmodium falciparum* proteome by high-accuracy mass spectrometry. *Nature* 2002, **419**, 537–542.

27. Francis, S.; Sullivan, D. J.; Goldberg, D. E. Hemoglobin metabolism in the malaria parasite *Plasmodium falciparum. Annu. Rev. Microbiol.* 1997, **51**, 97–123.

28. Klemba, M.; Goldberg, D. E. Biological roles of proteases in parsitic protozoa. *Annu. Rev. Microbiol.* 2002, **71**, 275–305.

29. Sullivan, D. J. Theories on malarial pigment formation and quinoline action. *Int. J. Parasitol.* 2002, **32**, 1645–1653.

30. Ross, R. In *Nobel Lectures, Physiology or Medicine*. Amsterdam: Elsevier, 1967, Vol. 1, pp. 25–116.

31. Sherman, I. W. In *Malaria: Parasite Biology, Pathogenesis, and Protection*. Sherman, I. W. (Ed.) Washington, DC: ASM Press, 1988, pp. 3–10.

32. Jungström, I.; Perlmann, H.; Schlichterle, M.; Scherf, A.; Wahlgren, M. *Methods in Malaria Research*, 4th ed. Manassas, VA: ATCC, 2004.

33. Centers for Disease Conrol, N. I. o. H. *Biosafety in Microbiological and Biomedical Laboratories*, 4th ed. Washington, DC: US Government Printing Office, 1999.

34. Feldman, A.; Antoine, M.; Lin, J.; Demirev, P. Covariance mapping in matrix-assisted laser desorption/ionization time-of-flight mass spectrometry. *Rapid Comm. Mass Spectrom.* 2003, **17**, 991–995.

35. Conzemius, R. J.; Capellen, J. M. A Review of the application to solids of the laser ion source in mass spectrometry. *Int. J. Mass Spectrom. Ion. Phys.* 1980, **34**, 197–271.

36. Cotter, R. J. Laser mass spectrometry—An overview of techniques, instruments and applications. *Anal. Chim. Acta.* 1987, **195**, 45–59.

37. Spengler, B.; Karas, M.; Bahr, U.; Hillenkamp, F. Excimer laser desorption mass spectrometry of biomolecules at 248 and 193 nm. *J. Phys. Chem.* 1987, **91**, 6502–6506.

38. Van Vaeck, L.; Struyf, H.; Van Roy, W.; Adams, F. Organic and inorganic analysis with laser microprobe mass spectrometry. *Mass Spectrom Rev.* 1994, **13**, 189–208.

39. Vertes, A.; Gijbels, R.; Adams, F. *Laser Ionization Mass Analysis*. New York: Wiley, 1993.

40. Brown, R. S.; Wilkins, C. L. Laser desorption Fourier transform mass spectrometry of synthetic porphyrins. *Anal. Chem.* 1986, **58**, 3196–3199.

41. Dale, M. J.; Costello, K. F.; Jones, A. C.; Langridge-Smith, P. Investigation of posphyrins and metalloporphyrins using two-step laser mass spectrometry. *J. Mass Spectrom.* 1996, **31**, 590–601.

42. Fenyo, D.; Chait, B. T.; Johnson, T.; Lindsey, J. S. Laser desorption mass spectrometry of synthetic multiporphyrin arrays. *J. Porph. Phthalocyan.* 1997, **1**, 93–99.
43. Jones, R. M.; Lamb, J. H.; Lim, C. K. Urinary porphyrin profiles by laser desorption ionization time-of-flight mass spectrometry without the use of classical matrices. *Rapid Comm. Mass Spectrom.* 1995, **9**, 921–923.
44. Quirke, J. M. *Mass Spectrometry of Porphyrins and Metalloporphyrins*. San Diego: Academic Press, 2000.
45. Tabet, J. C.; Jablonski, M.; Cotter, R. J.; Hunt, J. E. Time-resolved laser desorption: The metastable decomposition of Chlorpophyll-A and some derivatives. *Int. J. Mass Spectrom. Ion Phys.* 1985, **65**, 105–117.
46. Wagner, R. W.; Ciringh, Y.; Clausen, C.; Lindsey, J. S. Investigation and refinement of palladium-coupling conditions for the synthesis of diarylethyne-linked multiporphyrin arrays. *Chem. Mater.* 1999, **11**, 2974–2983.
47. Weeding, T. L.; Steenvoorden, R. J. M.; Kistenmaker, P. G.; Boon, J. J. Laser desorption multiphoton ionization mass spectrometry. *J. Anal. Appl. Pyrol.* 1991, **20**, 47–56.
48. Zhan, Q.; Voumard, P.; Zenobi, R. Chemical analysis of cancer therapy photosensitizers by 2-step laser mass spectrometry. *Anal. Chem.* 1994, **66**, 3259–3266.
49. Pandey, A.; Tekwani, B.; Pandey, V. Characterization of hemozoin from liver and spleen of mice infected with *Plasmodium yoelii*, a rodent malaria parasite. *Biomed. Res.* 1995, **16**, 115–120.
50. Ayanoglu, E.; Wegmann, A.; Pilet, O.; Marbury, G. D.; Hass, J. R.; Djerassi, C. Mass spectrometry of phospholipids—some applications of desorption chemical ionization and fast atom bombardment. *J. Am. Chem. Soc.* 1984, **106**, 5246–5251.
51. Haldar, K. Sphingolipid synthesis and membrane formation by *Plasmodium*. *Tr. Cell Biol.* 1996, **6**, 398–405.
52. Kirk, K. Membrane transport in the malaria-infected erythrocyte. *Physiolog. Rev.* 2001, **81**, 495–537.
53. Lauer, S. A.; Chatterjee, S.; Haldar, K. Uptake and hydrolysis of sphingomyelin analogues in *Plasmodium falciparum*-infected red cells. *Mol. Biochem. Parasitol.* 2001, **115**, 275–281.
54. Lauer, S. A.; Rathod, P. K.; Ghori, N.; Halder, K. A. A membrane network for nutrient import in red cells infected with the malaria parasite. *Science* 1997, **276**, 1122–1125.
55. Lingelbach, K.; Joiner, K. A. The parsitophorous vacuole membrane surrounding *Plasmodium* and *Toxoplasma*: An unusual compartment in infected cells. *J. Cell Sci.* 1998, **111**, 1467–1475.
56. Pessi, G.; Kociubinski, G.; Ben Mamoun, C. A pathway for phosphatidylcholine biosynthesis in *Plasmodium falciparum* involving phosphoethanolamine methylation. *Proc. Nat. Acad. Sci. USA* 2004, **101**, 6206–6211.
57. Vial, H. J.; Angelin, M. L. In *Malaria: Parsite Biology, Pathogenesis, and Protection*. Sherman, I. W. (Ed.) Washingtion, DC: ASM Press, 1988, pp. 159–175.
58. Aebersold, R.; Mann, M. Mass spectrometry-based proteomics. *Nature* 2003, **422**, 198–207.

59. Lay, J. O. MALDI-TOF mass spectrometry of bacteria. *Mass Spectrom. Rev.* 2001, **20**, 172–194.

60. Mann, M. Mass tool for diagnosis. *Nature* 2002, **418**, 731–732.

61. Dodd, R. Y.; Leiby, D. A. Emerging infectious threats to the blood supply. *Annu. Rev. Med.* 2004, **55**, 191–207.

9

BACTERIAL STRAIN DIFFERENTIATION BY MASS SPECTROMETRY

RANDY J. ARNOLD, JONATHAN A. KARTY, AND JAMES P. REILLY
Department of Chemistry, Indiana University, Bloomington, IN 47405

9.1 INTRODUCTION

The classification of microbes through analytical measurements of their chemical constituents, chemotaxonomy, has been an active field for more than two decades. Compared with microscopic observation of morphological characteristics, the use of analytical instrumentation to detect and identify bacteria offers the attractive prospects of higher speed, sensitivity, and accuracy. Mass spectrometry has been under investigation as a method for characterizing bacteria since the mid-1970s.[1] Pyrolysis mass spectrometry has been used to differentiate bacteria, but since the technique focuses on low molecular weight pyrolysis products, all information about the cellular macromolecules from which these products derived is lost.[2–4] Gas chromatography-mass spectrometry[5] and fast-atom bombardment mass spectrometry[6–8] have also been employed, and in these cases, ions having masses up to about 1000 amu have been detected. Non-mass spectrometric but otherwise standard methods of chemical analysis that have also been applied to the analysis of bacterial constituents include gas chromatography[9,10] high-performance liquid chromatography,[11,12] nuclear magnetic resonance,[13] and infrared spectroscopy.[14]

Identification of Microorganisms by Mass Spectrometry, Edited by Charles L. Wilkins and Jackson O. Lay, Jr.

Some of the most highly definitive methods of bacteria identification involve DNA probes, RNA probes, and the use of PCR for gene amplification.[15–17] A limitation of these methods is that mutations that occur in small, but critical parts of the genome can lead to false negative results. Extraction and sequencing of the 16s rRNA of bacteria is a method of growing popularity for identifying and classifying bacteria.[18] The complexity of ribosomal structure and the critical role that this organelle plays in protein synthesis insures that most random mutations in rRNA are selected against. Mutations that cells could survive generally occurred long ago and played a role in the development of different species. Thus rRNA sequencing is useful in species classification. On the other hand, 16s rRNA is only 1542 bases in length, and its variations are limited enough that it is *not* generally capable of distinguishing strains within a species. Pathogenic strains of bacteria are not typically identified through rRNA because the pathogen is usually a gene product that is completely independent of the ribosomal RNA. Instead, restriction endonucleases are employed to cleave the entire bacteria genome into a number of fragments and polyacrylamide gel electrophoresis is then used to resolve and detect these fragments.[19,20] A great deal of information, sufficient even to distinguish bacterial strains, is contained in the restriction fragment pattern. Unfortunately, the method suffers from being very slow (24–48 hours for the electrophoresis step, after typically a 24-hour bacteria culture). Furthermore there are reproducibility problems associated with the idiosyncracies of gel electrophoresis: results are sufficiently dependent on the method of gel manufacture and the experimental conditions under which the separation is performed that organizations employing this method often build their own libraries of gel data instead of importing data generated elsewhere.[21]

9.2 ANALYSIS OF CELLULAR PROTEINS BY MASS SPECTROMETRY

The analysis of cellular proteins by mass spectrometry is a new approach to bacterial identification that is rapidly gaining credibility. Although it does not have the sensitivity of a PCR-based DNA method it may be more universal. It is robust and can provide information about the *similarity* of two strains that may differ in only very subtle ways.[22] While the ribosomal RNA of a mutated bacteria strain rarely changes, its protein complement will likely be affected. Unlike nucleic acid-based methods, mass spectrometry has the potential for directly detecting toxins that a pathogenic strain might produce. Mass spectrometry of biological molecules has been facilitated by the development of two different ionization methods: electrospray (ESI)[23] and matrix-assisted laser desorption/ionization (MALDI).[24,25] Although each has its advantages, MALDI yields primarily singly charged ions while ESI creates a broad charge state distribution. Because MALDI mass spectra of complex samples such as cellular constituents tend to be substantially simpler than those produced by electrospray, it is the ionization method that we employ in our bacteria fingerprinting studies.

The combination of MALDI ionization with time-of-flight mass analysis (MALDI-TOF) yielded disappointingly poor mass resolution for several years. The source of the mediocre instrument performance was a combination of the broad velocity distribution with which MALDI ions are desorbed, the presence of unresolved adducts and the occurrence of metastable fragmentation processes.[24–28] In 1994 three groups independently demonstrated significant resolution improvements by employing pulsed ion extraction fields in linear TOF instruments.[29–32] The method was immediately commercialized and applied to reflectron-TOF analyzers as well.[33] As a result of the instrumentation advances that were achieved, it became possible to obtain MALDI-TOF mass spectral peaks that were 10 to 15 Da wide or narrower for essentially all proteins with masses below about 30 kDa. More important, protein masses could typically be assigned with an accuracy of ±1 Da. This kind of mass accuracy facilitated the production of highly unique fingerprints of bacteria and the capability of accurately comparing protein masses with predictions derived from proteomics databases.[22,34–37]

9.3 APPLICATION OF MALDI-TOF TO BACTERIA IDENTIFICATION

The application of MALDI-TOF mass spectrometry to the problem of bacteria identification has led to numerous encouraging results. Different species of bacteria yield mass spectra that display considerable variations.[22,38–46] Bacteria of the same genus but different species exhibit mass spectra having both similarities and differences, just as one might anticipate in a fingerprinting method. A few methods of sample handling and preparation, all of which yield ions, have thus far been described. Intact cells have been deposited directly onto the MALDI matrix and intense signals were observed.[40] Cells have also been sonicated and enzymatically digested and their protein contents extracted before mixing with the MALDI matrix.[38,42,47] This yielded somewhat different but nevertheless good results. In the latter case both fractions from the protein extraction step were found to yield strong mass spectral signals. The conclusion, which is corroborated by our own experience, is that simply recording mass spectra of bacteria is straightforward. However, the optimal strategies for cell handling in order to generate reproducibly strong signals and extract the maximum information from organisms have not been established. The sensitivity and selectivity of the mass spectral fingerprinting method depend on our ability to optimize various aspects of the experimental technique.

When our group began working with bacteria samples in 1996, the first question we were interested in was whether high-resolution MALDI-TOF mass spectrometry would enable us to resolve far more peaks in bacteria mass spectra than had appeared in previous low resolution data. Instead, we found that the limited number of peaks simply sharpened. We learned very quickly that the method of preparing MALDI sample spots affected the reproducibility and information content of the spectra. Three examples of whole cell *E. coli* mass spectra that demonstrate this reproducibility are displayed in Figure 9.1.

Encouraged by this spectral reproducibility, we focused our efforts on the particularly challenging problem of distinguishing bacterial strains by MALDI MS. We developed a modified correlation approach[22] that relies on two fundamental qualities of bacterial mass spectra. First, because different bacterial strains of the same species have substantial, if not complete, genetic overlap, most of the protein masses observed with two different strains will be identical. This feature limits the value of the "biomarker" approach that is commonly used to differentiate bacteria species. Second, as just noted, closely controlled sample preparation and mass analysis procedures can result in highly reproducible results.[22] The modified correlation approach takes advantage of subtle, yet reproducible, differences in mass spectra obtained from dif-

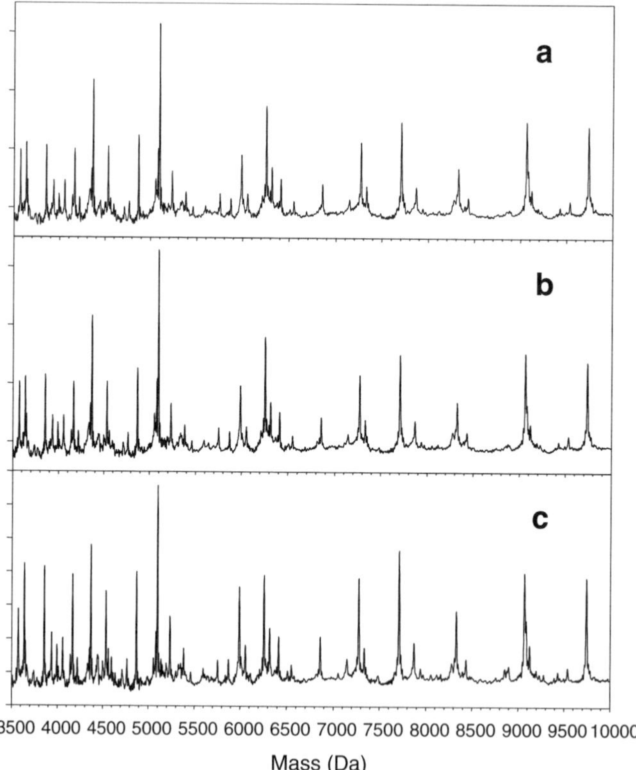

Figure 9.1 Comparison of mass spectra of three separate cultures of *E. coli* grown under essentially identical conditions.

ferent strains to produce an objective mathematical value that represents the similarity between two spectra. This approach removes the subjectivity inherent in biomarker identification and visual spectral matching that have been commonly employed for distinguishing bacterial species.

A thorough description of this approach has been previously presented.[22] Briefly, 25 common laboratory strains of *E. coli* bacteria, originally derived from the K-12 strain, were grown in Luria broth medium and harvested by pelleting after 18 to 22 hours. Pellets were washed, re-pelleted, and stored frozen prior to MALDI sample preparation. After thawing, the pellets were diluted with a volume of water proportional to the optical density of the original culture. The diluted sample was mixed with matrix, deposited on a stainless steel sample probe, and air dried. Positive ion MALDI mass spectra were acquired on a custom-built linear time-of-flight mass spectrometer using pulsed-ion extraction and 20 kV of total acceleration. A fast Fourier transform filter removed low and high frequencies and achieved baseline correction and smoothing. Spectra were externally calibrated with standard proteins.

The smoothed and calibrated spectra serve as fingerprints for the bacteria from which they were obtained. A computer program was written to compare pairs of these fingerprint files, each of which contains both mass and peak intensity information. The program's user selects the mass range over which spectra are to be compared and the number of intervals to divide this range into. In our *E. coli* studies the mass range from 3.5 to 10 kDa was chosen because of the richness of the data found in this region. This 6500 Da wide window was divided into 13 intervals of 500 Da each. The program calculates cross-correlation and auto-correlation values for each interval in the two spectra. These values correspond to the maxima that occur in the cross-correlation and auto-correlation functions.[48] To compensate for the fact that some mass regions contain intense features while others exhibit weaker peaks, each interval cross-correlation value is normalized by dividing it by the average of that interval's auto-correlation values from the two spectra. The composite correlation index for the two spectra is then defined as the product of all the normalized interval correlation values. These composite correlation indices can range from 0 (no match) to 1 (exact match).

As noted previously, identifying bacteria by their mass spectral fingerprints is possible only if the spectra are reproducible. Figure 9.2 shows the effect of one experimental parameter, laser light intensity, on spectral reproducibility. High laser intensity tends to broaden mass spectral peaks, causing subtle differences in the two spectra that result in a composite correlation index value of 0.573. When the light intensity is kept constant and duplicate spectra are acquired from the same spot, higher composite correlation indexes of 0.948 and 0.758 are obtained for high- and low-light conditions, respectively. As alluded to above, we demonstrated that reproducible MALDI spectra can be obtained from the same sample spot as well as from three different cultures of the same bacteria strain when sufficient care is taken in growing bacteria and recording mass spectra.[49]

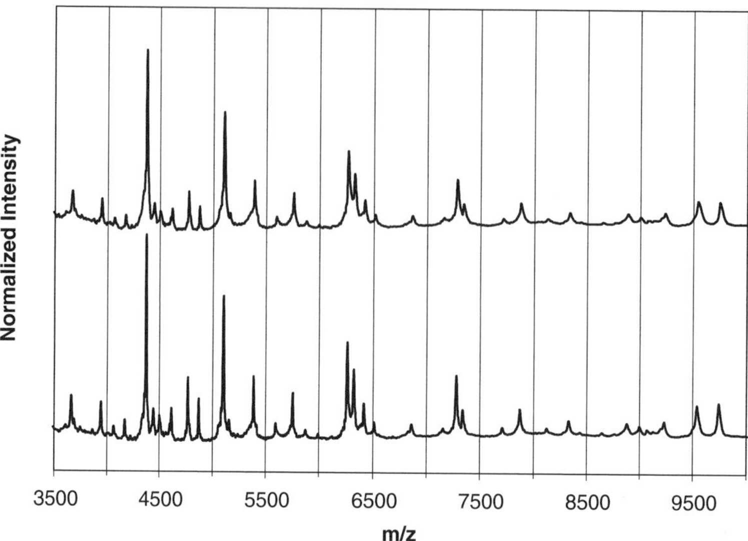

Figure 9.2 Mass spectra of the same E. coli strain, JM109, acquired under conditions of (*a*) high- (~12.5 μJ) and (*b*) low- (~7 μJ) light intensities. Interval correlation values are shown for the comparison of the two spectra. The composite correlation index value, the product of the 13 interval values, is 0.573.

Figure 9.3 displays what might be best described as typical variations among strains of *E. coli*. Peaks whose intensities vary significantly occur at masses of 3856, 4166, 4339, 4535, 5385, 5992, 7277, 7710, 8330, 9068, and 9743 Da. Table 9.1 presents a matrix of the composite correlation indexes obtained by comparing these four spectra. Numbers along the diagonal, shown in bold, are derived from a single comparison of two spectra of the same bacteria strain. Off-diagonal elements represent the average of four values obtained by comparing two spectra of one strain with two spectra of the other strain. Strains such as LE3 and DH5 appear to be quite different from strain AD4, as evidenced by composite correlation indices of only 2.27×10^{-9} and 4.76×10^{-8}. Strains whose spectra appear visually more similar, such as DH5 and Y10, yield a composite correlation index that is orders of magnitude larger (0.016). In our general experience, and certainly for all cases shown in Table 9.1, composite correlation indexes obtained from comparing different strains are substantially smaller than those derived from comparing two spectra of the same strain.

Figure 9.2 displays a division of the mass range into 13 intervals. After multiplication of all the individual normalized correlation values, the composite correlation index is smaller and the method achieves more selectivity than it would if a single correlation function were calculated over the full mass range. A lower limit of 0.0001 was placed on the value for each normalized interval comparison to prevent the method from generating absurdly low values. These

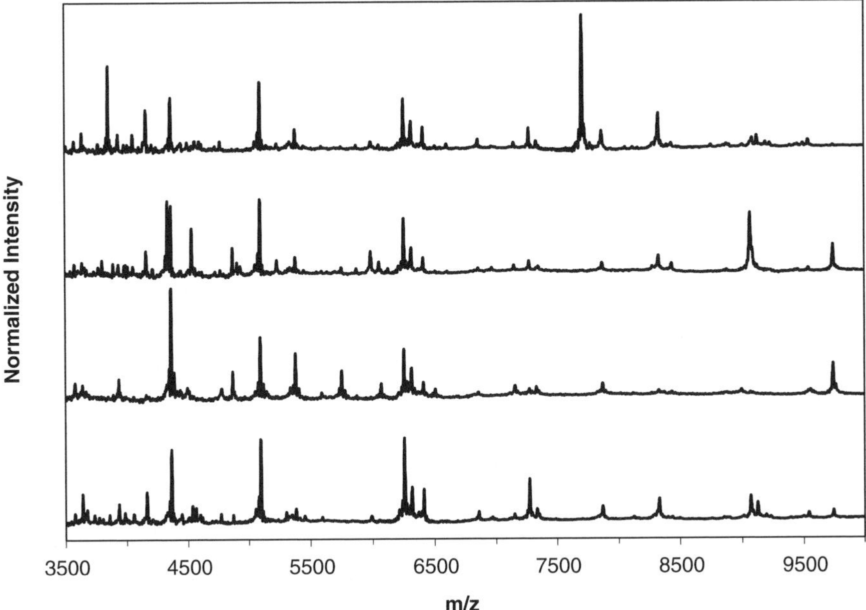

Figure 9.3 Mass spectra of four different *E. coli* strains: (*a*) AD4, (*b*) DH5, (*c*) LE3, and (*d*) Y10. Composite correlation indexes derived by comparing these spectra to one another are listed in Table 9.1.

TABLE 9.1 Composite Correlation Indexes for Modified Correlation of Spectra in Figure 9.3

	Y10	LE3	DH5	AD4
Y10	**0.933**	1.53×10^{-4}	0.016	4.58×10^{-4}
LE3	1.53×10^{-4}	**0.658**	2.76×10^{-4}	2.27×10^{-9}
DH5	0.016	2.76×10^{-4}	**0.828**	4.76×10^{-8}
AD4	4.58×10^{-4}	2.27×10^{-9}	4.76×10^{-8}	**0.749**

low-correlation intervals have a dominating effect on the overall composite correlation index. For this reason it is important to choose the mass range and number of intervals such that each interval has an appropriate number of significant peaks.

Application of the modified cross-correlation method was successfully demonstrated by comparing four independently recorded spectra with a library of spectra from 25 different strains of *E. coli*.[22] These four strains were chosen arbitrarily without prior knowledge of the results. Three of the spectra of these strains used to compare to the library were obtained from cultures other than those used to generate their corresponding library spectra. The fourth spectrum used to compare to the library was obtained from the same culture as the one used to generate the library spectrum. For each of the four strains the highest composite correlation index was obtained when comparing the spectrum of a particular strain with its library counterpart.

When applying our modified cross-correlation method, it is critical that multiple analyses of the same bacteria yield reproducible spectra. Indeed, large relative peak height variations for spectra derived from the same strain will cause composite correlation indexes to diminish, and could cause bacteria to be incorrectly identified. Spectra shown in Figure 9.2 suggest that even when MALDI mass spectra of whole bacteria cells are acquired using different light intensities, relative peak heights are relatively reproducible. The main difference between these two spectra is the diminished resolution under high-light intensities, a familiar experience with our instrumentation. A wide peak in one spectrum will not completely overlap a narrow peak in another even if they are both centered at the same mass, reducing the composite correlation index despite the otherwise indistinguishable spectra. When a constant light intensity is used, we obtain high composite correlation indexes for either low- or high-light conditions, indicating an acceptable degree of spectral reproducibility. This is consistent with previous results where all bacteria growth and mass spectral acquisition parameters were closely monitored.[22] Because two different instruments will generally not detect corresponding peaks with the exact same resolution, spectra recorded with different instruments may not compare as favorably as those recorded on the same one. Successful application of this modified correlation technique may require that both reference and unknown sample data be recorded with one type of mass spectrometer.

For the comparisons presented here, we have chosen the mass range from 3.5 to 10 kDa. With our sample preparation and instrument, we find the most information in this mass range, but there is certainly valuable information in other regions of the mass spectrum. For example, many peaks at smaller masses have been observed, some of which are very intense. Obviously many proteins with masses larger than 10 kDa are present in each *E. coli* cell.[50] We suspect that both reduced abundance and lower mass spectral sensitivity (relative to smaller masses) limit our ability to efficiently detect these large-mass signals. With modified sample preparation and experimental procedures, it is possible to compare bacteria using other mass ranges.[51,52] Nevertheless, this modified cross-correlation method is applicable over any mass range.

Different strains of *E. coli* have unique fingerprints when analyzed by MALDI mass spectrometry of whole cells, as illustrated by the spectra of Figure 9.3. Many peaks appear at the same masses in all of the spectra, with only modest intensity variations. A few peaks also vary dramatically in intensity for these four strains. We expect the spectra to be similar, since all the strains are laboratory derivatives of the single *E. coli* strain, K-12. Genetic modifications can, and most likely do, affect biological functions such as transcription of mRNA and subsequent translation of proteins. Some of the strains, for example, those with a single gene addition or deletion, are fairly similar, at least genetically. Others may have more extensive genetic modifications, but in all cases we should expect that protein masses will be rather similar for strains of the same species. The spectra in Figure 9.3 effectively demonstrate this. The similarity of protein masses means that the biomarker identification approach should have less utility in distinguishing strains than for identifying different species.

The real power of our approach is derived from performing comparisons of spectra over a number of intervals. Comparisons similar to those described here could have been done using the same mass range but only a single interval. If the calculations are performed with only one interval, the correlation index for a good match is only about twice as large as for a poor match. Dividing the spectrum into intervals and correlating these intervals yields correlation indices that are more sensitive to spectral differences. This greater sensitivity facilitates the process of distinguishing similar spectra, such as different strains of *E. coli* shown in Table 9.1.

To further demonstrate the capability of mass spectral fingerprinting to distinguish different strains of the same bacterial species, we investigated clinical isolates of methicillin-resistant *Staphylococcus aureus* (MRSA) bacteria. Eight samples were received from the laboratory of Dr. Alan Hartstein of the Indiana University School of Medicine, and a 25 ml starter culture of each was grown overnight in Luria broth. Aliquots of this culture were mixed 1:1 with glycerol prior to freezing at –80°C; these frozen cultures were used to produce all subsequent MRSA cultures. The bacteria analyzed by MALDI-TOF were grown for two different lengths of time. In one case a 125 ml flask containing 25 ml of LB media was inoculated with a small amount of the frozen glycerol stock. The flasks were shaken at 200 rpm for 24 hours at 37°C. Alternatively, 16 mm diameter test tubes containing 5 ml of Luria broth media were treated with an inoculum equivalent to 500 µl of a culture with an optical density at 600 nm (OD_{600}) of 1.0. These were then incubated for 5 hours at 37°C while being shaken at 200 rpm. All cells were harvested by centrifuging 1 ml of culture for 5 minutes at 7000 rpm in a bench top microcentrifuge. The OD_{600} of the cultures were measured with a Perkin Elmer 300 ultraviolet-visible spectrophotometer. Cultures with $OD_{600} > 1.4$ were diluted by a factor of 10 and their OD_{600} values re-measured.

The method that epidemiologists generally accept as the gold standard for distinguishing subtle variations in bacteria involves DNA analysis by pulsed field gel electrophoresis (PFGE).[53] In this approach bacterial DNA is digested with restriction endonucleases, and the restriction fragments are separated based on their gel migration propensities. Although it is very slow and has all of the reproducibility disadvantages that gel electrophoresis usually exhibits, this method is exquisitely sensitive to subtle variations in the genome of an organism and is therefore quite useful for distinguishing bacterial strains.

Application of this approach to the eight MRSA isolates of the present study led to the PFGE image displayed in Figure 9.4. It is quite apparent that the PFGE patterns of isolates 1 and 2 are nearly identical, as are those of isolates 3, 5, and 7. Likewise isolate 6 and the control are very similar. The PFGE pattern from isolate 4 identifies it as unique, though probably related to isolate 6 and the control.

Prior to MALDI-TOF analysis, cell pellets were suspended in 20 µl of distilled water per unit of OD_{600}. One µl of this suspension was added to 9 µl of 10 g/l α-cyano-4-hydroxycinnamic acid (Aldrich, Milwaukee, WI) in 50% v/v aqueous acetonitrile (EMD Chemicals, Gibbstown, NJ) with 0.1% v/v trifluoroacetic acid (Aldrich). One µl of the matrix-analyte mixture was deposited onto the target and allowed to air dry. Mass spectra were recorded with our standard home-built linear MALDI-TOF mass spectrometer.[22,34–37] employing pulsed ion extraction. Ions were generated by irradiating sample spots with a frequency tripled Nd:YAG laser (355 nm). The ions were accelerated to a kinetic energy of 20 keV and detected with a pulsed microchannel plate detector. A LeCroy 9300 digital oscilloscope digitized the ion current signal at a 1 GHz rate. Mass spectra were externally calibrated by fitting the masses and times of flight of singly, doubly, and triply charged ions of equine cytochrome C (MW = 12,360, Sigma, St. Louis, MO) and human ubiquitin (MW = 8565, Sigma) to a five term polynomial. The data were normalized such that the total integrated areas of all spectra were equal.

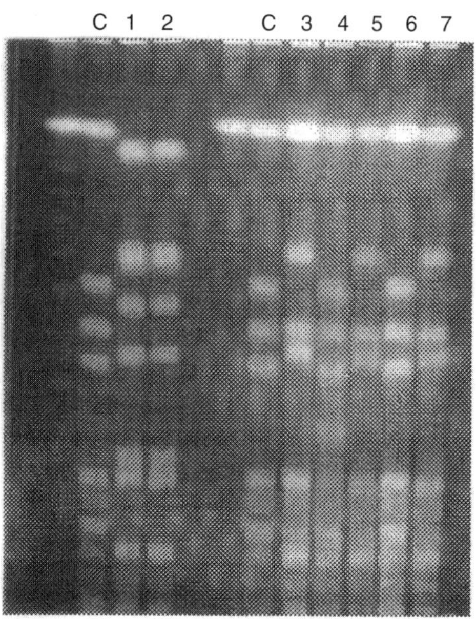

Figure 9.4 PFGE data for the eight MRSA isolates studied. C stands for control.

Mass spectra were compared using the modified cross-correlation algorithm discussed above.[22] Mass spectra from two of the 24-hour cultures were recorded from *m/z* 2600 to 7600 Da and divided into ten 500 Da wide sections. Corresponding mass intervals in the two mass spectra were compared. As in the earlier discussion, the cross-correlation values from all intervals were then multiplied together to generate final composite correlation indices. Data from the five-hour cultures were treated similarly except that mass spectra were recorded from 3000 to 8000 Da and divided into ten 500 Da wide regions.

Figure 9.5 contains representative MALDI-TOF mass spectra for 24-hour, flask-grown cultures of the eight strains of MRSA studied. Mass spectra from isolates 1 and 2 and from isolates 5 and 7 are visually very similar, consistent with the DNA analysis by PFGE. On the other hand, it would be difficult to argue that the spectra from isolates 4 and control are significantly different. The cross-correlation data shown in Table 9.2 demonstrate similar trends. The composite correlation indexes also suggest that isolates 1 and 2 are more similar to each other than to any other. However, considering the similarity of their spectra, the indices are rather small. Apparently the limited number of intense features above 3500 Da in the mass spectra of isolates 1 and 2 has an adverse impact on the correlation process. Cross-correlation of correspon-ding sections of spectra that contain insignificant information is basically a process of correlating random features. As discussed below, correlation indexes improved when more peaks were observed in mass spectra of isolates 1 and 2. When comparing the mass spectra from the other isolates, isolates 3 and 6 differ from all others, while isolates 4, 5, 7, and control all give correla-

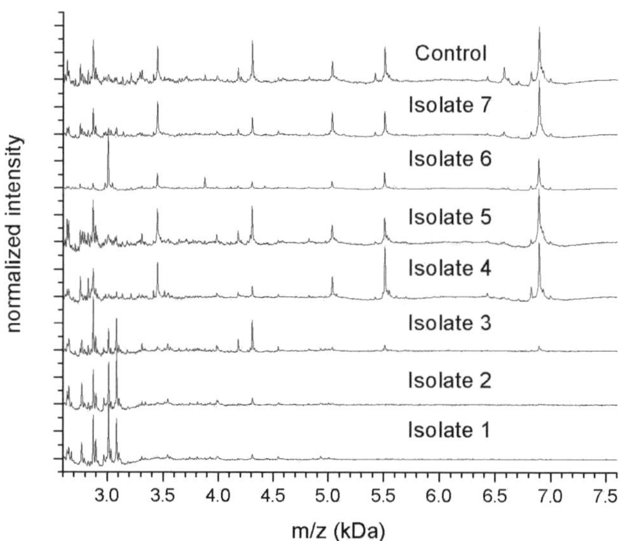

Figure 9.5 Representative MALDI-TOF mass spectra from each of the eight MRSA isolates.

TABLE 9.2 Composite Correlation Data for the Eight MRSA Isolates of Mass Spectra in Figure 9.5

Score	Isolate 1	Isolate 2	Isolate 3	Isolate 4	Isolate 5	Isolate 6	Isolate 7	Control
Isolate 1	**1.00**	3.39×10^{-3}	5.01×10^{-5}	3.47×10^{-9}	1.23×10^{-7}	2.24×10^{-8}	7.76×10^{-9}	9.77×10^{-12}
Isolate 2	3.39×10^{-3}	**1.00**	2.34×10^{-4}	5.75×10^{-8}	3.09×10^{-6}	9.77×10^{-8}	2.14×10^{-7}	1.05×10^{-7}
Isolate 3	5.01×10^{-5}	2.34×10^{-4}	**1.00**	1.86×10^{-4}	5.89×10^{-3}	1.07×10^{-4}	1.38×10^{-4}	2.45×10^{-4}
Isolate 4	3.47×10^{-9}	5.75×10^{-8}	1.86×10^{-4}	**1.00**	0.041	1.38×10^{-3}	0.158	0.026
Isolate 5	1.23×10^{-7}	3.09×10^{-6}	5.89×10^{-3}	0.041	**1.00**	5.37×10^{-3}	0.182	0.120
Isolate 6	2.24×10^{-8}	9.77×10^{-8}	1.07×10^{-4}	1.38×10^{-3}	5.37×10^{-3}	**1.00**	3.72×10^{-3}	3.72×10^{-3}
Isolate 7	7.76×10^{-9}	2.14×10^{-7}	1.38×10^{-4}	0.158	0.182	3.72×10^{-3}	**1.00**	0.209
Control	9.77×10^{-12}	1.05×10^{-7}	2.45×10^{-4}	0.026	0.120	3.72×10^{-3}	0.209	**1.00**

Note: Numbers in bold correspond to cross-correlation values comparing mass spectra of the same isolate.

tion scores > 0.01 when compared to each other. This indicates that the mass spectra from isolates 4, 5, 7, and the control are relatively similar. These observations differ slightly from results obtained by analyzing the DNA from each of these isolates by pulsed-field gel electrophoresis (PFGE).

Arnold and coworkers demonstrated that the mass spectra obtained from intact *E. coli* change considerably as a culture depletes its growth medium.[36] The OD_{600} versus time plot they reported demonstrated that an *E. coli* culture with an $OD_{600} > 4$ had passed from exponential growth (also known as log phase) into period of slow growth that gave rise to a nearly linear OD_{600} versus time plot (stationary phase). They also demonstrated that *E. coli* cultures yielded poorly reproducible mass spectra during the transition from log to stationary phase ($4 < OD_{600} < 6.5$). Six of the eight MRSA cultures analyzed above had OD_{600} values between 5.51 and 6.38, and this may have had an undesiredable effect on spectral reproducibility. With the goal of improving reproducibility and reducing analysis time, experiments on bacteria grown under different conditions were undertaken.

The second set of experiments featured very controlled growth conditions and relatively small (5 ml) cultures. All tubes were inoculated with a similar number of cells (based on OD_{600} measurements) and incubated for only 5 hours instead of 24. Their optical densities ranged from 1.33 (for the control) to 3.57 (for isolate 1). If MRSA is assumed to behave similar to *E. coli*, all eight MRSA cultures should have been in log phase when analyzed. Representative spectra from these eight cultures appear in Figure 9.6 and cross-correlation results appear in Table 9.3.

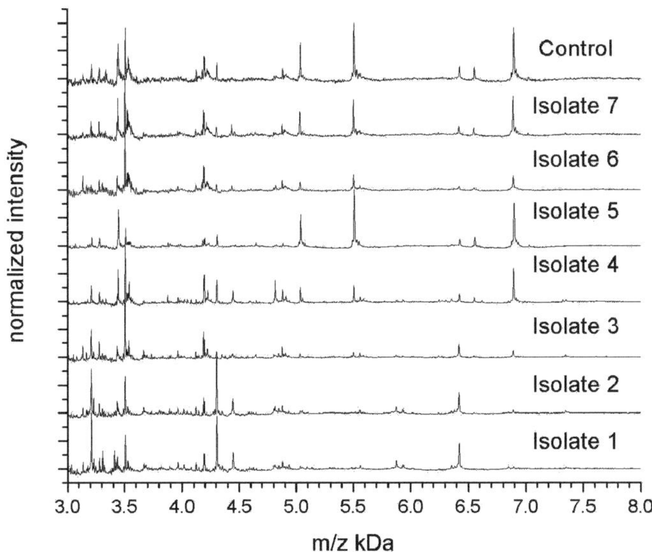

Figure 9.6 Representative spectra from the second set of MRSA cultures that were incubated for 5 hours.

TABLE 9.3 Composite Correlation Data for the Eight MRSA Isolates of Mass Spectra in Figure 9.3

Score	1	2	3	4	5	6	7	Control
1	**1.00**	0.182	8.71×10^{-4}	5.50×10^{-5}	2.69×10^{-10}	2.40×10^{-12}	8.91×10^{-11}	3.16×10^{-10}
2	0.182	**1.00**	5.01×10^{-3}	2.57×10^{-4}	6.46×10^{-10}	1.62×10^{-10}	5.25×10^{-10}	2.00×10^{-9}
3	8.71×10^{-4}	5.01×10^{-3}	**1.00**	0.016	2.82×10^{-6}	5.01×10^{-4}	1.17×10^{-4}	3.31×10^{-4}
4	5.50×10^{-5}	2.57×10^{-4}	0.016	**1.00**	4.57×10^{-3}	9.77×10^{-3}	0.030	0.017
5	2.69×10^{-10}	6.46×10^{-10}	2.82×10^{-6}	4.57×10^{-3}	**1.00**	1.32×10^{-3}	0.158	0.059
6	2.40×10^{-12}	1.62×10^{-10}	5.01×10^{-4}	9.77×10^{-3}	1.32×10^{-3}	**1.00**	0.018	0.028
7	8.91×10^{-11}	5.25×10^{-10}	1.17×10^{-4}	0.030	0.158	0.018	**1.00**	0.182
Control	3.16×10^{-10}	2.00×10^{-9}	3.31×10^{-4}	0.017	0.059	0.028	0.182	**1.00**

Note: Numbers in bold correspond to cross-correlation values comparing mass spectra of the same isolate.

Clearly, the new growth conditions improved the quality of mass spectra from isolates 1, 2, and 6 as all three now have several features above *m/z* 3.5 kDa. The new growth conditions did not radically alter the quality of the mass spectra from the other five isolates. The spectra can be differentiated by visual inspection. Mass spectra from isolates 1 and 2 are nearly identical, as are those from isolate 7 and the control. The mass spectrum from isolate 6 is similar to those from isolate 7 and the control at masses greater than 3.6 kDa, but differs somewhat between *m/z* 3 and 3.6 kDa. The spectra from isolates 3 and 4 are quite similar below 4.2 kDa. The mass spectra from isolates 4, 5, 6, 7, and the control are quite similar at $m/z > 6.4$ kDa. The isolate 5 mass spectrum is unique below *m/z* 5.5 kDa. The cross-correlation results did not completely agree with visual impressions.

Isolates 1 and 2 gave a very high composite index (0.182) when compared to each other, as did isolate 7 and the control. Interestingly isolate 3 appears to be somewhat similar (composite index 0.016) to isolate 4. This result is unexpected because the two spectra differ visually. Isolate 4 is also similar to isolates 6, 7, and control. Interestingly enough, isolate 5 is more similar to isolate 7 and the control than to isolates 4 or 6. Clearly, more of the cross-correlation parameters could be adjusted to enhance the algorithm's sensitivity to subtle differences in the mass spectra if it is to be employed as a routine screening technique. Alternatively, an algorithm that analyzes the spectra visually (similar to those used to interpret PFGE patterns[53] could be devised. Finally cultures could be grown to a predetermined OD_{600} to maximize reproducibility, which may increase the selectivity of any comparison scheme.

These results are in agreement with other published studies using MALDI-TOF analyses of whole cells to distinguish clinical isolates of *Staphylococcus aureus*.[54–57] Gordon and coworkers found that the vast majority of the mass spectral peaks derived from their intact *Staphylcoccus aureus* cells appeared below 4 kDa. Our results show a similar bias against features with high *m/z* ratios. It has generally been observed that strains with different PFGE patterns give rise to slightly different mass spectra and that those with similar PFGE patterns yield nearly identical mass spectra.[55] In our work, mass spectra from isolate 7 and the control yielded relatively high composite correlation indexes despite their having different PFGE patterns. Walker et al. demonstrated that careful control of MRSA growth conditions is required to ensure reproducible mass spectra.[57] In the study presented here, the mass spectra of isolates 1 and 2 changed dramatically when growth conditions were altered.

It is clear that both intact cell MALDI-TOF and PFGE have their limitations. PFGE analyses probes the chromosomal DNA of microorganisms for variations in the locations of specific restriction enzyme cleavage sites, while MALDI-TOF mass spectrometry of intact cells primarily examines *abundant* proteins such as ribosomal proteins[35] and those associated with or near bacterial cell walls.[58] In order for MALDI-TOF to detect a variation, a mutation must lead to noticeable changes in the expression of cell wall—associated and/or highly expressed soluble proteins whose molecular weights happen to fall in the mass range studied. Similarly PFGE differentiation requires that a mutation alter the location of a restriction enzyme cleavage site in a bacterial chromosome. PFGE offers more selectivity because it probes the entire genome of an organism as opposed to just those portions that encode abundant proteins with molecular weights that occur in the mass range examined by MALDI-TOF. On the other hand, MALDI-TOF analysis of whole cells is much more rapid than PFGE. Both methods require the initial culture of a clinical isolate prior to analysis, but radically different procedures follow this step, however. As demonstrated in this work, interpretable MALDI-TOF data can be generated just 5 hours after the initial culture is grown; PFGE analyses often take several days to complete.[53] Thus MALDI-TOF offers a rapid screening technique that could aid in the tracking of fast spreading outbreaks of MRSA. These results could then be confirmed by slower, but more selective PFGE.

In conclusion, MALDI-TOF mass spectrometry can discriminate between clinical isolates of methicillin-resistant *Staphylococcus aureus*. These results can be generated within 5 hours of the generation of the initial culture. Attention must be paid to growth conditions to ensure reproducible spectra. The mass spectra can be distinguished either by visual inspection or an automated cross-correlation algorithm. However, when employing an automated approach, care must be taken in selecting mass intervals over which spectra are compared since the cross-correlation algorithm can be frustrated by large featureless regions in the mass spectra.

While this chapter has described two examples of our efforts in bacterial strain identification, a number of other groups have contributed to this research area. It remains to be seen whether mathematical algorithms such as the modified correlation approach described herein are always more effective for strain assignments than the simple use of distinctive biomarker peaks. Nilsson reported MALDI mass spectra of *Helicobacter pylori* with three different matrices and solvent conditions.[59] She showed that some strains of this bacteria yield rather similar mass spectra while others are quite different. Nevertheless, each strain appears to exhibit some unique peaks that might be used for distinguishing strains.

Haag et al. found that several species of *Haemophilus* bacteria can be distinguished by MALDI-TOF mass spectrometry and further showed that different strains of *Haemophilus ducreyi* tend to exhibit specific unique biomarker peaks between 5000 and 15000 Da.[60] Lynn et al. observed a few peaks that could distinguish *E. coli* strains 0:157 and 7G1, though they specifically note that attempts to identify bacteria based on a limited number of biomarkers can lead to errors.[61]

Pearson and coworkers investigated several antibiotic-resistant strains of *E. coli* in an elegant and thorough study. Using a combination of MALDI-TOF and electrospray LC-MS they found that varying subtle mutations in ribosomal proteins occur in different strains that are resistant to various antibiotics.[62] Since ribosomal proteins are particularly easy to detect by mass spectrometry,[36] these observations may be quite useful in future strain identification efforts.

Along similar lines Li, Liu, and Chen[63] demonstrated that fungal spores from several strains of *Asperigillus* species yielded MALDI mass spectral features that were unique and characteristic. Likewise Fenselau and coworkers observed unique mass spectra from spores associated with *Bacillus* bacteria, indicating that different strains could be differentiated.[64]

9.4 CONCLUSIONS

An objective, mathematical approach for comparing mass spectra of bacteria has been developed and demonstrated. The use of this modified cross-correlation method eliminates most of the subjectivity in comparing two spectra and provides a numerical value for how well two spectra match. Our work confirms that MALDI mass spectrometry of whole cells is a powerful method for identifying bacteria. The modified cross-correlation method has impressive potential to provide accurate, strain-specific identification of bacteria. This approach will be most useful when analyzing clinical samples that can be reproducibly cultured and handled. The recognition of protein biomarkers may be helpful for identifying environmentally collected samples. The ability to distinguish and identify bacterial strains has many potential applications, including identifying biological warfare agents, food-borne pathogens, and in clinical analysis of infectious bacteria.

ACKNOWLEDGMENT

Funding in support of our research has been provided by the National Science Foundation and the National Institutes of Health.

REFERENCES

1. Anhalt, J. P.; Fenselau, C. Identification of bacteria using mass spectrometry. *Anal. Chem.* 1975, **47**, 219–225.
2. DeLuca, S.; Sarver, E. W.; Voorhees, K. J. Direct analysis of bacterial fatty acids by Curie-point pyrolysis tandem mass spectrometry. *Anal. Chem.* 1990, **62**, 1465–1472.
3. Snyder, A. P.; Smith, P. B. W.; Dworzanski, J. P.; Meuzelaar, H. L. C. In *ACS Symposium Series 541*. Fenselau, C. (Ed). Washington, DC,: American Chemical Society, 1994, p. 62.
4. Basile, F.; Voorhees, K. J.; Hadfield, T. L. Microorganism Gram-type differentiation based on pyrolysis–mass spectrometry of bacterial fatty acid methyl ester extracts. *Appl. Environ. Microbiol.* 1995, **61**, 1534–1539.

5. Fox, A.; Rosario, R. M. T.; Larsson, L. Monitoring of bacterial sugars and hydroxy fatty acids in dust from air conditioners by gas chromatography–mass spectrometry. *Appl. Environ. Microbiol.* 1993, **59**, 4354–4360.
6. Heller, D. N.; Cotter, R. J.; Fenselau, C.; Uy, O. M. Profiling of bacteria by fast atom bombardment mass spectrometry. *Anal. Chem.* 1987, **59**, 2806–2809.
7. Heller, D. N.; Murphy, C. M.; Cotter, R. J.; Fenselau, C.; Uy, O. M. Constant neutral loss scanning for the characterization of bacterial phospholipids desorbed by fast atom bombardment. *Anal. Chem.* 1998, **60**, 2787–2791.
8. Cole, M. J.; Enke, C. G. Direct determination of phospholipid structures in microorganisms by fast atom bombardment triple quadrupole mass spectrometry. *Anal. Chem.* 1991, **63**, 1032–1038.
9. Stead, D. E.; Sellwood, J. E.; Wilson, J.; Viney, I. Evaluation of a commercial microbial identification system based on fatty acid profiles for rapid, accurate identification of plant pathogenic bacteria. *J. Appl. Bacteriol.* 1992, **72**, 315–321.
10. Livesley, M. A.; Thompson, I. P.; Gern, L.; Nuttall, P. A. Analysis of intra-specific variation in the fatty acid profiles of *Borrelia burgdorferi. J. Gen. Microbiol.* 1993, **139**, 2197–2201.
11. Amadi, E. N.; Alderson, G. Menaquinone composition of some micrococci determined by high performance liquid chromatography. *J. Appl. Bacteriol.* 1991, **70**, 517–521.
12. Rönkkö, R.; Pennanen, T.; Smolander, A.; Kitunen, V.; Kortemaa, H.; Haahtela, K. Quantification of Frankia strains and other root-associated bacteria in pure cultures and in the rhizosphere of axenic seedlings by high-performance liquid chromatography-based muramic acid assay. *Appl. Environ. Microbiol.* 1994, **60**, 3672–3678.
13. de Ward, P.; van der Wal, G. N. M.; Huijberts, G. N.; Eggink, G. Heteronuclear NMR analysis of unsaturated fatty acids in oly(3-hydroxyalkanoates). *J. Biol. Chem.* 1993, **268**, 315–319.
14. Helm, D.; Labischinski, H.; Schallehn, G.; Naumann, D. Classification and identification of bacteria by Fourier-transform infrared spectroscopy. *J. Gen. Microbiol.* 1991, **137**, 69–79.
15. Cox, N.; Johnston, J.; Szarka, Z.; Wright, D. J. M.; Archard, L. C. Characterization of an rRNA gene-specific cDNA probe: Application in bacterial identification. *J. Gen. Microbiol.* 1990, **136**, 1639–1643.
16. Manz, W.; Amann, R.; Szewzyk, R.; Szewzyk, U.; Stenstrom, T.-A.; Hutzler, P.; Schleifer, K.-H. In situ identification of *Legionellaceae* using 16S rRNA-targeted oligonucleotide probes and confocal laser scanning microscopy. *Microbiology* 1995, **141**, 29–39.
17. Hurst, G. B.; Doktycz, M. J.; Vass, A. A.; Buchanan, M. V. Detection of bacterial DNA polymerase chain reaction products by matrix-assisted laser desorption/ionization mass spectrometry. *Rapid Comm. Mass Spectrom.* 1996, **10**, 377–382.
18. Podzorski, R. P.; Persing, D. M. In *Manual of Chemical Microbiology*. Murray, P. R.; et al. (Eds.). Washington, DC: ASM Press, 1995, pp. 130–150.
19. Persing, D. H.; Smith, T. F.; Tenover, F. C.; White, T. J. (Eds.). *Diagnostic Molecular Microbiology*. Washington, DC: American Society for Microbiology, 1993.

20. Arbeit, R. D. In *Manual of Chemical Microbiology*. Murray, P. R.; et al. (Eds.). Washington, DC: ASM Press, 1995, pp. 190–208.

21. Tenover, F. C. *ASM Conference on Molecular Diagnostics and Therapeutics*, August 1977, Kananaskis, Canada.

22. Arnold, R. J.; Reilly, J. P. Fingerprint matching of *E. coli* strains with MALDI-TOF mass spectrometry of whole cells using a modified correlation approach. *Rapid Comm. Mass Spectrom.* 1998, **12**, 630–636.

23. Fenn, J. B.; Mann, M.; Meng, C. K.; Wong, S. F.; Whitehouse, C. M. Electrospray ionization for mass spectrometry of large molecules. *Science* 1989, **246**.

24. Beavis, R. C.; Chait, B. T. Velocity distributions of intact high mass polypeptide molecule ions produced by matrix-assisted laser-desorption. *Chem. Phys. Lett.* 1991, **181**, 479–484.

25. Hillenkamp, F. In *Biological Mass Spectrometry: Present and Future*. Matsuo, T.; Capriolo, R. M.; Gross, M. L.; Seyama, Y. (Eds.). New York: Wiley, 1994, pp. 101–118.

26. Chien, B. M.; Michael, S. M.; Lubman, D. M. Enhancement of resolution in matrix-assisted laser desorption using an ion-trap storage/reflectron time-of-flight mass spectrometer. *Rapid Comm. Mass Spectrom.* 1993, **7**, 837–843.

27. Spengler, B.; Kirsch, D.; Kaufmann, R. Metastable decay of peptides and proteins in matrix-assisted laser-desorption mass spectrometry. *Rapid Comm. Mass Spectrom.* 1991, **5**, 198–202.

28. Chan, T.-W. D.; Thomas, I.; Colburn, A. W.; Derrick, P. J. Initial velocities of positive and negative protein molecule-ions produced in matrix-assisted ultraviolet laser desorption using a liquid matrix. *Chem. Phys. Lett.* 1994, **222**, 579–585.

29. Colby, S. M.; King, T. B.; Reilly, J. P. Improving the resolution of matrix-assisted laser desorption/ionization time-of-flight mass spectrometry by exploiting the correlation between ion position and velocity. *Rapid Comm. Mass Spectrom.* 1994, **8**, 865–868.

30. Brown, R. S.; Lennon, J. J. Mass resolution improvement by incorporation of pulsed ion extraction in a matrix-assisted laser desorption/ionization linear time-of-flight mass spectrometer. *Anal. Chem.* 1995, **67**, 1998–2003.

31. Brown, R. S.; Lennon, J. J. Sequence-specific fragmentation of matrix-assisted laser-desorbed protein/peptide ions. *Anal. Chem.* 1995, **67**, 3990–3999.

32. Whittal, R. M.; Li, L. High-resolution matrix-assisted laser desorption/ionization in a linear time-of-flight mass spectrometer. *Anal. Chem.* 1995, **67**, 1950–1954.

33. Vestal, M. L.; Juhasz, P.; Martin, S. A. Delayed extraction matrix-assisted laser desorption time-of-flight mass spectrometry. *Rapid Comm. Mass Spectrom.* 1995, **9**, 1044–1050.

34. Arnold, R. J.; Karty, J. A.; Ellington, A. D.; Reilly, J. P. Monitoring the growth of a bacteria culture by MALDI-MS of whole cells. *Anal. Chem.* 1999, **71**, 1990–1996.

35. Arnold, R. J.; Polevoda, B.; Reilly, J. P.; Sherman, F. The action of *N*-terminal acetyltransferases on yeast ribosomal proteins. *J. Biol. Chem.* 1999, **274**, 37035–37040.

36. Arnold, R. J.; Reilly, J. P. Observation of *Escherichia coli* ribosomal proteins and their posttranslational modifications by mass spectrometry. *Anal. Biochem.* 1999, **269**, 105–112.

37. Arnold, R. J.; Reilly, J. P. Observation of tetrahydrofolyl-poly-glutamic acid in bacteria cells by matrix assisted laser desorption/ionization mass spectrometry. *Anal. Biochem.* 2000, **281**, 45–54.

38. Cain, T. C.; Lubman, D. M.; W.J. Weber, J. Differentiation of bacteria using protein profiles from matrix-assisted laser desorption/ionization time-of-flight mass spectrometry. *Rapid Comm. Mass Spectrom.* 1994, **8**, 1026–1030.

39. Krishnamurthy, T.; Ross, P. L. Rapid identification of bacteria by direct matrix-assisted laser desorption/ionization mass spectrometric analysis of whole cells. *Rapid Comm. Mass Spectrom.* 1996, **10**, 1992–1996.

40. Holland, R. D.; Wilkes, J. G.; Rafii, F.; Sutherland, J. B.; Persons, C. C.; Voorhees, K. J.; Lay, J. O. Rapid identification of intact whole bacteria based on spectral patterns using matrix-assisted laser desorption/ionization with time-of-flight mass spectrometry. *Rapid Comm. Spectrom.* 1996, **10**, 1227–1232.

41. Dai, Y.; Li, L.; Roser, D. C.; Long, S. R. Detection and identification of low-mass peptides and proteins from solvent suspensions of *Escherichia coli* by HPLC fractionation and MALDI mass spectrometry. *Rapid Comm. Mass Spectrom.* 1999, **13**, 73–78.

42. Liang, X.; Zheng, K.; Qian, M. G.; Lubman, D. M. Determination of bacterial protein profiles by matrix-assisted laser desorption/ionization mass spectrometry with high-performance liquid chromatography. *Rapid Comm. Mass Spectrom.* 1996, **10**, 1219–1226.

43. Lay, J. O, Jr. MALDI-TOF mass spectrometry and bacterial taxonomy. *Trends Anal. Chem.* 2000, **19**, 507–516.

44. Lay, J. O. MALDI-TOF mass spectrometry of bacteria. *Mass Spectrom. Rev.* 2001, **20**, 172–194.

45. Fenselau, C.; Demirev, P. A. Characterization of intact microorganisms by MALDI mass spectrometry. *Mass Spectrom. Rev.* 2001, **20**, 157–171.

46. van Baar, B. L. M. Characterization of bacteria by matrix-assisted laser desorption/ionization and electrospray mass spectrometry. *FEMS Microbiol. Rev.* 2000, **24**, 193–219.

47. Krishnamurthy, T.; Ross, P. L.; Rajamani, U. Detection of pathogenic and non-pathogenic bacteria by matrix-assisted laser desorption/ionization time-of-flight mass spectrometry. *Rapid Comm. Mass Spectrom.* 1996, **10**, 883–888.

48. Ng, R.; Horlick, G. *Spectrochim. Acta* 1981, **36**B, 529–542.

49. Arnold, R. J.; Reilly, J. P. High resolution time of flight mass spectra of alkylthiolate-coated gold nanocrystals. *J. Am. Chem. Soc.* 1998, **120**, 1528–1532.

50. Van Bogelen, R. A. In *E. coli and Samonella*, 6 ed. Neidhardt, F. C.; et al. (Eds.). Washington, DC: ASM Press, 1996, pp. 2067–2117.

51. Ogorzalek Loo, R.; Mitchell, C.; Loo, J. A.; VanBogelen, R.; Moldover, B.; Cavalcoli, J.; Stevenson, T.; Andrews, P. C. *Proc. 45th American Society for Mass Spectrometry Conference*, Palm Springs, CA 1997; 311.

52. Chong, B. E.; Wall, D. B.; Lubman, D. M.; Flynn, S. J. Rapid profiling of *E. coli* proteins up to 500 kDa from whole cell lysates using matrix-assisted laser desorption/ionization time-of-flight mass spectrometry. *Rapid Comm. Mass Spectrom.* 1997, **11**, 1900–1908.

53. Struelens, M. J.; De Ryck, R.; Deplano, A. In *New Approaches for the Generation and Analysis of Microbial Typing Data*. Dijkshoorn, L.; Towner, K. J.; Struelens, M. (Eds.). Amsterdam: Elsevier, 2001, pp. 159–176.

54. Edwards-Jones, V.; Claydon, M. A.; Evason, D. J.; Walker, J.; Fox, A. J.; Gordon, D. B. Rapid discrimination between methicillin-sensitive and methicillin-resistant *Staphylococcus aureus* by intact cell mass spectrometry. *J. Med. Microbiol.* 2000, **49**, 295–300.

55. Bernardo, K.; Pakulat, N.; Macht, M.; Krut, O.; Seifert, H.; Fleer, S.; Hunger, F.; Kronke, M. Identification and discrimination of *Staphylococcus aureus* strains using matrix-assisted laser desorption/ionization-time of flight mass spectrometry. *Proteomics* 2002, **2**, 747–753.

56. Du, Z.; Yang, R.; Guo, Z.; Song, Y.; Wang, J. Identification of *Staphylococcus aureus* and determination of its methicillin resistance by matrix-assisted laser desorption/ionization time-of-flight mass spectrometry. *Anal. Chem.* 2002, **74**, 5487–5491.

57. Walker, J.; Fox, A. J.; Edwards-Jones, V.; Gordon, D. B. Intact cell mass spectrometry (ICMS) used to type methicillin-resistant *Staphylococcus aureus*: Media effects and inter-laboratory reproducibility. *J. Microbiol. Meth.* 2002, **48**, 117–126.

58. Evason, D. J.; Claydon, M. A.; Gordon, D. B. Exploring the limits of bacterial cell identification by intact cell-mass spectrometry. *J. Am. Soc. Mass Spectrom.* 2000, **12**, 49–54.

59. Nilsson, C. Fingerprinting of *Helicobacter pylori* strains by matrix assisted laser desorption/ionization mass spectrometric analysis. *Rapid Comm. Mass Spectrom.* 1999, **13**, 1067–1071.

60. Haag, A. M.; Taylor, S. N.; Johnston, K. H.; Cole, R. B. Rapid identification and speciation of *Haemophilus* bacteria by matrix-assisted laser desorption/ionization time-of-flight mass spectrometry. *J. Mass Spectrom.* 1998, **33**, 750–756.

61. Lynn, E. C.; Chung, M.-C.; Tsai, W.-C.; Han, C.-C. Identification of *Enterobacteriaceae* bacteria by direct matrix-assisted laser desorption/ionization mass spectrometric analysis of whole cells. *Rapid Comm. Mass Spectrom.* 1999, **13**, 2022–2027.

62. Wilcox, S. K.; Cavey, G. S.; Pearson, J. D. Single ribosomal protein mutations in antibiotic-resistant bacteria analyzed by mass spectrometry. *Antimicrob. Agents Chemother.* 2001, **45**, 3046–3055.

63. Li, T.-Y.; Liu, B.-H.; Chen, Y.-C. Characterization of *Asperigillus* spores by matrix-assisted laser desorption/ionizaton time-of-flight mass spectrometry. *Rapid Comm. Mass Spectrom.* 2000, **14**, 2393–2400.

64. Ryzhov, V.; Hathout, Y.; Fenselau, C. Rapid characterization of spores of *Bacillus cereus* group bacteria by matrix-assisted laser desorption-ionization time-of-flight mass spectrometry. *Appl. Environ. Microbiol.* 2000, **66**, 3828–3834.

10

BACTERIAL PROTEIN BIOMARKER DISCOVERY: A FOCUSED APPROACH TO DEVELOPING MOLECULAR-BASED IDENTIFICATION SYSTEMS

TRACIE L. WILLIAMS, STEVEN R. MONDAY, AND STEVEN M. MUSSER

Center for Food Safety and Nutrition, Food and Drug Administration, College Park, MD 20740

10.1 INTRODUCTION

Human illness, as a result of microbial infection, is an ever-increasing public health crisis. Children often represent the most at-risk population. Worldwide every year billions of cases of microbial infection occur and millions of deaths can be directly attributed to microbial pathogens.[1] The Centers for Disease Control estimates that microbial infections are the fourth leading cause of death in the United States, with approximately 75,000 deaths attributable to microbial pathogens annually.[2] Not only are there serious physical health effects associated with microbial infections, but the cost of health care associated with microbial infections is shocking. For example, in the United States alone, upper respiratory tract infections in children under the age of 15 account for more than 50 million visits to a doctor's office each year.[3]

Despite the widespread availability of antibiotics and modern health care practices, microbial pathogens continue to be a serious health issue. Numer-

Identification of Microorganisms by Mass Spectrometry, Edited by Charles L. Wilkins and Jackson O. Lay, Jr.

ous factors are contributing to this problem, including newly acquired antibiotic resistance profiles, production of new and more potent toxins, and the emergence of new pathogenic organisms.[4,5] The acquisition of these traits occurs very rapidly and may be acquired via several mechanisms, including direct transfer of genetic material and naturally occurring mutation. As a case in point, between 1997 and 2003 clinical isolates of vancomycin-resistant *enterococci* have increased 27.5% and cases of methicillin-resistant *Staphylococcus aureus* have increased 57.1%.[6] In addition new strains of multidrug-resistant *Streptococcus*[7] and Shigatoxin-producing strains of *Escherichia coli*[8–10] are appearing in human outbreaks of disease every year.

Regardless of the method by which new pathogenic organisms appear in the environment, an urgent need exists to rapidly identify markers that can be used in the identification and diagnosis of the new pathogen. This information is necessary to improve drug therapy, provide for contamination trace-back, and to develop better food handling and clinical practices. Ideally the development of specific diagnostic methods for new pathogenic strains would be accomplished following sequencing of the entire genome of the new strains. While this is technically feasible, given the large number of strains to be sequenced along with the rapid rate of mutation, it would be prohibitively expensive and time-consuming. Therefore other analytical approaches are being investigated, among them are methods based on mass spectrometry.

Mass spectrometry has been used for investigations into microbial identification for more than three decades. Much of this research involved developing methods for the rapid identification and characterization of bacteria that may be important for advance warning systems in military applications. These methods have used fatty acid analysis,[11,12] pyrolysis,[13,14] and more recently matrix-assisted laser desorption ionization (MALDI) mass spectrometry[15,16] to obtain rapid identification information. In the case of pyrolysis MS and fatty acid GC/MS analysis, heavy reliance on pattern recognition software is required, and many times the markers for specific traits are unknown. While MALDI MS of whole bacteria has proved to be a reliable, robust technique for rapidly identifying various species of bacteria, it often does not supply enough information to identify strain or phenotypic markers of pathogenic microorganisms. A more promising approach to identifying specific protein markers related to a phenotypic trait is to use MS methods developed for studying proteomics.

There are several practical approaches, based on proteomics, that have been used to investigate differences among closely related species of bacteria. One of the most frequently used methods for comparing protein expression among bacteria is two-dimensional gel electrophoresis of the extracted proteins, followed by in-gel digestion of the protein spots with a protease, MS sequencing of resulting peptides and protein identification based on the resulting peptide sequences.[17,18] This approach is particularly useful for comparison of protein expression between a small numbers of closely related strains. Due to difficulties associated with gel reproducibility and the inherent low mass resolution of the gels, examining large numbers of closely related strains is an extremely cumbersome and difficult task. Another approach is to simply extract all the proteins from the cell, digest with a protease (usually trypsin), separate the resulting peptide mixture on a HPLC column, follow by MS/MS of the peptides, and finally identify the marker via database searching. This technique is often referred to as a shotgun approach to the analysis of the proteome.[19] A modification of this approach, automated multidimensional protein identification technology (MudPIT),[20] is now commonly used to improve the total number of proteins identified. This technique can obtain sequences of thousands of peptides and identify hundreds of proteins in a single analysis using readily available and highly reproducible protocols and instrumentation.

Both approaches have several drawbacks. In each case they can easily fail to identify small differences, such as small substitutions and deletions in the expressed protein. Since the MS/MS of tryptic peptides does not always produce readily interpretable spectra, a number of mistakes may result. Many times only a partial sequence is observed or the spectrum is of poor quality and the wrong sequence is assigned. In both cases misidentification of the protein occurs. A more damaging problem associated with these two techniques occurs when the peptide sequence information is of very good quality, but a match for the parent protein is not found in the database. This rarely occurs in eukaryotes but often occurs in bacteria due to the high mutation rate and relatively small genetic diversity found in the genomic databases. When the purpose of the research is to identify new protein markers, these peptides may be very significant. Nevertheless, it is impossible to determine the identity of the intact protein, and consequently this very important piece of information is often discarded.

For these reasons we have developed a different approach that measures differential expression of intact proteins.[21] In this approach the proteins are extracted from the cell, separated on an HPLC column, ionized via electrospray, and automatically deconvoluted into their respective uncharged nominal masses. By this methodology it is then possible to obtain accurate, intact protein profiles of the individual strains of bacteria. Because the masses of the detected proteins are accurate to ±2 Da from run to run, it is possible to subtract protein profiles from known strains to quickly identify differences in protein expression among newly mutated strains.

Additionally it has been our experience that mass spectrometry as a routine detection/identification technique for bacteria is not well received by microbiologists and clinicians who prefer less expensive, less complicated approaches to bacterial typing and identification, such as methods based on polymerase chain reaction (PCR) and enzyme-linked immunosorbent assays (ELISA). For that reason we have adapted our MS approach to serve as a means of biomarker discovery that feeds candidate proteins or leads into development as PCR targets or other immunoassay techniques.

10.2 PROTEIN EXTRACTION METHODS

Sample preparation used to extract proteins from cells prior to analysis is an important step that can have an effect on the accuracy and reproducibility of the results. Proteins isolated from bacterial cells will have co-extracted contaminants such as lipids, polysaccharides, and nucleic acids. In addition various organic salts, buffers, detergents, surfactants, and preservatives may have been added to aid in protein extraction or to retain enzymatic or biological activity of the proteins. The presence of these extraneous materials can significantly impede or affect the reproducibility of analysis if they are not removed prior to analysis.

Ultrasonic cell disruption, homogenization (French press), freeze/thaw cycles, glass bead vortexing, and enzymatic lysis (lysozyme) are the most common means of preparing cell lysates from bacterial cells.[22,23] A common feature of many studies is simplification of the proteome by division into extracts of differential solubility representative of the cytosolic and cell wall protein fractions.[24–28] Organic solvents have been used in an attempt to selectively extract the hydrophobic subset of proteins from *E. coli*.[29] However, most of the means to enhance protein solubility involve detergents[30–32] and chaotropes (urea, thiourea)[26] that are amenable to analysis by two-dimensional gel electrophoresis but require complete removal before injection onto a mass spectrometer. The method of detergent removal depends on the chemical properties of the detergent.[33]

Dialysis and gel filtration are size exclusion means by which salts, chaotropes, and detergents with high critical micelle concentration values (i.e., CHAPS) can be removed from the cell lysate prior to analysis via LC/MS. These cleanup methods, while simple to perform, are time-consuming and require a concentration step to compensate for the sample dilution that occurs upon processing the samples in this manner. Efforts have been made to speed up sample cleanup methods, reduce sample handling by doing the cleanup on line, and permitting the processing of smaller sample amounts.[34,35]

Ion exchange chromatography is another means to remove detergents and chaotropes from protein samples. This is one rarely mentioned, but well understood benefit of using ion exchange chromatography as a first dimension separation step in a two-dimensional LC experiment.

10.3 PROTEIN CHROMATOGRAPHY

Most comparative analyses of proteins expressed by closely related biological samples have used two-dimensional polyacrylamide gels electrophoresis (2D PAGE).[17,36] However, this method is time-consuming and difficult to reproduce. More recent efforts have focused on using liquid chromatography/mass spectrometry (LC/MS) with either a top-down[37,38] or bottom-up[39–41] approach. The top-down approach looks at the overall protein expression profile to single out unique proteins that are then sequenced and identified. The bottom-up approach involves proteolytic digestion of all the proteins in the sample and relies on database analysis or de novo sequencing of the individual peptides to identify the source proteins in the sample. With the current chromatography technology peptides are easier to separate and quantify[42–46] than whole proteins, and therefore the bottom-up approach has been the method of choice for many comparative protein expression studies. This approach however, generates enormous amounts of data that requires efficient and accurate software to derive meaningful correlations. More important, small changes in proteins can be easily overlooked during the analysis of large amounts of data.

Reverse phase chromatography is suitable for low molecular weight proteins (<50 kDa) using a polymeric C4-like stationary phase. In our laboratory we have had success with POROS R1 (4000 Å pore size) and R2 (2000 Å pore size), a 20 μ Poly (Styrene-Divinylbenzene) particle produced by Applied Biosystems (Foster City, CA). This stationary phase is quite easily packed into columns with inner diameters as small as 320 μm using an HPLC with the maximum pressure set at 2000 psi to eliminate the possibility of crushing the particles. Agilent (Palo Alto, CA) sells 500 μm inner diameter and larger columns packed with Zorbax Poroshell 300 Å 5 μ particles with C3 and C8 stationary phases. Both the POROS and the Poroshell packings require fast flow rates (250 μl/min and above), which necessitates splitting prior to injection into the mass spectrometer. For lower flow rates, Jupiter 300 Å 5 μ particles with a C4 stationary phase (Phenomenex, Torrance, CA) provide for efficient protein separations.

The mobile phase used for separating whole proteins by reverse phase chromatography is the same as that used for peptides save for one difference: an increase in acid composition in the mobile phase. Typically 0.1% acid is added to the mobile phase to improve the chromatographic peak shape and to provide a source of protons in reverse phase LC/MS. The most common acids used are acetic, formic, and trifluoroacidic acid (TFA). TFA-containing mobile phases have an ion suppression effect, thereby reducing the sensitivity and analytical reliability of LC/MS techniques.[47] For this reason the percentage of TFA is often lower to 0.05% or even 0.002% without significant loss in chromatographic efficiency. A small percentage of heptafluorobutyric acid is sometimes added to acetic acid solvents or low TFA containing solvents to help improve peak shape.

Many proteins will precipitate out of solution upon encountering the acetonitrile typically used for gradient elutions. However, by increasing the acid composition from 0.1% acetic or formic acid as is usually used for peptide separations to 5.0% acetic or formic acid, the solubility of many proteins is increased. As shown in Figures 10.1 and 10.2, not only does the change in percent composition of acid change the retention time, but there is trade-off between solubility and peak resolution. Increasing the amount of acid above 5.0% of the mobile phase substantially decreases the chromatographic peak

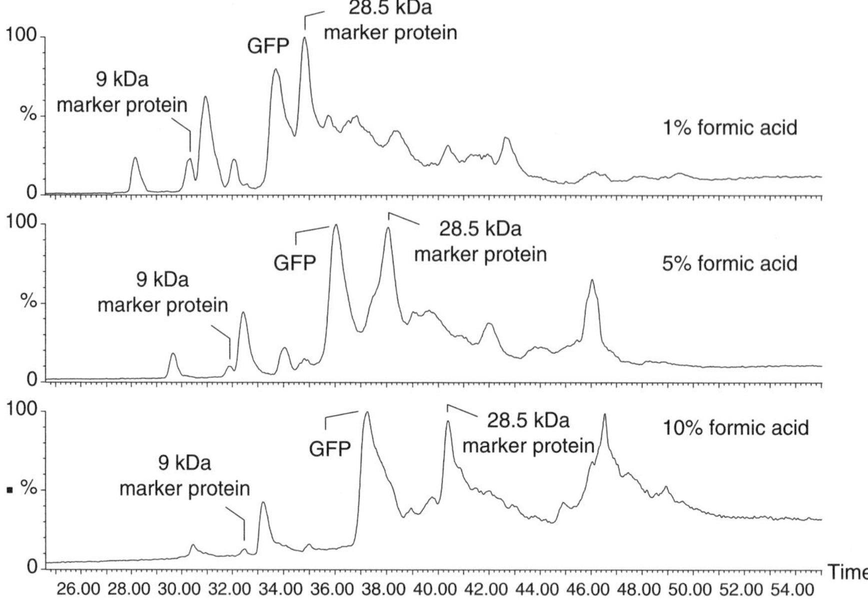

Figure 10.1 Proteins extracted from *E. coli* DH5α strain were separated by LC/MS using different concentrations of formic acid in the mobile phase.

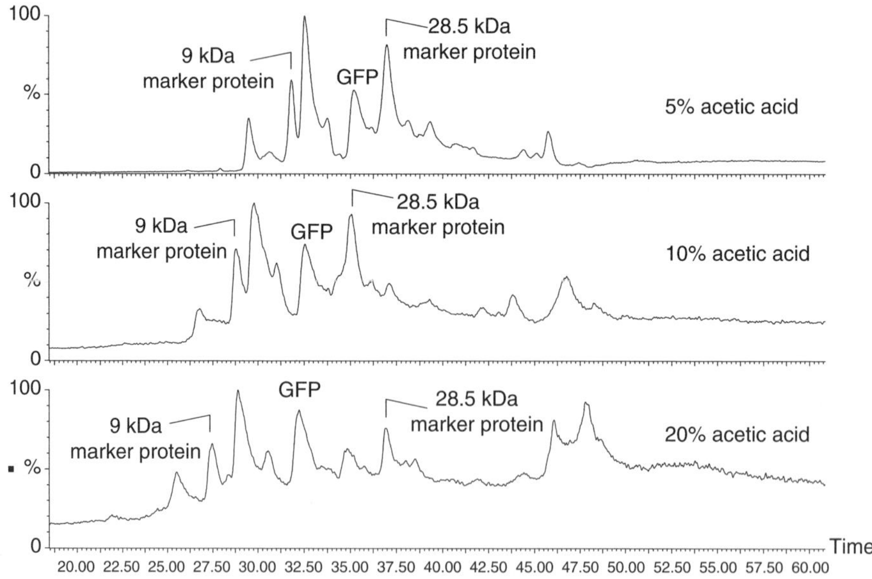

Figure 10.2 Proteins extracted from *E. coli* DH5α strain were separated by LC/MS using different concentrations of acetic acid in the mobile phase.

resolution. For this reason we typically use 5.0% acidic acid or 2.0% to 5.0% formic acid in the mobile phase. Even though this helps with protein solubility, significant fouling of the column does occur, and occasionally the frits at the head of the column need to be replaced.

Methanol, isopropanol, and 50 mM hexafluoroisopropanol were also evaluated for use as organic modifiers in the mobile phase. Little improvement to the solubility of whole proteins or their chromatographic separation was observed, however.

Regardless of the type of column, organic modifiers, or ion-pairing reagents, we have not yet found a set of parameters that allows for sufficient chromatographic resolution for direct comparison of the chromatograms to determine differences in the whole protein profiles of closely related bacterial samples. Figure 10.3 shows the LC chromatograms of two strains of *Escherichia coli*. Due to the lack of baseline resolution and the co-elution of proteins, only the most abundant proteins could be discerned in a simple comparison of the chromatograms. Therefore simply comparing chromatograms without associated mass information does not distinguish between very similar, but not identical proteins. These proteins are likely to elute at the same

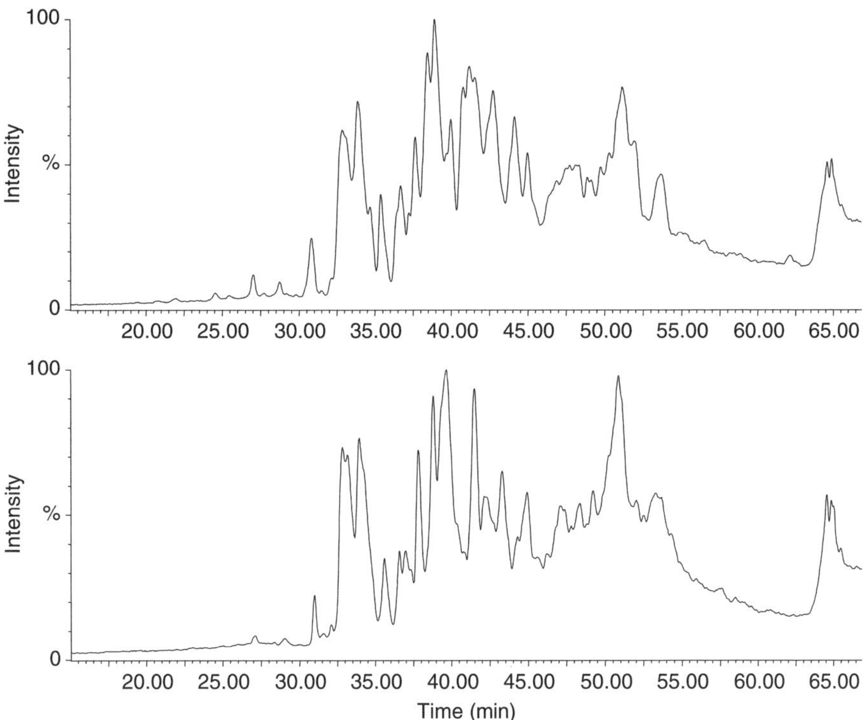

Figure 10.3 Chromatograms of proteins extracted from *E. coli* K12 (*top*) and E. coli 493/89 (*bottom*).

retention time because there is little difference in their chemical properties. Thus chromatographic information alone is not a sufficient means of determining the presence of proteins that are unique to a bacterial strain.

Two-dimensional (2D) LC methods have been demonstrated as useful for comparative proteomic studies. These include ion exchange chromatography followed by reverse phase chromatography,[48] size-exclusion chromatography followed by reverse phase chromatography,[18,49] and liquid phase isoelectric focusing (IEF) followed by reverse phase chromatography.[50–52] While 2D separations do improve the chromatographic resolution and provide more information about the whole protein (namely pI) over simple single-dimensional chromatography, small differences in protein structure (i.e., single amino acid substitutions, posttranslational modifications) are not chemically altered to the extent that protein's pI or chromatographic retention time will be measurably different.

10.4 MASS SPECTROMETRY

The high resolving power of mass spectrometry nevertheless does provide for observing the differences in small mass changes between proteins. Even with what is typically considered a "low" resolution mass spectrometer (nominal mass over the accessible *m/z* range),[53] proteins with different masses that indicate different proteins, amino acid substitutions, or posttranslational modifications can usually be easily distinguished. For this reason we use chromatography to grossly separate the proteins of the extract and simplify the mass analysis, but we rely on the molecular weight measurements to differentiate proteins of closely related bacterial species. In our laboratory a Waters (Manchester, UK) QTOF II was used to acquire the data in full scan continuum mode with an *m/z* range from 500 to 2000. Data were processed using Masslynx v. 4.0 software, and the multiply-charged protein spectrum was deconvoluted into a molecular weight spectrum using maximum entropy deconvolution software, MaxEnt 1.

To obtain a high-quality final spectrum, it is essential that spectra be combined in short intervals, and subsequently deconvoluted, rather than summing the scans for the entire chromatographic run. To illustrate this point, all mass spectra obtained from 20 to 65 minutes of the *E. coli* K12 strain (top chromatogram, Figure 10.3) were summed and combined into a single spectrum and deconvoluted using MaxEnt 1. The resulting deconvoluted spectrum (Figure 10.4) appears noisy, contains a small number of well-defined protein masses, and makes it impossible to distinguish between less abundant proteins and noise. The poor quality of the final spectrum can be attributed to the great number of multiply-charged ions obtained in the summed spectrum, creating a peak at every mass. As the number of successive scans added to the spectrum increases, so does the overall noise. The result is the successful deconvolution of only the most intense proteins present in the sample.

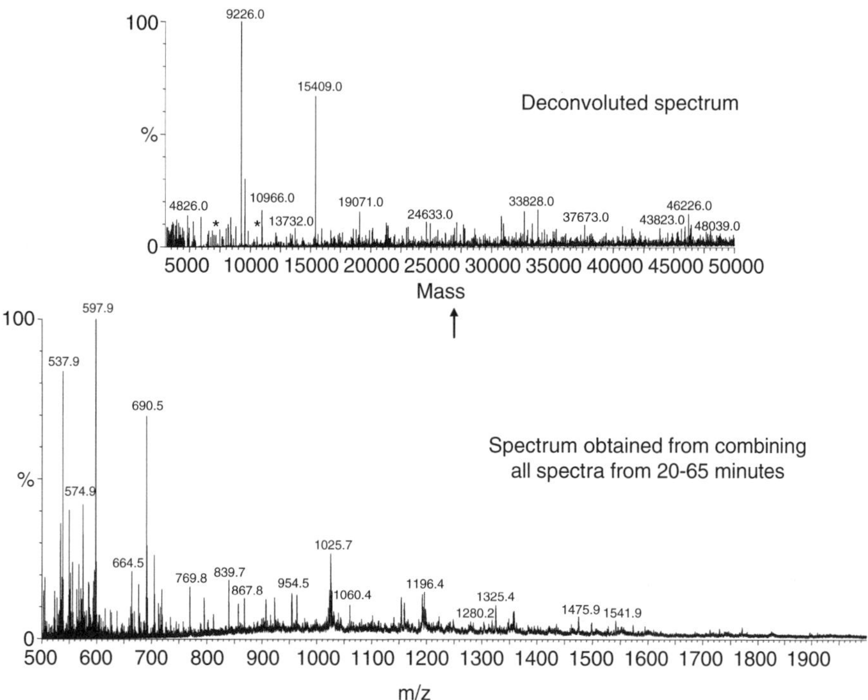

Figure 10.4 All spectra obtained between 20 and 66 minutes are combined into a single spectrum and then deconvoluted using MaxEnt 1. The resulting molecular weight spectrum is noisy and hampers the identification of low abundance proteins. The asterisks indicate where proteins of mass 7274 and 10652 should be observed but are not.

Analyzing smaller time segments of the LC analysis (0.5 or 1 min intervals) provides for an unequivocal identification of the protein, as there are fewer multiply-charged ions present in the summed spectrum and improved signal to noise. Figure 10.5 shows the spectrum obtained from combining all spectra between 31.5 and 32.0 minutes and the resulting deconvolution spectrum showing the presence of proteins with masses of 7274 and 10652 Da. An asterisk is placed in the region where they should be observed in Figure 10.4. However, when the entire experiment is combined into one spectrum and then deconvoluted, proteins such as these are lost in the noise. For this reason the spectra must be combined in smaller intervals and the separate deconvolution processes performed on the individual intervals.

Manually processing each chromatographic peak is not only time and labor intensive but difficult to reproduce. To overcome these problems and to provide a consistent data format that was independent of retention time, a number of data-processing subroutines were automated to produce a single representative cellular protein spectrum.

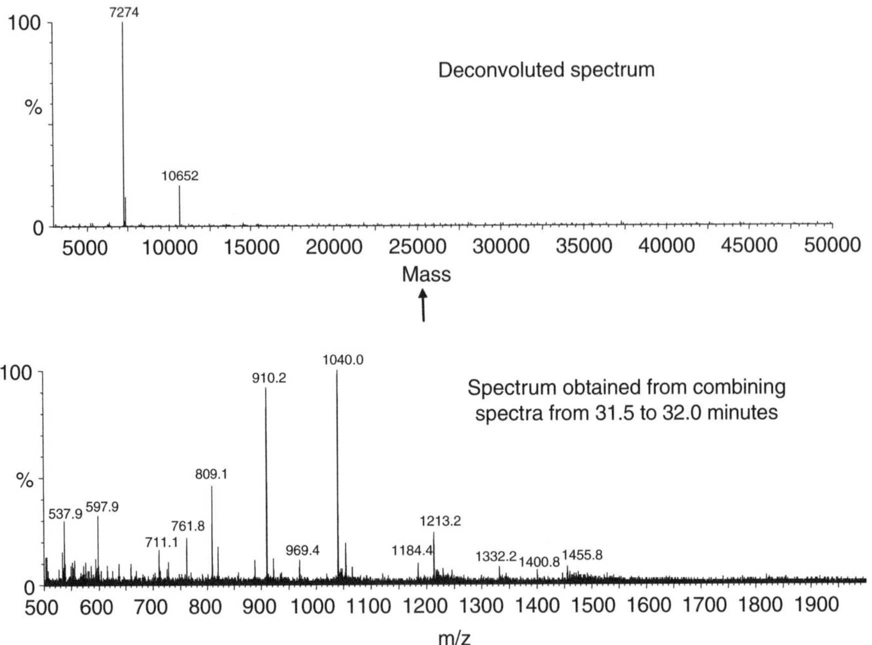

Figure 10.5 Spectra obtained between 31.5 and 32.0 minutes are combined into a single spectrum and then deconvoluted using MaxEnt 1. Proteins with a molecular weight of 7274 Da and 10,652 Da are easily observed. These proteins are not observed in Figure 10.4.

10.5 AUTOMATING THE PROCESS

Automated analysis of the full scan (MS) data was performed with ProteinTrawler, custom software written for this purpose by BioAnalyte, Inc. (Portland, ME). The function of this program is to automate data processing subroutines within the data-processing program and to produce a combined time and intensity text output file. A detailed explanation of this program has been published.[21] Briefly, the program sums all data within a specified time interval, uses MaxEnt 1 to deconvolute the multiply-charged ions, centers the result, performs a threshold selection, and reports the mass, intensity, and retention time of the protein in a text file. It continues this process across sequential portions of the chromatogram. All aspects of the subroutines, including retention times, mass windows, number of MaxEnt 1 iterations, and spectra to combine, can be controlled through ProteinTrawler.

Upon completion of the ProteinTrawler program, the text file contains a cumulative list of all the protein masses that were observed upon deconvolution of the individual summed spectra. This text file records mass, intensity, and retention time. The retention time information is held in the text file for the user to reference if a protein is singled out or deemed significant for further study, and thereby facilitates the isolation and purification process. It can also be used to verify that proteins of the same mass are actually two unrelated proteins as indicated by their different retention times.

Advanced Chemistry Development (ACD, Inc., Toronto, Ontario, Canada) software package version 8.0 was used to display the data as reconstructed mass spectra (Figure 10.6), as well as reflect spectra for comparison purposes (Figure 10.7) and display peaks that were not common between two spectra (Figure 10.8).

The presence or absence of a protein is not the only possible indicator of pathogenicity. Instead, the amount of a particular protein in a sample may serve as a trigger for a signaling process to turn on or off that which makes the strain pathogenic. Therefore, comparing the protein peak intensities of the two bacterial strains may identify a characteristic marker of bacterial pathogenicity. If the spectra of the two *Vibrio parahaemolyticus* are normalized, the peak at 9465 Da dominates the spectrum of Bac-98-3547, and it is difficult to compare the less intense peaks. However, if the spectrum of Bac-98-3547 is normalized to the second largest peak in the display (Figure 10.9), significant changes in the intensities of only a few peaks are observed. There are some differences in the intensities of the protein peaks in the range of 7000 to 8000 Da. Again, since ProteinTrawler records retention time, these proteins

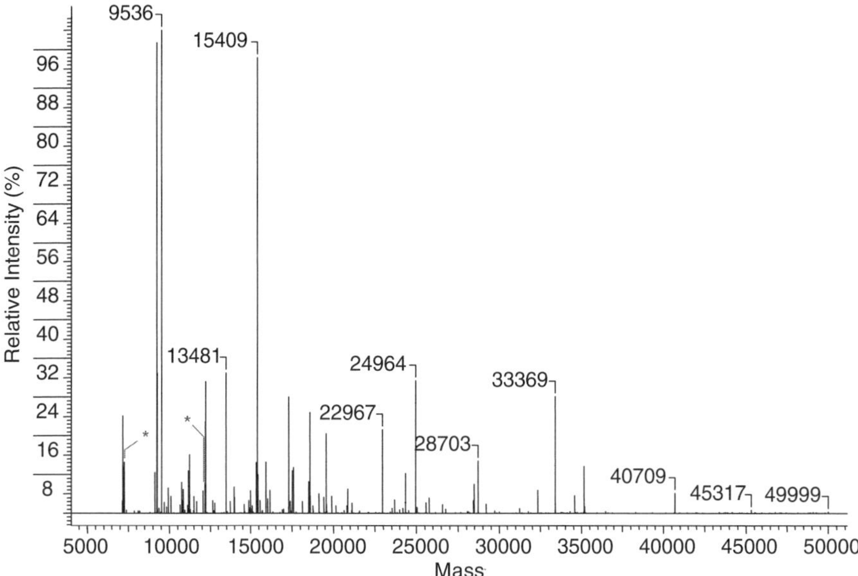

Figure 10.6 Single representative cellular protein spectrum of *E. coli* K12 generated by deconvoluting the spectra in 30-second intervals from 20 minutes to 66 minutes. The asterisks indicate where proteins of mass 7274 and 10,652 are observed.

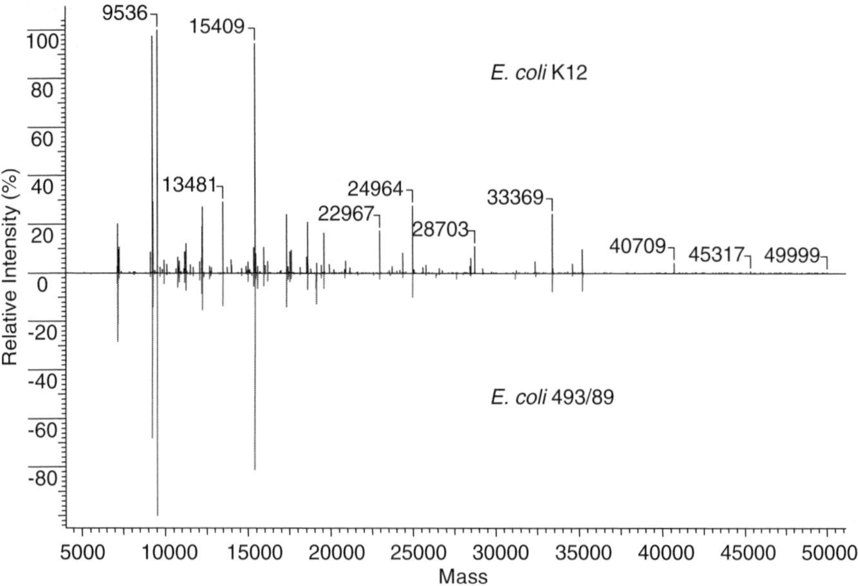

Figure 10.7 Reconstructed spectra of the two *E. coli* strains are reflected across the *x*-axis.

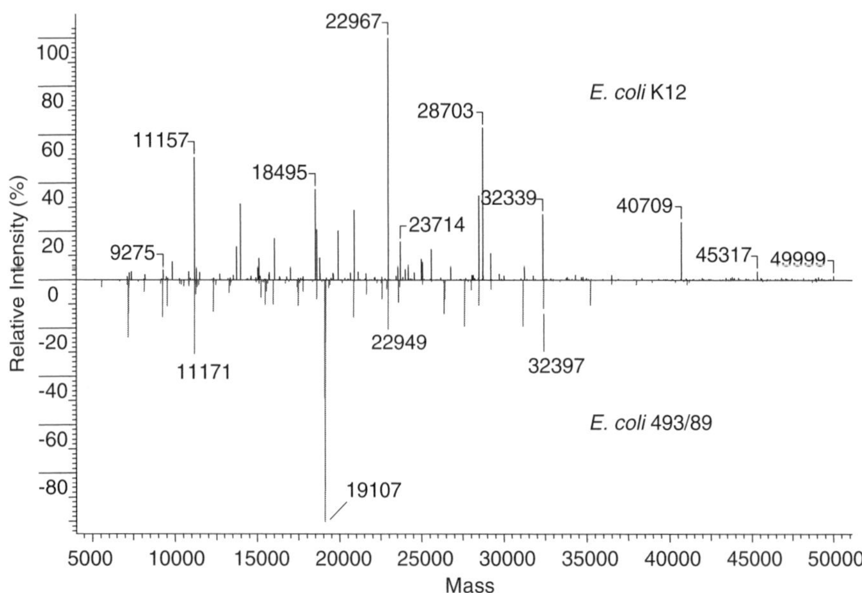

Figure 10.8 Proteins unique to each strain of *E. coli* are shown.

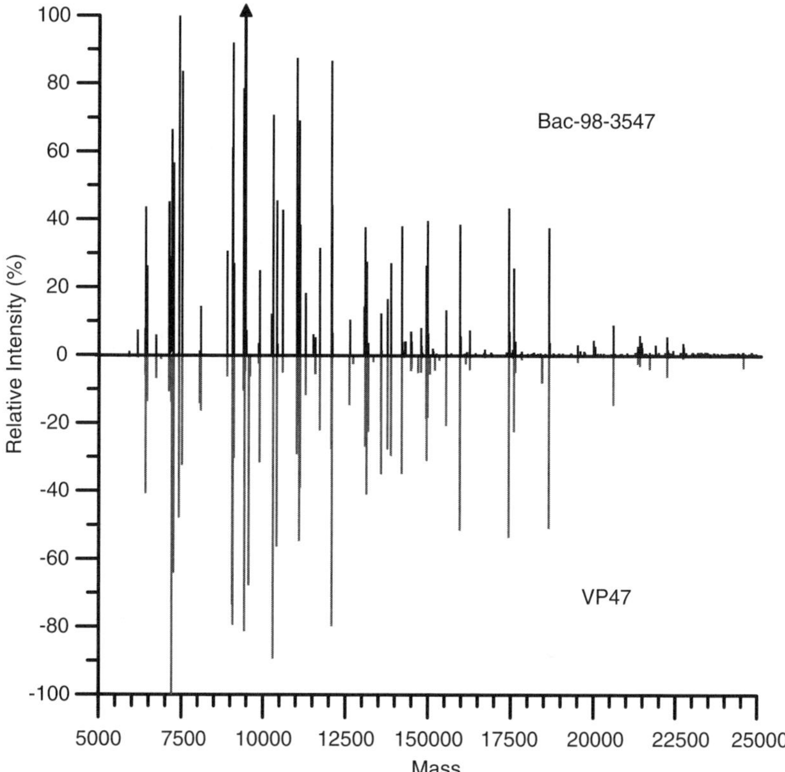

Figure 10.9 When the spectra of Bac-98-3547 is normalized to the second largest peak in the display, it is evident that there is little difference in the relative intensities of the majority of the proteins in the sample.

can be easily isolated and further studied. Re-analysis of samples prepared on different days, demonstrates that the protein ratios are highly reproducible. Therefore, by normalizing the final profile spectrum to proteins with the same mass in several different strains, it is possible to identify not only proteins of different mass but also significant changes in level of protein expression.

The feasibility of this approach to not only differentiate pathogenic and nonpathogenic strains of bacteria based on significant differences in protein mass but also on the basis of variations in levels of protein expression was demonstrated using a method for quantitating protein expression by LC/MS of whole proteins.[54] This method is based on the fact that some proteins present in cells are abundant universal proteins whose expression levels exhibit little variation. This method demonstrates that these co-extracted proteins can be used as internal standards to which the other proteins in the sample can be compared. By comparing the intensities of a selected protein to a marker protein, or internal standard, a relative ratio is obtained. This ratio can then be used to determine the relative amount of protein expression between cellular extracts.

10.6 COLLECTING AND SEQUENCING PROTEINS

Since ProteinTrawler records the retention time of each protein mass, it is a simple endeavor to maintain chromatographic conditions, split the flow that exits the LC column with a small portion set to the mass spectrometer to monitor for assurance that there were no changes in the retention time that would hinder the pooling of fractions from multiple runs, and to facilitate the determination of which fractions contained the desired proteins. In our experimental setup, the flow was split after the column with 25% of the flow going to the mass spectrometer while the remaining diverted to an HP1100 fraction collector. The fraction collector was used to collect fractions at 1.0-minute intervals.

The fraction containing the protein of interest was evaporated to dryness and reconstituted in 50 μL of Rapigest (Waters, Bedford, MA). The protein was incubated at 37° with 1 μmole of modified trypsin (Promega, Madison, WI) for 2 hours for complete protein digestion. The protein digest (8 μl) was injected onto a Symmetry300™ (Waters, Bedford, MA) C18 column with dimensions of 150 mm × 0.320 mm i.d. Chromatography was completed using the HP1100 with the same mobile phase and gradient as was used in the whole protein separation but with a reduced flow rate of 20 μl/min.

The PepSeq program of Micromass's ProteinLynx software package was used for de novo analysis of the sequence (MS/MS) data. MS-Pattern of ProteinProspectror[55] used the sequence tag determined from PepSeq to search the nonredundant database of the National Center for Biotechnology Information (NCBI) for protein identification.

10.6.1 *Salmonella enterica* Serovar *Newport* Isolated from Different Animals

An example of how this method could be used to find biomarkers to differentiate closely related species is described in our efforts to find a biomarker of *Salmonella enterica* serovar *Newport* that would indicate from which animal the strain originated. This would be useful in food-borne illness to quickly identify and isolate the source of the infection. The top spectrum in Figure 10.10 shows only the proteins that are unique to the strain isolated from cattle while proteins unique to the strain isolated from chicken is shown in the negatively reflected spectrum. We focused on proteins with a molecular weight of 11,219 Da that had been isolated from cattle and proteins of 11,232 Da that had been isolated from chicken. Considering that the retention time of the two proteins was similar, our first assumption was that this protein was identical except that the protein isolated from chickens had undergone a post-

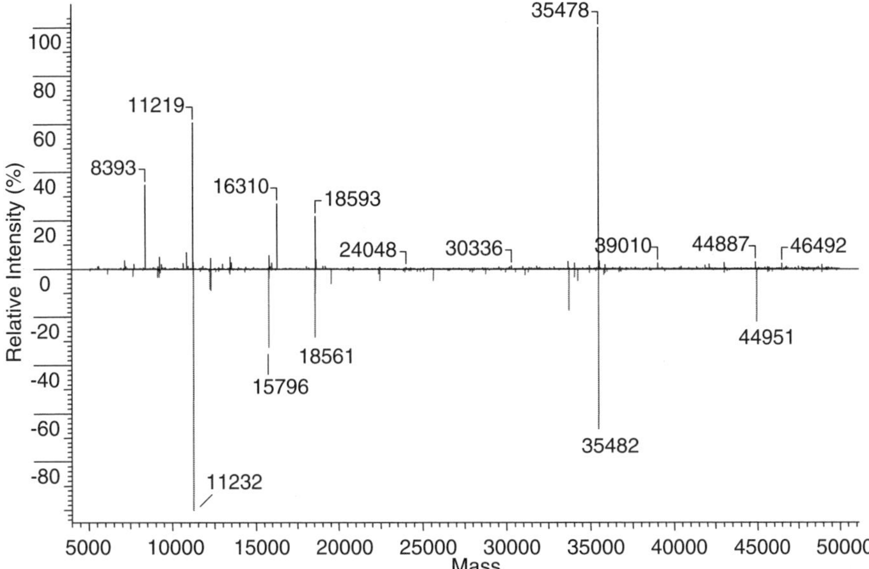

Figure 10.10 Proteins unique to each strain of *Samonella enterica* serovar *Newport.* The top spectrum is that which had been isolated from cattle (20769), while the bottom is from a strain that had been isolated from chicken (21541).

translational modification, namely methylation.[56,57] When the protein was purified and sequenced, it was identified as homologous to hypothetical protein STY2542 of *Salmonella enterica* serovar *Typhi* strain CT18 (Database entry NP_456854) with the amino acid sequence

MRMSYQFGESR VDDDLTLLSETLEEVLR SSGDPADQK
YIELK AR AEQALEEVK NR VSHASDSYYYR AK QAVYK
ADDYVHEKPWQGI GVGAAVGLVLGLLLAR R

The underlined amino acids were identified using tryptic digestion and MS/MS analysis. Interesting to note here is that the start terminal MRM sequence is not observed in the experimentally derived sequence. If the both methionines had been present, the second would have been retained on the tryptic piece SYQFGESR. As it was, the second methionine likely was actually the terminal methionine that was lost, based on the fact that the second amino acid was an uncharged serine residue.[58,59] Since these protein sequences are translated from the genomic sequences, such discrepancies are likely as the sequence is simply read from the nucleotides that designate a start codon (which translates as a methionine) until it reaches a stop codon. While there were two start codons in near proximity in the nucleotide sequence, the true start of the open reading frame of this protein was the second methionine. This clearly demonstrates how an experimentally measured molecular weight of a protein may differ from that found in the database and how, currently, sequencing must be conducted in order to positively identify the protein.

The second and most important finding of this sequencing project was that the difference of 14 Da between the two proteins was not because of a methylation of the protein isolated from chicken but because of a substitution of valine for leucine. The MS/MS analysis of the tryptic pieces in which the substitution occurred is shown in Figure 10.11. Had this been a bottom-up

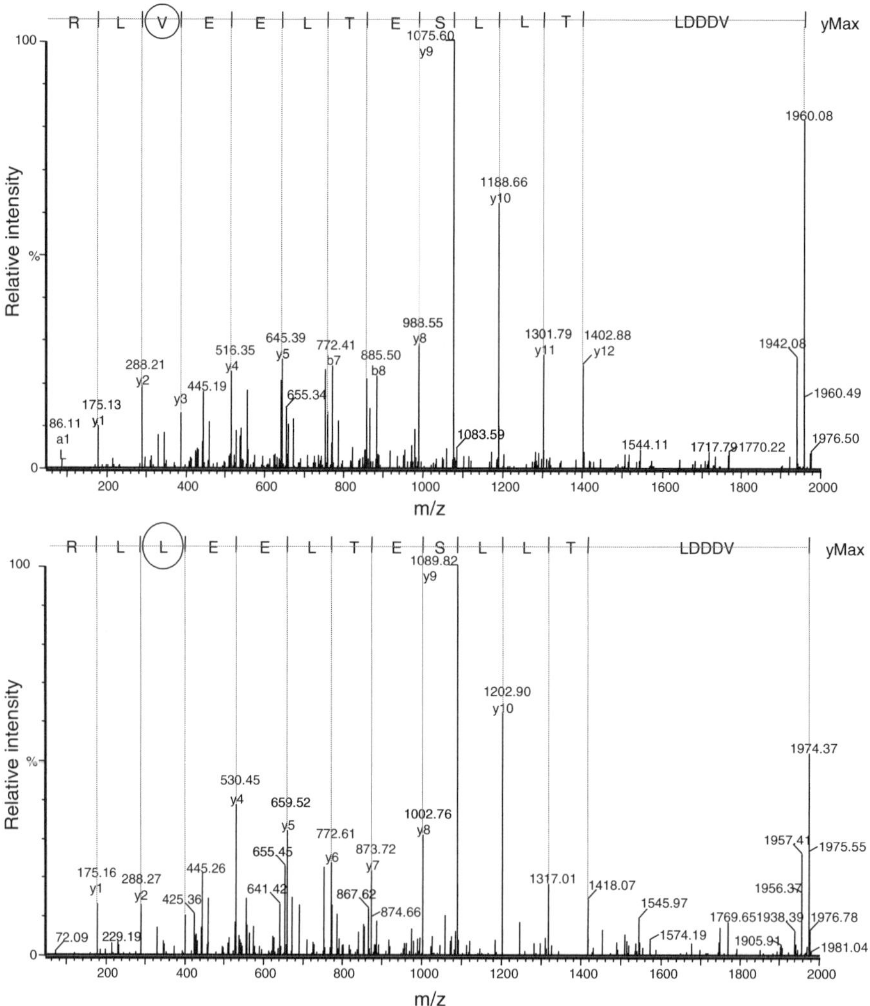

Figure 10.11 MS/MS spectra of a tryptic piece of the 11,222 Da protein (*top spectrum*) and the 11,236 Da protein (*bottom spectrum*). Both proteins were identified as homologous to a conserved hypothetical protein of *Salmonella enterica* serovar *Typhi*. The only difference observed in the sequence is that of a substitution of valine with leucine, which increases the molecular weight of the protein by 14 Da.

sequencing project, this protein would likely not have been observed as unique because three of the four tryptic pieces that we identified would have indicated the identical protein for both strains. Such small differences in protein sequence can only be observed if the molecular weight of the intact protein is known prior to digestion and MS/MS analysis. This point was also illustrated by our discovery of a marker protein for the pandemic O3:K6 clone of *Vibrio parahaemolyticus* that differed from the protein in environmental/nonpathogenic strains by merely five amino acids at the carboxyl-terminus.[60] Thus, for comparative proteomic analysis of closely related bacterial strains, a top-down approach greatly simplifies the amount of data that must be analyzed and quickly focuses efforts on proteins that could potentially be used as biomarkers to differentiate closely related strains.

10.6.2 Phenotypic Marker of Thermal Resistant Strains of *Enterobacter sakazakii*

Enterobacter sakazakii is a rare, but significant cause of life-threatening neonatal meningitis, necrotizing enterocolitis, and sepsis.[61] The association of severe *E. sakazakii* infections in neonates with milk-based powdered infant formulas has led to intensified research into the characterization and control of this microorganism.[62–64] Since thermal treatments have long been used to control pathogenic microorganisms, both at manufacture and preparation, several studies have examined the microorganism's thermal resistance.[65–69] These studies found that some strains of *E. sakazakii* were substantially more thermotolerant, although standard pasteurization practices were effective for the inactivation of all strains in their investigations.[70] To date, no other specific genotypic or phenotypic characteristic has been linked to the strains with increased thermotolerance. A means for characterizing and classifying these isolates would be important for understanding the ecology of this microorganism, including the emergence of phenotypic traits that allow it to survive better in the manufacturing environment.

To identify a biomarker specific for this phenotypic trait, intact protein expression profiles of 12 strains of *E. sakazakii* were obtained using LC/MS followed by automated processing of the data. From the individual expression profiles it was possible to identify numerous unique proteins corresponding to strains with thermal resistance (data not shown). Because the genome sequence of *E. sakazakii* is largely unknown, de novo sequencing of any candidate marker protein would likely be required, thus involving a laborious fractionation and purification of the extracted proteins. Since this is a time-consuming process, we tried to select and sequence an abundant protein, rather than attempt to sequence all the unique proteins. In this case the best candidate protein present in the heat-tolerant strains of *E. sakazakii* but absent in the heat-sensitive strains had molecular weight of 29,563 Da (Figure 10.12).

Because the genome of *E. sakazakii* has not yet been sequenced and deposited into the National Center for Biotechnology Information (NCBI)

Strain	Lab ID	D_{58}-value (seconds)	29,563 Da protein (MS detection)	Hypothetical protein Mflag020121 (PCR detection)
ATCC 51329	701	30.5 ± 0.1	-	-
NQ2[c]	703	31.5 ± 1.8	-	-
NQ3[c]	704	34.4 ± 4.1	-	-
LCDC 674[a]	707	36.9 ± 4.6	-	-
CDC A3 (10)[a]	706	37.5 ± 8.2	-	-
NQ1[c]	702	47.9 ± 1.0	-	-
EWFAKRC11NNV1493[a]	712	307.8 ± 6.7	+	+
ATCC 29544	700	367.1 ± 23.4	+	+
SK 90[a]	708	465.4 ± 15.3	+	+
LCDC 648[a]	709	540.9 ± 20.7	+	+
4.01C[b]	713	571.9 ± 23.1	+	+
607[d]	705	591.9 ± 49.9	+	+

[a]Original strains provided by J.M. Farber, Health Canada
[b]Isolated from dried infant formula by S.Edelson-Mammel, FDA
[c]Environmental samples obtained from M. Kotewicz, FDA
[d]Clinical isolate obtained from F. Khambaty, FDA

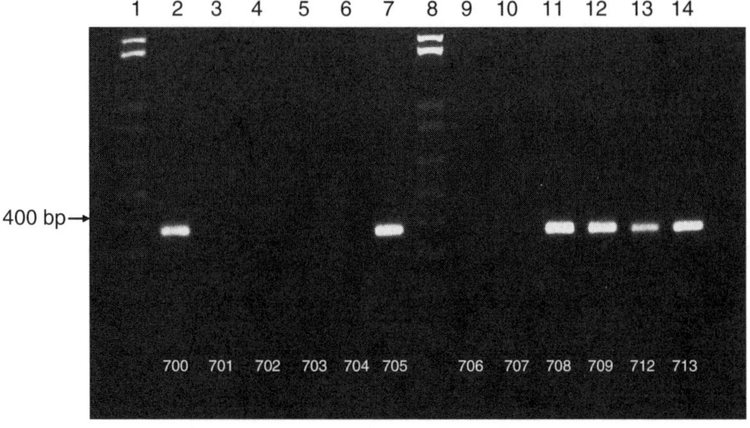

Figure 10.12 *E. sakazakii* strains used in the study are listed in the table. PCR results for *M. flagellatus* hypothetical protein ORF in *E. sakazakii* strains are shown below. Ln 1 and 8, 1 kb Plus MW marker. Ln 2, 700; Ln3, 701; Ln4, 702; Ln5, 703; Ln6, 704, Ln7, 705; Ln9, 706; Ln10, 707; Ln11, 708; Ln 12, 709; Ln13, 712; and Ln14, 713.

database, it was clear that an exact identification of the protein may not be possible (at the time of this writing, only 10 protein entries for *E. sakazakii* were available). However, it was hoped that the tryptic pieces that were identified would be conserved among other closely related species (i.e., *Enterobacter cloacae*, *Salmonella enterica*, *Escherichia coli*) such that the protein identity could be deduced. It was clear that the amino acid sequence of the tryptic fragments were correct based on the quality of the MS/MS data (data not shown). Initially the tryptic pieces that were identified did not match any entries in the NCBI database.

Recently the genome sequence of *Methylobacillus flagellatus KT* was deposited into the NCBI database and provided a match to our sequence data with a database entry (ZP_00173833). The protein was identified as a hypothetical protein Mflag020121. The sequence of this protein along with the sequences matching the tryptic fragments from our *E. sakazakii* marker protein (shown underlined) is

MTLLDAKKTKEALAALELASGKLELVLARDAKLALAPVDVRVITH
DIHAN**VESVKK**AVKLSRELLGDGEVQKARPIVANLASEIVIETDN
LPMATYPAAIKSAARLIDSGKIDNAKAELARALNTLVVTSVAFPLP
MLRAEAAMEKAEKLAETDKRDAKQNEELS**TLLSSV**RTEIELAQI
LGYGKKEDFKPIFDQVKSIEQKSAGGKSGKGWFDELKTRLQKLF

Methylobacillus flagellatus KT is an obligate methylotroph able to use only methanol and methylamine as the sole carbon and energy sources.[71] Unlike the majority of the known obligate methylotrophs, *M. flagellatum* grows at relatively high temperatures.[72] It is unclear what, if any, role this hypothetical protein plays in the thermal tolerance of either *M. flagellatum* or *E. sakazakii*. However, the protein marker clearly distinguished the thermal tolerant strains of *E. sakazakii* from the thermal sensitive strains (Figure 10.12).

To verify that this protein is only present in the thermal-tolerant strains of *E. sakazakii* and would make a good marker for thermotolerant strains, PCR primers were made for a portion of the gene sequence that encodes the open reading frame (ORF) of the protein. The amino acids that were targeted are in bold in the above amino acid sequence of the hypothetical protein. As shown in Figure 10.12, only the thermal-tolerant strains of *E. sakazakii* had the 364 bp amplification product of the gene sequence within the ORF of the protein. This indicated that both the protein and its corresponding gene sequence were unique to thermotolerant strains of *E. sakazakii*.

We were fortunate that the nucleotide sequence of our biomarker had been deposited in GenBank permitting straightforward primer development. However, it would have been possible to obtain this information experimentally using readily available molecular biology techniques. In this particular case, had the database sequence not been available, degenerative primers could be designed from the experimentally determined amino acid sequence using an *E. sakazakii* codon bias table (www.kazusa.or.jp/codon/). These primers could be used in PCR with *E. sakazakii* DNA to generate probes for use in Southern blot analyses to clone and sequence the DNA encoding the protein biomarker detected using our technique. Although this approach would have required more time, it is a proven technique from which the correct nucleotide sequence can be determined from an experimentally derived amino acid sequence. Thus our approach allows this type of highly directed sequence analysis to be conducted in the absence of complete genetic information.

10.7 CONCLUSIONS

We have described the use of a top-down MS based proteomics approach for the discovery of bacterial protein biomarkers. The method is specific, reproducible, and accurately identifies differences in large sets of closely related strains of the same species of bacteria. Although this is a semiquantitative approach to biomarker discovery, it is not complicated and does not rely on the use of special reagents or isotopic labels. Furthermore the technique delivers highly specific molecular level targets that can be exploited for the purpose of microbial pathogen identification or phenotypic matching. Another important and often overlooked aspect related to the usefulness of specific biomarkers in microbial identification systems is the ability to positively identify one species in the presence of another. This is particularly useful in clinical diagnosis and tracing the source of infection. Perhaps the most useful information specific microbial pathogen biomarkers can provide is allowing the rapid selection of the appropriate antibiotic treatment. While custom, patient-tailored therapeutic response is not often used, methods such as we describe provide the tools that will allow the more widespread use of this type of therapy. Finally, although this method has been designed for use in microbiological applications, it could easily be applied to other areas of life sciences research such as early disease diagnosis and treatment.

ACKNOWLEDGMENTS

We would like to thank our FDA collaborators Dr. Peter C. H. Feng for providing the *E. coli* cultures, Dr. Robert Buchanan and Ms. Sharon Edelson Mammel for providing the *E. sakazakii* cultures, and Dr. Ruby Singh, Dr. Heather Harbottle, and Dr. Robert Walker for providing the *Salmonella enterica* cultures.

REFERENCES

1. King, C. K.; Glass, R.; Bresee, J. S.; Duggan, C. Managing acute gastroenteritis among children: Oral rehydration, maintenance, and nutritional therapy. *MMWR Recomm. Rep.* 2003, **52**, 1–16.
2. Mokdad, A. H.; Marks, J. S.; Stroup, D. F.; Gerberding, J. L. Actual causes of death in the United States, 2000 *Jama* 2004, **291**, 1238–1245.
3. Freid, V. M.; Makuc, D. M.; Rooks, R. N. Ambulatory health care visits by children: Principal diagnosis and place of visit. *Vital Health Stat.* 1998, **13**, 1–23.

4. Stratton, C. W. Antimicrobial resistance in respiratory tract pathogens. *Expert Rev. Anti. Infect. Ther.* 2004, **2**, 641–647.

5. Bielecki, J. Emerging food pathogens and bacterial toxins. *Acta Microbiol. Pol.* 2003, **52**(suppl), 17–22.

6. National Nosocomial Infections Surveillance (NNIS) System Report, data summary from January 1992 through June 2003, issued August 2003. *Am. J. Infect. Control* 2003, **31**, 481–498.

7. Jones, M. E.; Blosser-Middleton, R. S.; Thornsberry, C.; Karlowsky, J. A.; Sahm, D. F. The activity of levofloxacin and other antimicrobials against clinical isolates of *Streptococcus pneumoniae* collected worldwide during 1999–2002. *Diagn. Microbiol. Infect. Dis.* 2003, **47**, 579–586.

8. Beutin, L.; Krause, G.; Zimmermann, S.; Kaulfuss, S.; Gleier, K. Characterization of Shiga toxin-producing *Escherichia coli* strains isolated from human patients in Germany over a 3-year period. *J. Clin. Microbiol.* 2004, **42**, 1099–1108.

9. de Castro, A. F.; Guerra, B.; Leomil, L.; Aidar-Ugrinovitch, L.; Beutin, L. Multidrug-resistant Shiga toxin-producing *Escherichia coli* 0118:H16. *Lat. Am. Emerg. Infect. Dis.* 2003, **9**, 1027–1028.

10. Strauch, E.; Schaudinn, C.; Beutin, L. First-time isolation and characterization of a bacteriophage encoding the Shiga toxin 2c variant, which is globally spread in strains of *Escherichia coli* O157. *Infect. Immun.* 2004, **72**, 7030–7039.

11. Larsson, L.; Saraf, A. Use of gas chromatography-ion trap tandem mass spectrometry for the detection and characterization of microorganisms in complex samples. *Mol. Biotechnol.* 1997, **7**, 279–287.

12. Moss, C. W.; Dees, S. B. Identification of microorganisms by gas chromatographic-mass spectrometric analysis of cellular fatty acids. *J. Chromatogr.* 1975, **112**, 594–604.

13. Barshick, S. A.; Wolf, D. A.; Vass, A. A. Differentiation of microorganisms based on pyrolysis-ion trap mass spectrometry using chemical ionization. *Anal. Chem.* 1999, **71**, 633–641.

14. Basile, F.; Beverly, M. B.; Abbas-Hawks, C.; Mowry, C. D.; Voorhees, K. J.; Hadfield, T. L. Direct mass spectrometric analysis of in situ thermally hydrolyzed and methylated lipids from whole bacterial cells. *Anal. Chem.* 1998, **70**, 1555–1562.

15. Welham, K. J.; Domin, M. A.; Scannell, D. E.; Cohen, E.; Ashton, D. S. The characterization of micro-organisms by matrix-assisted laser desorption/ionization time-of-flight mass spectrometry. *Rapid Comm. Mass. Spectrom.* 1998, **12**, 176–180.

16. Holland, R. D.; Wilkes, J. G.; Rafii, F.; Sutherland, J. B.; Persons, C. C.; Voorhees, K. J.; Lay, J. O., Jr. Rapid identification of intact whole bacteria based on spectral patterns using matrix-assisted laser desorption/ionization with time-of-flight mass spectrometry. *Rapid Comm. Mass. Spectrom.* 1996, **10**, 1227–1232.

17. Cordwell, S. J.; Nouwens, A. S.; Walsh, B. J. Comparative proteomics of bacterial pathogens. *Proteomics* 2001, **1**, 461–472.

18. Opiteck, G. J.; Ramirez, S. M.; Jorgenson, J. W.; Moseley, M. A., 3rd. Comprehensive two-dimensional high-performance liquid chromatography for the isolation of overexpressed proteins and proteome mapping. *Anal. Biochem.* 1998, **258**, 349–361.

19. Link, A. J.; Eng, J.; Schieltz, D. M.; Carmack, E.; Mize, G. J.; Morris, D. R.; Garvik, B. M.; Yates, J. R., 3rd. Direct analysis of protein complexes using mass spectrometry. *Nat. Biotechnol.* 1999, **17**, 676–682.

20. Wolters, D. A.; Washburn, M. P.; Yates, J. R., 3rd. An automated multidimensional protein identification technology for shotgun proteomics. *Anal. Chem.* 2001, **73**, 5683–5690.

21. Williams, T. L.; Leopold, P.; Musser, S. Automated postprocessing of electrospray LC/MS data for profiling protein expression in bacteria. *Anal. Chem.* 2002, **74**, 5807–5813.

22. *2-D Proteome Analysis Protocols.* Totowa, NJ: Humana Press, 1999.

23. Kaufman, P. B. *Handbook of Molecular and Cellular Methods in Biology and Medicine.* Boca Raton, FL: CRC Press, 1995.

24. Coldham, N. G.; Woodward, M. J. Characterization of the *Salmonella typhimurium* proteome by semi-automated two dimensional HPLC-mass spectrometry: Detection of proteins implicated in multiple antibiotic resistance. *J. Proteome Res.* 2004, **3**, 595–603.

25. Molloy, M. P.; Herbert, B. R.; Walsh, B. J.; Tyler, M. I.; Traini, M.; Sanchez, J. C.; Hochstrasser, D. F.; Williams, K. L.; Gooley, A. A. Extraction of membrane proteins by differential solubilization for separation using two-dimensional gel electrophoresis. *Electrophoresis* 1998, **19**, 837–844.

26. Herbert, B. Advances in protein solubilisation for two-dimensional electrophoresis. *Electrophoresis* 1999, **20**, 660–663.

27. Dreger, M. Subcellular proteomics. *Mass Spectrom. Rev.* 2003, **22**, 27–56.

28. Chaffin, W. L.; Lopez-Ribot, J. L.; Casanova, M.; Gozalbo, D.; Martinez, J. P. Cell wall and secreted proteins of Candida albicans: Identification, function, and expression. *Microbiol. Mol. Biol. Rev.* 1998, **62**, 130–180.

29. Molloy, M. P.; Herbert, B. R.; Williams, K. L.; Gooley, A. A. Extraction of *Escherichia coli* proteins with organic solvents prior to two-dimensional electrophoresis. *Electrophoresis* 1999, **20**, 701–704.

30. Perdew, G. H.; Schaup, H. W.; Selivonchick, D. P. The use of a zwitterionic detergent in two-dimensional gel electrophoresis of trout liver microsomes. *Anal. Biochem.* 1983, **135**, 453–455.

31. Chevallet, M.; Santoni, V.; Poinas, A.; Rouquie, D.; Fuchs, A.; Kieffer, S.; Rossignol, M.; Lunardi, J.; Garin, J.; Rabilloud, T. New zwitterionic detergents improve the analysis of membrane proteins by two-dimensional electrophoresis. *Electrophoresis* 1998, **19**, 1901–1909.

32. Molloy, M. P.; Herbert, B. R.; Slade, M. B.; Rabilloud, T.; Nouwens, A. S.; Williams, K. L.; Gooley, A. A. Proteomic analysis of the *Escherichia coli* outer membrane. *Eur. J. Biochem.* 2000, **267**, 2871–2881.

33. Swiderek, K. M.; Alpert, A. J.; Heckendorf, A.; Nugent, K.; Patterson, S. D. Structural analysis of proteins and peptides in the presence of detergents: Tricks of the trade. *ABRF News* 1997.

34. Liu, C.; Hofstadler, S. A.; Bresson, J. A.; Udseth, H. R.; Tsukuda, T.; Smith, R. D.; Synder, A. P. On-line dual microdialysis with ESI-MS for direct analysis of complex biological samples and microorganism lysates. *Anal. Chem.* 1998, **70**, 1797–1801.

35. Xiang, F.; Anderson, G. A.; Veenstra, T. D.; Lipton, M. S.; Smith, R. D. Characterization of microorganisms and biomarker development from global ESI-MS/MS analyses of cell lysates. *Anal. Chem.* 2000, **72**, 2475–2481.

36. Li, C.; Chen, Z.; Xiao, Z.; Wu, X.; Zhan, X.; Zhang, X.; Li, M.; Li, J.; Feng, X.; Liang, S.; Chen, P.; Xie, J. Y. Comparative proteomics analysis of human lung squamous carcinoma. *Biochem. Biophys. Res. Comm.* 2003, **309**, 253–260.

37. Kettman, J. R.; Frey, J. R.; Lefkovits, I. Proteome, transcriptome and genome: Top down or bottom up analysis? *Biomol. Eng.* 2001, **18**, 207–212.

38. Hamler, R. L.; Zhu, K.; Buchanan, N. S.; Kreunin, P.; Kachman, M. T.; Miller, F. R.; Lubman, D. M. A two-dimensional liquid-phase separation method coupled with mass spectrometry for proteomic studies of breast cancer and biomarker identification. *Proteomics* 2004, **4**, 562–577.

39. Wu, C. C.; MacCoss, M. J. Shotgun proteomics: tools for the analysis of complex biological systems. *Curr. Opin. Mol. Ther.* 2002, **4**, 242–250.

40. Bodnar, W. M.; Blackburn, R. K.; Krise, J. M.; Moseley, M. A. Exploiting the complementary nature of LC/MALDI/MS/MS and LC/ESI/MS/MS for increased proteome coverage. *J. Am. Soc. Mass. Spectrom.* 2003, **14**, 971–979.

41. McDonald, W. H.; Yates, J. R., 3rd. Shotgun proteomics and biomarker discovery. *Dis. Markers* 2002, **18**, 99–105.

42. Griffin, T. J.; Han, D. K.; Gygi, S. P.; Rist, B.; Lee, H.; Aebersold, R.; Parker, K. C. Toward a high-throughput approach to quantitative proteomic analysis: Expression-dependent protein identification by mass spectrometry. *J. Am. Soc. Mass. Spectrom.* 2001, **12**, 1238–1246.

43. Gygi, S. P.; Rist, B.; Gerber, S. A.; Turecek, F.; Gelb, M. H.; Aebersold, R. Quantitative analysis of complex protein mixtures using isotope-coded affinity tags. *Nat. Biotechnol.* 1999, **17**, 994–999.

44. Chakraborty, A.; Regnier, F. E. Global internal standard technology for comparative proteomics. *J. Chromatogr. A* 2002, **949**, 173–184.

45. Ji, J.; Chakraborty, A.; Geng, M.; Zhang, X.; Amini, A.; Bina, M.; Regnier, F. Strategy for qualitative and quantitative analysis in proteomics based on signature peptides. *J. Chromatogr. B Biomed. Sci. Appl.* 2000, **745**, 197–210.

46. Regnier, F. E.; Riggs, L.; Zhang, R.; Xiong, L.; Liu, P.; Chakraborty, A.; Seeley, E.; Sioma, C.; Thompson, R. A. Comparative proteomics based on stable isotope labeling and affinity selection. *J. Mass. Spectrom.* 2002, **37**, 133–145.

47. Apffel, A.; Fischer, S.; Goldberg, G.; Goodley, P. C.; Kuhlmann, F. E. Enhanced sensitivity for peptide mapping with electrospray liquid chromatography-mass spectrometry in the presence of signal suppression due to trifluoroacetic acid-containing mobile phases. *J. Chromatogr. A* 1995, **712**, 177–190.

48. Opiteck, G. J.; Lewis, K. C.; Jorgenson, J. W.; Anderegg, R. J. Comprehensive on-line LC/LC/MS of proteins. *Anal. Chem.* 1997, **69**, 1518–1524.

49. Opiteck, G. J.; Jorgenson, J. W. Two-dimensional SEC/RPLC coupled to mass spectrometry for the analysis of peptides. *Anal. Chem.* 1997, **69**, 2283–2291.

50. Kachman, M. T.; Wang, H.; Schwartz, D. R.; Cho, K. R.; Lubman, D. M. A 2-D liquid separations/mass mapping method for interlysate comparison of ovarian cancers. *Anal. Chem.* 2002, **74**, 1779–1791.

51. Wall, D. B.; Kachman, M. T.; Gong, S.; Hinderer, R.; Parus, S.; Misek, D. E.; Hanash, S. M.; Lubman, D. M. Isoelectric focusing nonporous RP HPLC: A two-dimensional liquid-phase separation method for mapping of cellular proteins with identification using MALDI-TOF mass spectrometry. *Anal. Chem.* 2000, **72**, 1099–1111.

52. Zhu, Y.; Lubman, D. M. Narrow-band fractionation of proteins from whole cell lysates using isoelectric membrane focusing and nonporous reversed-phase separations. *Electrophoresis* 2004, **25**, 949–958.

53. Balogh, M. P. Debating resolution and mass accuracy in mass spectrometry. *Spectroscopy* 2004, **19**, 34–40.

54. Williams, T. L.; Callahan, J. H.; Monday, S. R.; Feng, P. C.; Musser, S. M. Relative quantitation of intact proteins of bacterial cell extracts using coextracted proteins as internal standards. *Anal. Chem.* 2004, **76**, 1002–1007.

55. Clauser, K. R.; Baker, P.; Burlingame, A. L. Role of accurate mass measurement (±10 ppm) in protein identification strategies employing MS or MS/MS and database searching. *Anal. Chem.* 1999, **71**, 2871–2882.

56. Strader, M. B.; Verberkmoes, N. C.; Tabb, D. L.; Connelly, H. M.; Barton, J. W.; Bruce, B. D.; Pelletier, D. A.; Davison, B. H.; Hettich, R. L.; Larimer, F. W.; Hurst, G. B. Characterization of the 70S Ribosome from *Rhodopseudomonas palustris* using an integrated "top-down" and "bottom-up" mass spectrometric approach. *J. Proteome Res.* 2004, **3**, 965–978.

57. Wittmann, H. G. Components of bacterial ribosomes. *Annu. Rev. Biochem.* 1982, **51**, 155–183.

58. Giglione, C.; Boularot, A.; Meinnel, T. Protein *N*-terminal methionine excision. *Cell. Mol. Life Sci.* 2004, **61**, 1455–1474.

59. Giglione, C.; Vallon, O.; Meinnel, T. Control of protein life-span by *N*-terminal methionine excision. *Embo. J.* 2003, **22**, 13–23.

60. Williams, T. L.; Musser, S. M.; Nordstrom, J. L.; DePaola, A.; Monday, S. R. Identification of a protein biomarker unique to the pandemic O3:K6 clone of *Vibrio parahaemolyticus*. *J. Clin. Microbiol.* 2004.

61. Nazarowec-White, M.; Farber, J. M. *Enterobacter sakazakii*: A review. *Int. J. Food. Microbiol.* 1997, **34**, 103–113.

62. van Acker, J.; de Smet, F.; Muyldermans, G.; Bougatef, A.; Naessens, A.; Lauwers, S. Outbreak of necrotizing enterocolitis associated with *Enterobacter sakazakii* in powdered milk formula. *J. Clin. Microbiol.* 2001, **39**, 293–297.

63. Biering, G.; Karlsson, S.; Clark, N. C.; Jonsdottir, K. E.; Ludvigsson, P.; Steingrimsson, O. Three cases of neonatal meningitis caused by *Enterobacter sakazakii* in powdered milk. *J. Clin. Microbiol.* 1989, **27**, 2054–2056.

64. Simmons, B. P.; Gelfand, M. S.; Haas, M.; Metts, L.; Ferguson, J. *Enterobacter sakazakii* infections in neonates associated with intrinsic contamination of a powdered infant formula. *Infect. Control Hosp. Epidemiol.* 1989, **10**, 398–401.

65. Muytjens, H. L.; Kollee, L. A. Enterobacter sakazakii meningitis in neonates: Causative role of formula? *Pediatr. Infect. Dis. J.* 1990, **9**, 372–373.

66. Jaspar, A. H.; Muytjens, H. L.; Kollee, L. A. [Neonatal meningitis caused by *Enterobacter sakazakii*: Milk powder is not sterile and bacteria like milk too!] *Tijdschr. Kindergeneeskd* 1990, **58**, 151–155.

67. Kindle, G.; Busse, A.; Kampa, D.; Meyer-Konig, U.; Daschner, F. D. Killing activity of microwaves in milk. *J. Hosp. Infect.* 1996, **33**, 273–278.

68. Edelson-Mammel, S. G.; Buchanan, R. L. Thermal inactivation of *Enterobacter sakazakii* in rehydrated infant formula. *J. Food. Prot.* 2004, **67**, 60–63.

69. Nazarowec-White, M.; Farber, J. M. Thermal resistance of *Enterobacter sakazakii* in reconstituted dried-infant formula. *Lett. Appl. Microbiol.* 1997, **24**, 9–13.

70. Nazarowec-White, M.; McKellar, R. C.; Piyasena, P. Predictive modelling of *Enterobacter sakazakii* inactivation in bovine milk during high-termperature short-time pasteurization. *Food. Res. Int.* 1999, **32**, 375–379.

71. Govorukhina, N. I.; Kletsova, L. V.; Tsygankov, Y. D.; Trotsenko, Y. A.; Netrusov, A. I. Characteristics of a new obligate methylotroph. *Mikrobiologiya* (in Russian) 1987, **56**, 849–854.

72. Tsygankov, Y. D.; Kazakova, S. M.; Serebrijski, I. G. Genetic mapping of the obligate methylotroph *Methylobacillus flagellatum*: Characteristics of prime plasmids and mapping of the chromosome in time-of-entry units. *J. Bacteriol.* 1990, **172**, 2747–2754.

11

HIGH-THROUGHPUT MICROBIAL CHARACTERIZATIONS USING ELECTROSPRAY IONIZATION MASS SPECTROMETRY AND ITS ROLE IN FUNCTIONAL GENOMICS

Seetharaman Vaidyanathan and Royston Goodacre
School of Chemistry, The University of Manchester, Manchester M60 1QD

11.1 INTRODUCTION

Postgenome science is driven by the need to evaluate functional aspects of genes and gene products for cellular characterization. Understanding the link between genomic information (which is increasingly becoming available for several organisms) and cellular activities is important in appreciating cellular function or dysfunction. Several processes mediate cellular activities, often in parallel and in short time scales. Consequently there is a paradigm shift in the emphasis of biochemical research from the traditional reductionist approach of evaluating cellular processes individually, and often independent of each other, toward a more global approach of analysing cellular compositions in parallel and in its entirety in order to get a holistic picture. While the genetic makeup of a cell is helpful in characterizing an organism to some extent, there are questions that remain unanswered by analyzing at the genetic level alone, the physiological response of a cell to its environment being the most obvious

Identification of Microorganisms by Mass Spectrometry, Edited by Charles L. Wilkins and Jackson O. Lay, Jr.

case in point. In such instances analysis at the level of gene products, such as mRNAs, proteins, and metabolites could be of greater relevance[1] as we strive to understand and perhaps even define the organism's phenotype.

Developments in analytical techniques are progressing to enable simultaneous high-throughput measurements of several analytes, at the level of the transcripts (transcriptomics), proteins (proteomics), and metabolites (metabolomics). Mass spectrometry (MS) offers valuable potential for such analyses, especially in analyzing proteomes and metabolomes. Electrospray (ES) as a method of generating gas-phase ions for mass spectrometric investigations has sprung into life over the last decade or so, thanks to the demonstration of its utility in detecting high molecular weight cellular components, such as proteins.[2] Its development parallels that of its contemporary, matrix-assisted laser desorption ionization (MALDI), and both methods of analysis occupy center-stage in the functional genomics arena. The rationale of characterizing microbes beyond the genomic level, and the characteristics of ESMS and its application in the analysis of whole cells and crude cell extracts for high-throughput functional genomic investigations are discussed below.

11.2 MICROBIAL CHARACTERIZATIONS BEYOND THE GENOMIC LEVEL: FUNCTIONAL GENOMICS

Microbial characterizations serve the purpose of identifying microorganisms for taxonomic classification, in medical diagnosis, for ecological purposes, or for detecting biohazards in food and the environment. At a fundamental level it serves in unraveling the functions of orphan genes, understanding cellular biochemistry and physiology with respect to identifying biosynthetic bottlenecks and regulatory networks, for growth media optimization, and in screening for production of bioactive molecules. Depending on the analytical objective, microbial characterizations may involve (1) identification of a specific biomarker or a set of biomarkers, for instance, in screening samples to detect the presence of a known microorganism in a sample matrix, or identify a condition or physiological state of a microorganism; (2) isolation and enrichment of an uncharacterised microorganism of interest from a population before making elaborate biochemical characterizations to look for characteristic fingerprints; or (3) assessment of the biochemical makeup of known microorganisms under sets of conditions to screen for novel metabolites, pathways, and so forth, for instance, in analysing genetic mutants to assign function to "orphan" genes.

Genetic methods based on DNA characterizations have been (and still are) the mainstay of microbial identifications for many years. As there are conserved regions in the genetic makeup of a living organism that can be used for reproducible characterizations, it makes sense to use such genetic elements for identification purposes. Consequently cataloguing the genetic makeup of several microorganisms is underway. Since the advent of DNA sequencing technology in 1977, and the first genomic sequencing of a free-living organism in 1995, that of *Haemophilus influenzae*, a bacterium of relevance to human health and disease,[3] over 150 microbial genomes have been sequenced (http://www.genomesonline.org/), and many more are in progress. The availability of the basic genetic makeup of several microbes is paving the way for comparative assessments between and within microbial species (comparative genomics) to gain insight into evolutionary traits, and to identify potential disease causing genes in pathogens and understand pathogenic basis, for instance.[4] However, DNA sequence information alone does not tell one how cells work, or explain complications resulting from cellular function or dysfunction. It is now known that the total number of human genes does not differ substantially from the number of genes of the nematode worm *Caenorhabditis elegans*, and that there are considerable similarities in the genetic makeup of diverse species, suggesting that contextual combination of gene products confers complexity and diversity to the functional genome. Therefore the goal in genomics cannot be the simple provision of a catalog of all the genes and information about their function but also to generate an understanding of how the components come together to comprise a functional entity such as a cell or organism. Consequently in the "post-genomic" era there is greater emphasis on technologies that would elucidate the functional aspects of genes and gene products. The objectives of microbial characterizations discussed in the previous paragraph are also better served by assessing at the functional level. There is therefore greater emphasis in looking for chemical or biochemical markers that will help in identifying microbial strains and define its physiological state.

Since mRNAs, proteins and metabolites effect the flow of information from gene to function/phenotype (Figure 11.1), "omic" technologies that enable large-scale assessment at the levels of mRNAs (transcriptomics), proteins (proteomics), and metabolites (metabolomics) have emerged, and these form the basis of functional genomics. Initial efforts were expended on technologies that would enable large-scale assessment of gene expression via the transcriptome (the spatial and temporal accumulation patterns of mRNAs). Subsequently array-based methodologies emerged for gene expression monitoring. Nucleic acid arrays produced by the robotic deposition of PCR products, plasmids, or oligonucleotides onto a glass slide or in situ synthesis of oligonucleotides by photolithography have been used. A key advantage of using arrays, especially those that contain probes for tens of thousands of genes, is that a holistic (rather than targeted and potentially biased) view of cellular response can be obtained, as compared to looking for specific genes, without a priori knowledge of which genes or mechanisms are important.[5] These and other tools, such as SAGE (sequential analysis of gene expression), can be used for monitoring differences in gene expression that are responsible for morphological and physiological/phenotypical differences, and can be indicative of cellular response to environmental stimuli and perturbations. However, mRNA is only an intermediate in translating the genetic informa-

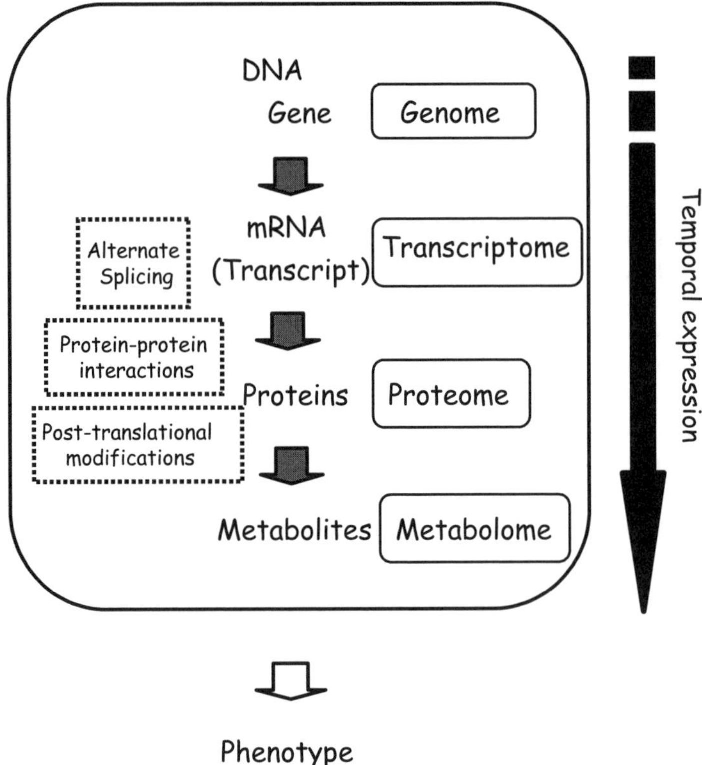

Figure 11.1 Schematic of "omic" expression, showing some prominent events that accompany the respective "omic" component of the cell.

tion to cellular response and function, the business end of which is enacted by proteins and metabolites (Figure 11.1). While the genome sequence can give an idea of which proteins the cell has a potential to make, and the mRNA profiles can indicate which proteins are being produced, analyses beyond the genomic and transcriptomic levels are clearly required to gain information on what the expressed proteins do to carry out cellular function at the molecular level and how metabolites are involved in effecting cellular function. Changes in the temporal expression and accumulation patterns of these latter entities would therefore throw more light on the phenotypic responses of the cell.

Several observations lend credence to the rationale of monitoring cellular activities at the level of the proteins (proteomes) and metabolites (metabolomes). Some are listed below:

- The existence of an ORF does not necessarily imply the existence of a functional gene.

- It is now recognized that the relationship between genes and gene products (i.e., proteins) is not necessarily linear. A given gene can express more than one protein. In fact, in eukaryotes, 6 to 8 proteins are expressed per gene. The human genome itself contains around 40,000 genes, yet the human proteome is predicted to comprise more than 10^6 proteins.[6]
- Expression profiles at the protein level may throw more light on function, than those at the transcript level, as mRNA levels do not necessarily correlate with protein levels.[7]
- Cellular activities are mediated by complex networks of interactions in response to physiological signals, and the nature of the response is dependent on the cell type and states. These aspects cannot be accounted for by investigating at the genomic or transcriptomic levels alone, as genomes and transcriptomes are fairly conserved between cell states and types within a system.
- Several biochemical events occur posttranscriptionally that define the response of cells to stimuli. For instance, alternative splicing, posttranslational modifications, regulation of enzyme activities, distribution of metabolites between cellular compartments, necessitate analysis at the level of the proteome and the metabolome.
- At a molecular level, cellular function is closely associated with activities of the proteins and metabolites. They are instrumental in translating the genetic information to phenotypic function and thus offer an ideal platform for characterizing cellular activities.

Thus there is sufficient ground to cover at the functional level even after successful genetic characterizations that justifies the development of technologies for functional genomics, especially for high-throughput monitoring of proteomes and metabolomes.

11.3 ELECTROSPRAY (IONIZATION) MASS SPECTROMETRY (ESMS)

Strategies based on the application of MS are widely being investigated for functional genomic characterizations, thanks to developments in ionization techniques and improvements with mass analyzers. ES is an ionization technique that has generated considerable interest in functional genomic initiatives. It is a method of generating a fine spray of highly charged droplets that can lead to the release of ions (charged analytes) in the gas phase. The amount of internal energy imparted to generated ions is the lowest of all mass spectrometric ionization techniques,[8] and consequently it is the "softest" method of generating ions in the gas phase to date that is useful for characterizing intact macromolecular assemblies. Initial experimentation on the application of electrospray as a method of generating gas-phase macromolecular ions for mass spectrometric investigations is generally attributed to Malcom Dole and coworkers.[9] However, the demonstration of the capability of electrospray mass spectrometry (ESMS) to analyze large molecular weight compounds such as proteins is credited to Fenn and coworkers,[2] an effort that released the floodgates of biological and biochemical applications of ESMS and won Fenn a Nobel prize in 2002. Less well known is the parallel efforts of (former) Soviet researchers who investigated the phenomenon at the same time.[10]

An electrospray is generally produced by the application of an electric field to a small flow of liquid from a capillary tube toward a counter electrode. The principles of electrospray as applicable to mass spectrometry and the mechanisms involved have been a subject of intense debate over the last decade and have been addressed even before that. This is evident from the discussions in the 2000 issue of the *Journal of Mass Spectrometry* (e.g., Mora[11]), the book by Cole,[12] and several reviews.[8,10,13,14] Here we present a summary encapsulating the relevant observations and direct the readers to the above articles for a more elaborate account.

In ESMS (Figure 11.2), a dilute solution of the analyte is pumped through the capillary at a flow rate of typically 0.1 to 10 μl/min. A high voltage (typically 2–5 kV) is applied to a narrow bore sample injection tube (capillary) relative to its surrounding, which generally includes a counter electrode that is

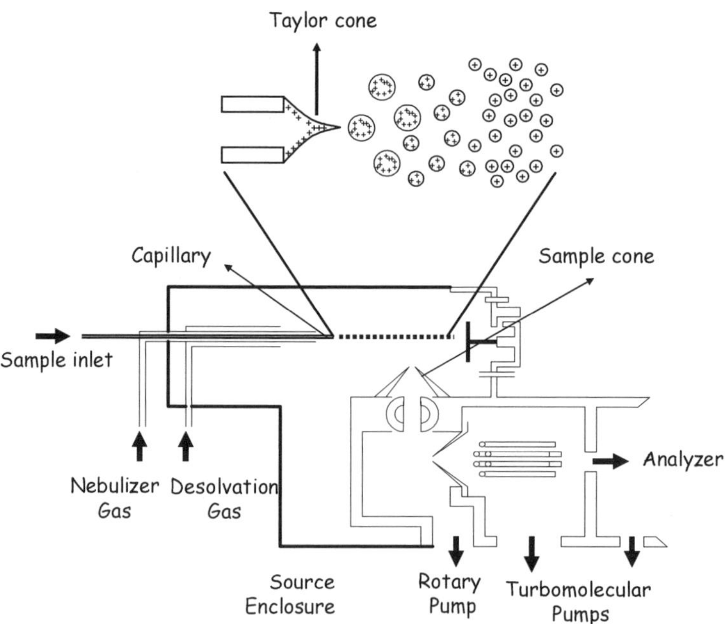

Figure 11.2 A schematic of the electrospray process, showing the release of charged droplets from the Taylor cone and the Z-spray arrangement with respect to the sample inlet, sample cone, and the subsequent path of the ions into the analyzer.

held at ground potential and situated a few centimeters away. The potential difference between the capillary and the counterelectrode creates an intense electric field at the tip of the capillary, resulting in the dispersal of the emerging liquid as a fine spray of charged droplets, usually into the ambient gas at or near atmospheric pressure. The intensity of the electric field is dependent on the potential applied, the radius of the capillary, and the distance between the capillary and the counterelectrode.

As ionic species in the sample solution emerge from the capillary they undergo electrophoretic movement in response to the imposed electric field. When the capillary is not a conducting metal and is made of a dielectric, such as fused silica, the potential is conducted through the liquid to where the liquid is in contact with the source of its potential. The applied voltage can be either positive or negative, depending on the type of analysis, and accordingly cations or anions are respectively analyzed. The high electric potential between the capillary and the counterelectrode give rise to redox reactions at the metal solution interface of the capillary and the counterelectrode. Excess charges accumulate near the end of the capillary and are electrically attracted to the counterelectrode, causing the emerging liquid to elongate in the direction of the counterelectrode. Under the influence of the intense electric field at the capillary tip the emerging liquid forms a so-called Taylor cone. When the solution that comprises the Taylor cone reaches the point where Coulombic repulsion of the surface charge equals the surface tension of the solution—the Rayleigh limit, droplets that contain excess of positive or negative charges detach from the cone tip and are released into the atmosphere. Just prior to reaching the Rayleigh limit the droplets undergo "Coulomb fission," a process that leads to the production of smaller "offsring" droplets. The charged droplets then migrate toward the entrance of the mass spectrometer, releasing ions into the gas phase that enter the mass spectrometer, and are influenced by the electric potential gradient and the pressure differential between the interior and the exterior of the mass spectrometer.

Two principal mechanisms have been put forward to explain the process by which gas-phase ions originate from small highly charged droplets. The charged residue model (CRM), proposed by Dole and coworkers,[15] depicts a scenario in which a series of fission events leads to the formation of small droplets that eventually consist only of single ions. As the last few solvent molecules evaporate, the excess charge present will reside on the sites affording the most stable gas-phase ion. The ion evaporation mechanism (IEM) due to Iribarne and Thomson[16,17] is envisioned in a scenario where direct emission of ions to the gas phase begins to occur when solvent evaporation and Coulomb droplet fissions have reduced the size of the charged droplets to a certain radius (~10–20 nm). In such a case it is assumed that the increased charge density that results from solvent evaporation eventually causes Coulombic repulsion to overcome the liquid's surface tension and releases ions directly from the droplet surface. While there is general consensus that ionic forms of macromolecules, including multiply charged varieties, such as proteins are formed in ESMS by the CRM, it is more likely that small ions are produced by the IEM.[14] Regardless of the mechanism, the process generates vapor phase ions that can be analyzed in the mass spectrometer.

In addition to ions, the electrospray process also generates neutrals and clusters of ions with neutrals. In "conventional" electrospray where sample flow rates are on the order of μl/min, the electrospray process is not very efficient and only a fraction of the charged analytes are transferred to the gas phase. However, such flow rates enable high-throughput autosampling[18] and the online combination of ESMS with analyte separation techniques such as liquid chromatography (LC). In order to minimize formation of clusters and the entry of neutrals into the mass spectrometer and to facilitate desolvation of the charged droplets at high flow rates, a co-axial flow of a neutral gas around the needle tip is usually resorted to. Heating the capillary and/or the desolvation gas can also be used to assist the desolvation process. In addition an off-axis positioning of the capillary assists in maximizing the entry of desolvated analytes into the mass spectrometer over unevaporated droplets. In one configuration the capillary is positioned orthogonal to the mass spectrometer entrance, as is the sampling and the extraction cones, in a Z-spray arrangement (Figure 11.2) that is even better in minimizing the entry of neutrals into the mass analyzer.

Improved ion-source efficiency is achieved by reducing the sample flow rate to the order of a few nl/min. The technique termed nanospray[19] allows improved sensitivity to be achieved, resulting in the detection of analytes at amol levels. Typically such flow rates are achieved by allowing the electrospray process to pull the solution to the capillary tip at the appropriate flow rate. Since the flow is driven by the electrospray process itself and does not require the assistance of temperature or co-axial gas flows to assist in the desolvation process, the process can be more efficient. Smaller droplets characterized by higher surface-to-volume ratios permit higher desorption efficiencies of charged analyte molecules, and the smaller aperture capillaries employed (1–2μm inner diameter) allow narrower dispersion of sprayed droplets; hence sample transfer into the mass spectrometer is more efficient. Furthermore at such low flow rates the sample consumption is minimal, and spectral acquisition can be carried over extended time periods, leading to improved signal-to-noise ratios by signal averaging over extended acquisitions. Nanospray is also more tolerant to salt concentrations than conventional electrospray, as the charged droplets are promptly decomposed to smaller droplets from which the ions are released, leaving less chance for any influence of increase in salt concentration due to solvent evaporation.[20]

A characteristic feature of ESMS is the detection of multiply charged analytes. Macromolecules, such as proteins have multiple sites where protonation or deprotonation (the two most common charge inducing mechanisms in electrospray—other routes to charge induction include, ionization through adduct formation, through gas-phase reactions, and through electrochemical oxidation or reduction) occur. These are desorbed effectively in ESMS and detected, allowing for high molecular weight analytes to be detected at low mass-to-charge values (m/z). Several factors influence the magnitude and detectability of signals in ESMS,[21] including analyte characteristics such as concentration and nature of analyte, solvent characteristics such as pH and ionic strength, and instrumental parameters. Differences in analyte structure can induce variations in ESMS response among charged species. For instance, the maximum positive charge state observed in the ES mass spectra for peptides and proteins has been correlated to the number of basic amino acid residues.[22] It is generally true that analytes with high surface activity (high affinity for the surface of the ES droplet) have higher ESMS response. Accordingly Enke and coworkers[23] observed that peptides with more extensive non-polar regions have better ESMS response than those with polar residues and that for simple singly charged anaytes, a relationship between ESMS response and HPLC retention times exists.[24] With proteins, solvent and pH-induced unfolding can determine the charge state distribution,[25–27] as will inter- and intramolecular interactions in the proteins.[28] Apart from solution-based characteristics, gas-phase processes significantly influence the maximum charge state detected, the charge state distribution, and the intensity of analyte signals.[29]

Different mass analysers can be combined with the electrospray ionization source to effect analysis. These include magnetic sector analysers, quadrupole filter (Q), quadrupole ion trap (QIT), time of flight (TOF), and more recently the Fourrier transform ion cyclotron resonance (FTICR) mass analysers. Tandem mass spectrometry can also be effected by combining one or more mass analysers in tandem, as in a triple quadrupole or a QTOF. The first analyzer is usually used as a mass filter to select parent ions that can be fragmented and analyzed by subsequent analysers.

11.4 ESMS OF MICROBES

Mass spectrometry has been in the realms of chemical and biochemical analysis of low molecular weight compounds for several years, albeit usually coupled to analyte separation techniques, such as gas chromatography (GC) or LC. The objective in most of its traditional applications has been the targeted analysis of small molecular weight cellular components after appropriate sample preparation and cleanup operations. Electron impact or chemical ionization techniques served the purpose in such analyses. The idea of using signals registered from chemical/biochemical species of microbial origin as biomarkers that are unique and representative of individual microorganisms was mooted in 1975 when it was demonstrated that mass spectra characteristic of different species of gently heated bacteria could be obtained.[30] The electron impact spectra contained biomarker ions assigned to volatile metabolites. Subsequently Fenselau and coworkers studied other emerging ionization techniques, such as fast-atom bombardment (FAB), laser desorption, and plasma desorption, for the purpose and demonstrated the detection of lipids and other metabolites that were already known to be biomarkers for bacteria. Much of the work in the area at that time was summarized in a book edited by Fenselau in 1994.[31] The use of pyrolysis mass spectrometry with the combination of pattern recognition techniques has also been the basis for bacterial identifications, as discussed in Chapter 15 by Goodacre and Timmins. However, the highest m/z value reproducibly attainable in the commonest pyrolysis MS instruments is very small (typically $< m/z$ 200). Moreover, due to the in vacuo thermal degradation step, essentially all information on the structure or identity of the molecules producing the pyrolysate is lost,[32] unless the metabolite is thermally stable and of very low molecular weight.[33]

With the advent of "soft" ionization techniques, such as MALDI and ES in the late 1980s, the application of MALDI MS and ESMS for the detection of larger molecular weight compounds as biomarkers suitable for microbial characterizations took a leap in the mid- to late 1990s.[10,34–38] Since most microbial components have now become amenable to mass analysis with these gentler ionization techniques, analysis of whole cells or crude cell extracts using a mass spectrometer, without the use of sample cleanup operations in a nontargeted manner to obtain information rich spectra, became practical. Short and simple analytical protocols can be useful for such purposes as routine monitoring of biohazards in the food and the environment, including the detection of biological warfare agents and for preliminary screening in medical diagnosis, if they can still provide the required information. In this regard microbial fingerprinting using mass spectrometry of crude cell extracts and whole cells offers potential. Little sample preparation is involved in such methods. Nevertheless, the information provided can be sufficient for discrimination purposes. With appropriate validation such information can be invaluable as is or as a preliminary screening tool prior to more extensive investigations in functional genomics.

ESMS has been shown to be a valuable tool for the reproducible analysis of diverse analytes, such as phospholipids,[39,40] glycolipids,[41] lipopolysaccharides,[42] lipooligosaccharides,[43] muramic acids,[44] carbohydrates,[45] and proteins,[46,47] from complex biological samples, but in most of these investigations the analysis has been via specific cell fractions or lysates. Moreover these have usually been presented to the ESMS after liquid chromatographic (LC),[40,45,46] capillary electrophoretic (CE)[42] separations, or online microdialysis sample cleanup.[47] Intact phospholipid profiles obtained using LC-ESMS have been shown to be useful for characterizing methanotrophic bacteria.[48] Unlike lipids, which usually constitute less than 10% of bacterial dry cell weight, proteins can constitute over 50% and therefore have a greater potential to yield unique biomarkers that will be useful for microbial characterizations. Indeed, proteomic information is invaluable from a functional genomics viewpoint. The combination of LC separations with the gentle ionization of ESMS has been shown to be useful in detecting several proteins from bacterial cell lysates.[46,49,50] LC-ESMS profiles of peptides from global protein digests[51] can be put to use for microbial characterizations. Even without the application of analyte separation steps, it has been possible to detect muramic acid,[44] glycolipids,[41] and phospholipids[52] by direct infusion of unfractionated bacterial cell extracts that can be used as biomarkers for microbial characterizations.

The possibility of detecting signals from noncovalent complexes of small molecule ligands bound to proteins and of multiprotein complexes in ESMS has prompted researchers to investigate the feasibility of analyzing "intact" ribosomes[53] and viral particles[54,55] using ESMS. Robinson and coworkers[53] were able to detect *E. coli* 30S and 50S subunits, the earlier containing 21 proteins and the latter 33 proteins with masses in the range of 1 to 1.5 MDa including the associated RNAs. They were also able to detect the "intact" 70S particle, and were able to demonstrate that component proteins, even in the absence of cross-linking, do not dissociate from the 70S ribosome under the mass spectrometric conditions used. However, controlled dissociation of the 50S subunits using collision energies of varying intensities enabled the researchers to identify the subunit components. ESMS in conjunction with MALDI MS has been used to characterize viral subunits, to probe viral dynamics, and even characterize "intact" viruses.[55] Transmission electron microscopy of the electrosprayed viral ions collected on a brass plate revealed that the examined viruses retained their shape and were even viable (as found on infecting tobacco plants), demonstrating that native biomolecular structures can be conserved through the ESMS process.

ESMS is well suited for analyzing liquid samples, and can be employed for detecting and characterizing microorganisms present in liquid matrices, in their native state. For instance, it could be useful as a method of directly detecting and identifying bacteria in diseased body fluids for medical diagnosis. Besides, such a method would be useful as a preliminary screening tool in high-throughput rapid characterizations for functional genomic applications. Toward the development of such a tool, direct infusion of whole bacterial cell suspensions in ESMS (DIESMS) for rapid microbial characterizations was investigated.[56] In order to minimize blockage of the capillary, dilute cell suspensions with relatively high flow rates (μl/min) were employed in a pneumatically assisted electrospray process. Both positive and negative ion modes were found to yield spectra with the potential for microbial characterizations.[56]

Subsequently 36 strains of aerobic endospore-forming bacteria, consisting of six *Bacillus* species and one *Brevibacillus* species could be discriminated using cluster analysis of ESMS spectra acquired in the positive ion mode (m/z 200–2000).[57] The analysis was carried out on harvested, washed bacterial cells suspended in aqueous acidic acetonitrile. The cell suspensions were infused directly into the ionization chamber of the mass spectrometer (LCT, Micromass) using a syringe pump. Replicates of the experiment were performed over a period of six months to randomize variations in the measurements due to possible confounding factors such as instrumental drift. Principal components analysis (PCA) was used to reduce the dimensionality of the data, followed by discriminant function analysis (DFA) of the PC scores to generate class-based clusters that were then classified by hierarchical cluster analysis (HCA) based on Euclidian distances of the DF clusters. The species level discrimination that resulted had similarities to results obtained from biochemical tests[58–60] and 16S rDNA sequence analysis[61] but were much more rapidly effected. Even isolates that belonged to the same species could be discriminated using the technique when the spectra of seven *B. subtilis* strains were analyzed in a separate experiment. It was thus demonstrated that the spectral information obtained is sufficient for the discrimination of bacteria even at subspecies level (Figure 11.3).

Fungal characterizations have also been carried out using ESMS, where ESMS of cellular fractions containing metabolites,[62,63] lipids,[64] and proteins[65] have been shown to be useful. ESMS profiles of crude ethylacetate/methanol/chloroform fungal extracts have been shown to yield secondary metabolite signals, even without the application of any prior fractionation step, providing useful information for taxonomy and secondary metabolite profiling.[37]

11.4.1 Direct Infusion ESMS of Cell Extracts in the Flow Injection Mode (FIESMS)

The spectral information of whole cells and cell-free supernatants were found to be similar when the ESMS spectra of crude bacterial extracts were examined for discriminatory purposes.[57] It was also found that whole cells and cell clusters tend to block the thin electrospray capillary, when multiple measurements were attempted. Consequently the possibility of using cell-free supernatants in an automated mode via FIESMS was investigated to facilitate high-throughput analysis. Such analysis has been proven useful in fungal characterizations.[37] In bacterial investigations,[18] five bacterial strains (two *E. coli*, two *Bacillus* spp., and one *Br. laterosporus*) were axenically grown, harvested, washed, and resuspended in an acidic organic solvent. The cells were centrifuged, and the cell-free extracts were sequentially injected into a solvent flow stream that was sprayed into the ionisation chamber of the ESMS. The spectra produced contained reproducible information that was useful for discrimination purposes. Bacterial extracts stored under different conditions gave very similar mass spectra for each of the five bacterial strains, indicating that the extracts were stable even at room temperatures for up to 24 hours without loss of information content, and this has obvious implications for automated high-throughput analysis. An analysis of the components of the extracting solvent mixture and their effects on the spectral information showed that acetonitrile contributes most significantly to the extraction process and hence to the information content of the spectra. This approach has potential application within high-throughput screening (HTS) programs for microbial characterisations (as well as in the analysis of plant and animal systems), since it requires little sample preparation, is automated, and is rapid.

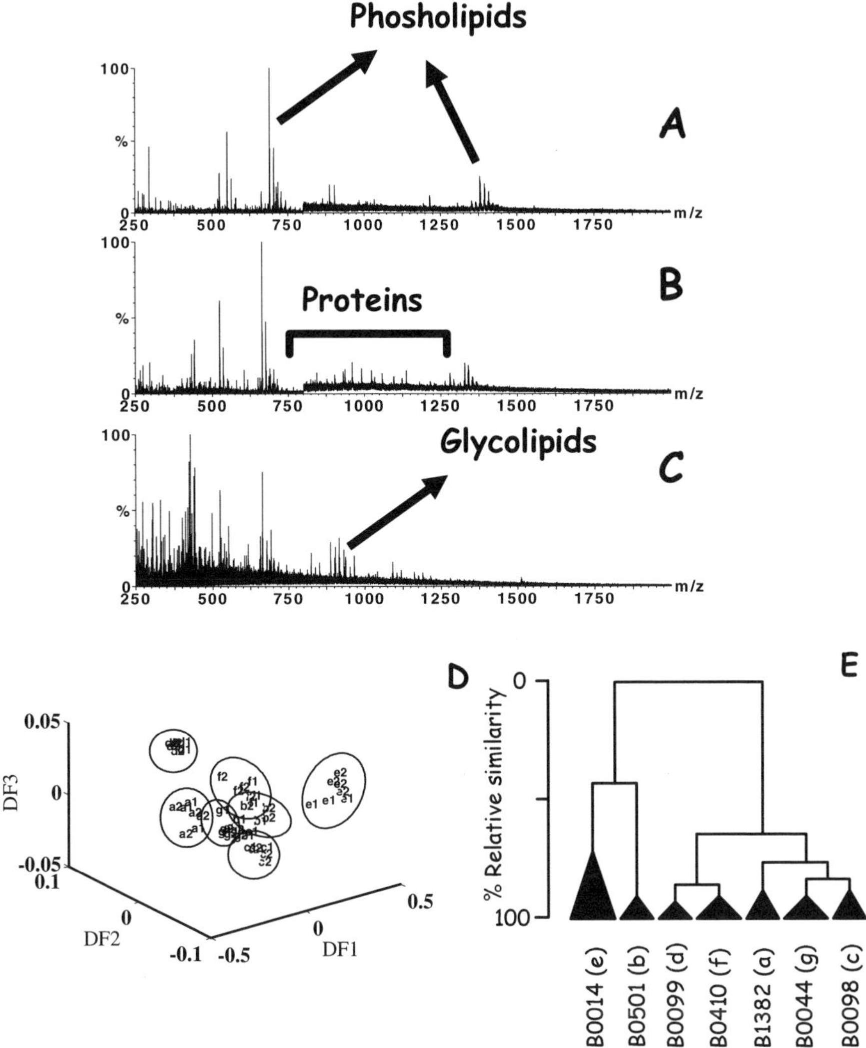

Figure 11.3 Positive ion FIESMS spectra of crude cell extracts from *Escherichia coli* HB101 (*A*), *Bacillus sphaericus* DSM 28 (*B*), and *Bacillus licheniformis* NTCC 10341 (*C*). (*D*) A pseudo-3D plot of the first three discriminant functions (DF1–3) obtained from positive ion whole-cell DIESMS spectra of seven *Bacillus subtilis* strains (*a–g*); (*E*) the corresponding abridged dendrogram obtained from the same information as in *D*. (Adopted from Vaidyanathan et al.[57])

11.4.2 Biochemical Basis of the Mass Spectral Information in Bacterial ESMS

The biochemical basis for the most prominent peaks in the direct infusion ESMS spectra of the whole cells and crude extracts has been studied using tandem mass spectrometry.[18,57] In these studies three classes of macromolecules, namely phospholipids, glycolipids, and proteins were found to contribute most to the bacterial mass spectral information (Figure 11.3). In most cases phospholipid peaks were found to dominate the spectral information. The dominance of phospholipids is understandable from the nature of the analyte. Phospholipids are highly surface active. They are composed of long chains of fatty acids that give them a hydrophobic character and a relatively small polar head group that is hydrophilic in nature. Under the electrospray ionization conditions charge is stabilized in such analytes, as has been demonstrated by Enke and coworkers,[13,23,24,66] who show that analytes with high surface activity are more likely to sequester available charge. Although phospholipids have been detected in direct infusion ESMS,[39,48,52,64] this has been in extractions using nonpolar solvents, usually chlororfom/methanol mixtures. Their spectral dominance, even in a polar solvent such as aqueous acidic acetonitrile,[18] is remarkable and can be attributed to the amphipathic nature of the analyte, which possibly helps in forming stable miscelles that in polar solvents have their hydrophobic regions folded inward with the polar head groups on the surface for sequestering available charge, under the electrospray conditions (Figure 11.4). Peaks that are possibly due to lipopeptides were also detected in *B. amyloliquifaciens*.[57]

At the low mass end (<500 *m*/*z*) several peaks that correspond to metabolites could be detected that were shown to be of discriminatory value.[57] Fungal investigations[37,63,64] have also shown the utility of direct ESMS on crude cell extracts in detecting secondary metabolites and producing mass profile useful for fungal characterizations. These studies suggest that under appropriate conditions, it is possible to obtain metabolic information from ESMS of cell extracts with minimal sample preparations that will be useful for high-throughput metabolomic investigations. In addition a charge distribution of peaks in the *m*/*z* 750 to 1250 region was observed for *B. sphaericus*.[57] The "base" MaxEnt mass underlying this distribution was found to be 32 kDa. MALDI MS of the bacterial cells also showed a peak at 32 kDa, as did SDS-PAGE analysis of the cell extracts. SwissProt searches indicated that this was an *S*-layer protein, which is found in *B. sphaericus*.[67] In order to examine the role of the cell surface in generating spectral patterns, crude cell extracts from 20 strains that have previously been identified to belong to *B. sphaericus*[68–70] were analyzed by FIESMS. As observed previously *S*-layer proteins were found in the spectra. These were different depending on the *B. sphaericus* strain analyzed, and thus suggesting the utility of the technique in proteomic investigations. Optimizing the extraction conditions to counter the dominance of phospholipids signals in the bacterial spectral information resulted in the

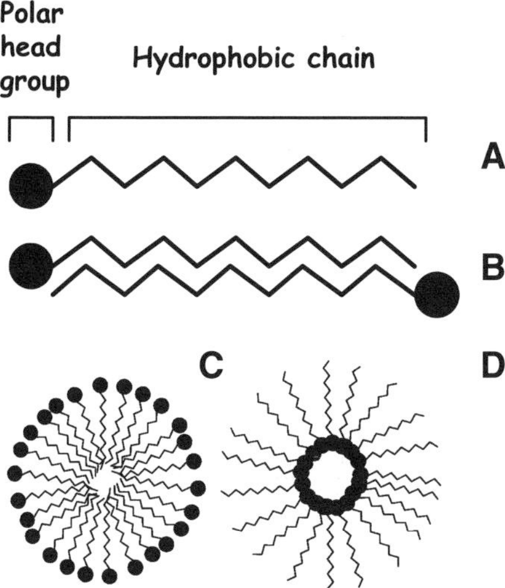

Figure 11.4 Schematic of a phospholipids unit (*A*) showing the polar head group that holds the charge in the electrospray; (*B*) a phospholipids bilayer that is the possible origin of the signals at around 1400 m/z in the DIESMS spectra of bacteria; (Figure 11.3) (*C*) the micellar arrangement that is stabilized in a polar solvent; (*D*) the reverse micellar arrangement that is likely to be encountered in nonpolar environments.

detection of different proteins in *E. coli* K12, under different extraction conditions (vide infra), and under one condition the MaxEnt of the observed charge state distributions showed the presence of more than four proteins in the mass range 5 to 30 kDa (Figure 11.5). It is noteworthy that we were able to observe peaks attributable to proteins, even from such crude preparations of the cells. Moreover it was observed that these reproducible signals differed between bacterial species, such that they can be used as chemotaxonomic biomarkers for proteomic characterization purposes. Investigations are underway to assess the utility of such information in bacterial characterizations with respect to its physiological status.

11.4.3 Conditions for Rapid Extractions

The suspension of microbial cells in a solvent such as aqueous acidic acetonitrile is a procedure used routinely in "soft" ionization mass spectrometric investigations of microorganisms. This is particularly the case in MALDI-MS studies where whole-cell suspensions have been analyzed directly without separating the cellular residue. By contrast, ESMS is usually carried out with cell-free supernatants after analyte separation by LC. Some workers[71] report that partial lysis of the cells occurs due to the acidic conditions employed in such techniques and that this results in the release of proteins and peptides from

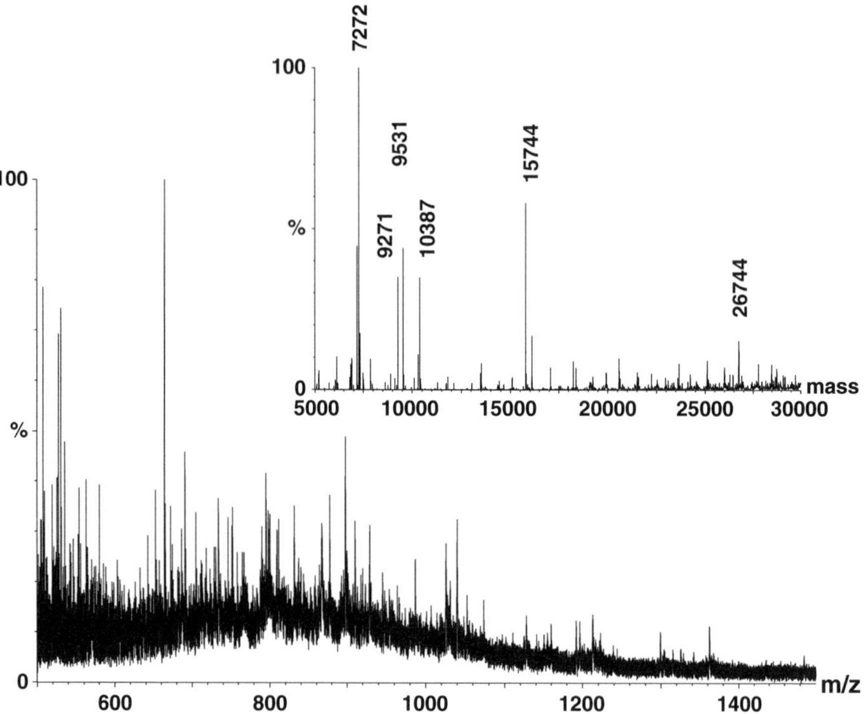

Figure 11.5 Positive ion DIESMS spectra of the crude cell extract of an *E. coli* strain that expresses the GFP protein, showing protein peaks. The inset is the MaxEnt deconvoluted spectrum that shows more than three peaks attributable to proteins from the extract.

the cells that are being detected on MS analysis. Protein database search has revealed that the proteins exuded from such treatments, at least with MALDI MS of *E. coli* K12, can be traced to the cytosol, suggesting cell lysis.[72] Other investigations,[73,74] however, have shown that there is little evidence of cellular damage by lysis, even after such a treatment, and that the signals observed are those of cell wall and/or membrane associated proteins.[35,75] Our own studies on MALDI MS of bacterial cells[76] revealed that although cell lysis during the MALDI process cannot be ruled out, as under SEM a majority of the bacterial cells appeared to be intact, the detection of proteins by SDS-PAGE in the extracts prior to MALDI MS suggests that some lysis occurs when the cells are suspended in the extraction solvent. Although the origins of the spectral contributors in such techniques are inconclusive and remain an issue to be resolved, the approach of suspending the cells in weak acidic solvents offers a more rapid method compared to the laborious, elaborate approach of cell lysis, and it is an attractive option for rapid microbial characterization using mass spectrometry.

Aqueous acidic acetonitrile is an ideal solvent for ESMS; it is known to have good electrospray properties,[2] and has been widely employed in the ESMS research community. In order to assess the influence of the solvent composition on the bacterial mass spectral signal, solvent compositions with different aqueous:organic ratios were investigated at pH ranging from 1 to 10, adjusted using formic acid or ammonia. Four bacterial strains (two *E. coli*, one *B. sphaericus*, and one *B. subtilis*) were studied. It was observed that in the positive ion mode, lower organic content minimizes the phospholipid signals, in relation to others at all pH values tested, and at lower pH values (<4) the protein signals are quite prominent. Higher organic content appears to solubilize phospholipids better, increasing their signal content, and the overall information content is better in the positive than in the negative ion mode. In an attempt to minimize the dominance of phospholipids and to see other species in the spectrum, hexane and Folch (chloroform/methanol) extraction of the crude extracts were performed prior to ESMS. Solvents with different ratios of chloroform, methanol, and formic acid were also tested. Phospholipid signals were suppressed and protein signals were enhanced with Folch extraction. Under one solvent condition more than four protein peaks (in maximum entropy deconvolution (MaxEnt, Micromass, UK)) could be detected for an *E. coli* strain (Figure 11.5).

11.4.4 Influence of Instrumental Parameters

Preliminary investigations on the influence of instrumental parameters on the bacterial ESMS signals showed that the skimmer potential (sample cone voltage) significantly influenced the bacterial mass spectral signals.[57] Skimmer potential is an instrumental parameter that is known to influence ES mass spectra in polymer and protein analyses.[77,78] Whole-cell ESMS of type strains from seven *Bacillus* species were examined at different cone potentials. Variations in the spectral information were observed with respect to its discriminatory utility with different cone potentials, implying, on the one hand, that such parameters need to be optimized to obtain "generally applicable ESMS conditions" and, on the other hand, that such differences could be put to use by combining the information obtained from two or more conditions, such that the composite information is of discriminatory value.

Since proteins form a major component in bacterial cells and were found to contribute to the mass spectral signal of whole cells and crude cell extracts, the influence of the instrumental parameters were investigated at a greater detail for a mixture of five standard proteins.[79] The relative signals of the individual proteins in the mixture were influenced by the instrumental settings, with a defined set of conditions giving signals that had more or less equal responses from all the proteins, and that resulted in the detection of a maximum number of charge states. It was also noted that there were sets of conditions in which one or more of the proteins were preferentially detected, enabling selective detection of proteins at appropriate settings (Figure 11.6).

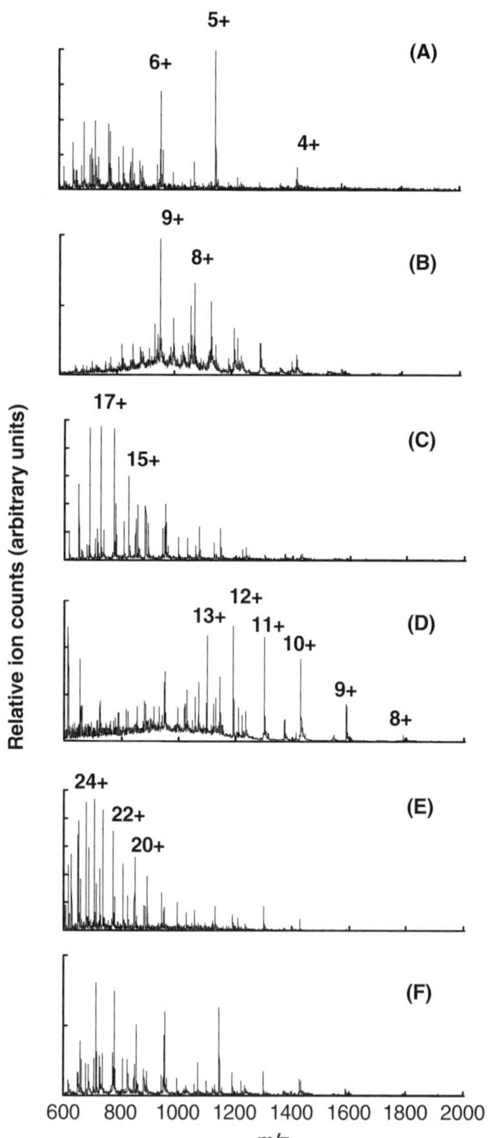

Figure 11.6 Positive ion electrospray mass spectra of an equimolar mixture of five standard proteins, under different instrumental settings, showing cases where prominent signals for the different charge states of (*A*) insulin, (*B*) ubiquitin, (*C*) cytochrome c, (*D*) lysozyme, and (*E*) myoglobin were preferentially observed, and (*F*) where signals for all the proteins were more uniformly detected.

Fourteen instrumental settings were studied, and these were optimized using genetic search methods. It was possible in six generations with a total of <500 experiments out of some 10^{14} to find good combinations of experimental variables (electrospray ionisation mass spectral settings) that yielded a set of conditions giving signals that had more or less equal responses from all the proteins.

11.5 DIRECT INFUSION ESMS OF CRUDE CELL EXTRACTS FOR HIGH-THROUGHPUT CHARACTERIZATIONS—METABOLIC FINGERPRINTING AND FOOTPRINTING

Functional genomics represents a systematic approach to elucidating the function of the novel genes revealed by complete genome sequences.[1] As discussed earlier, it incorporates analysis at the level of the transcriptome, proteome, and the metabolome. As genome sequences of several microbes are revealed, functional annotation of the genes constituting the sequenced genomes follow. Since homologies exist in the genetic information of different biological species (e.g., at least 40% of single-gene determinants of human heritable diseases find homologues in yeast), inferences drawn from one species can be usefully translated in annotating genetic information from others. In this regard, since the genome sequence of yeast is well characterized, it serves as an ideal model system for assessing functional genomic characterizations.[1,80,81] Consequently, as a part of the European Functional Analysis Network (EUROFAN) in which the function of novel *S. cerevisiae* genes revealed by genome sequencing is being systematically analyzed, trascriptomic analysis has revealed or confirmed useful information about the general function of several ORFs.[82–85] Proteomic investigations have also revealed useful information on gene function, and the utility of assessing protein turnover rather than mere analysis of protein accumulation profiles has been discussed with an application in the analysis of abundant proteins in glucose-limited yeast cells grown in aerobic chemostat culture at steady state.[86]

Although transcriptomic and proteomic investigations are at the center stage of current efforts in functional genomics, the utility and scope of metabolomic investigations is nevertheless beginning to appear in a few investigations, more so with model organisms, such as yeast. A large proportion of the 6000 odd ORFs present in the *S. cerevisiae* genome encode proteins of unknown function, for which a library of mutants (each deleted for a protein encoding gene) has been constructed.[80] Many of these mutants are "silent" (i.e., show no overt phenotype for measurement when deleted from the genome) and can be assessed using comparative metabolomic approaches such as functional analysis of co-response in yeast (FANCY), where "metabolic snapshots" from strains deleted for unstudied genes and those deleted for known genes are compared.[87] In such cases the intracellular metabolites, at a standard point in time of the growth phase, provide a "metabolic snapshot" for comparisons. While strategies that help in detailing the intracellular compositions will undoubtedly be useful, they will, with current technology, inevitably involve analyte separation steps that can prolong analysis times and are cumbersome for high-throughput investigations.

Rapid methods such as analysis of crude extracts with NMR, FT-IR, or DIESMS that involve simple sample preparation protocols but nevertheless provide reproducible discriminatory information on intracellular metabolites offer the potential for providing metabolic "fingerprints" that can be put to use for functional genomic characterizations. However, there are several considerations in developing such strategies. Intracellular metabolic reactions can be fast, with rapid turnover of metabolite concentrations necessitating rapid quenching of metabolism prior to analysis, so as to make meaningful observations from such investigations. Changes in the intracellular milieu should be reproducibly detected with the information content robust to variations in analytical factors, such as instrumental drifts. The brief sample preparation protocols or the instrumental measurements should be devoid of analytical artifacts that could result in incidental measurements that do not reflect true changes in the system investigated. Rapid extraction protocols are seldom comprehensive, albeit they can potentially preserve sample information better than more comprehensive analysis, due to shorter analysis times. A corollary of the fact above is that extraction protocols should be such that they preserve analyte information while still being effective in extracting sufficient information. The method of extraction directed at functional genomic investigations should maximize the proportion of analytes representing all functional categories that can be extracted simultaneously. Finally it is often imperative to isolate cells from bulk solutions before carrying out analysis in order to minimize contamination of information from the extracellular milieu due to secreted analytes and media components.

Freeze clamping in liquid nitrogen, treatment with strong acids such as perchloric or nitric acids, dousing in cold organic solvents, and the application of heat are some of the methods employed to quench enzyme activities and extract intracellular metabolites. A comparison of six different methods that included extractions with hot (90°C) ethanol, hot (70°C) methanol, cold (–40°C) methanol, perchloric acid, alkali (KOH), and methanol/chloroform showed that the methods resulted in different profiles and that cold methanol is effective in rapidly quenching metabolism and is ideally suited for investigating intracellular metabolites.[88] The influence of cultivation conditions, sampling, quenching, extraction, concentration, and storage of samples have been investigated for rapid profiling of intracellular metabolites using DIESMS.[89] Since maintaining the pH of the extracting medium at or near the physiological levels is important for preserving the extracted metabolites in their native state, buffers can be used. But to minimize ion suppression effects in ESMS, non–salt based buffers are better suited. In this regard tricine has been shown to be useful. Our own investigations discussed earlier (Section 11.4.4) shows the importance of rapid extraction conditions on the analytical information.

The aforementioned concerns with measuring intracellular metabolites for obtaining metabolic fingerprints have prompted the development of an alternative strategy termed metabolic "footprinting"[90] that involves assessing the extracellular matrix for secreted metabolites. The strategy employs FIESMS to profile extracellular metabolites, and has been shown to be useful in discriminating information from mutant yeast strains. The potential for the application of machine learning techniques, such as genetic programming to extract rules from such profiles that will be useful for functional genomic assignments, was also discussed in this report. In another investigation,[91] FIESMS employed to classify *E. coli* mutants in the tryptophan metabolism according to their metabolic footprints and showed information on the potential for such techniques in high-throughput functional genomic investigations.

11.6 CONCLUSIONS

Strategies for the analysis of proteomic and metabolomic compositions are increasingly becoming relevant for microbial characterizations, especially with respect to functional genomic investigations. Methods based on the analysis and assessment of gene deletion mutants under various conditions necessitate that these strategies be rapid, have high-throughput, and be amenable to automation. DIESMS is one of the techniques that can be a valuable component in the development of such strategies. Electrospray is the softest ionization technique available to date, and is useful in generating gas-phase ions of virtually most macromolecular species within the cell. Coupled with high-resolution mass spectrometry such as FTICR or even moderate resolution analyzers such as quadrupole and TOF, it offers a powerful platform for functional genomic investigations. The sensitivities attainable through nanospray operations enable the assessment of minor changes within biological systems, unparalleled by contemporary techniques. However, as with any modern technique there are limitations and considerations to bear in mind while employing the technique in the development of rapid strategies to extract useful information. The ionization mechanism is not entirely understood and changes in the sample matrix can lead to differences in the ionization profiles of analytes and hence in the mass spectral information due to differential ion suppression effects that cannot always be explained. Salt concentrations can significantly influence spectral information, often leading to the complete masking of analyte signals. Signals due to clusters and adduct formations are not uncommon, and these may reduce the utility of the mass spectral information. Multiple charging of analytes results in the detection of several peaks for a single analyte that albeit increases the accuracy of prediction, leads to redundancy of spectral information. Since it is a mass (and charge) based technique, its application is limited to changes in the system associated with a change in mass-to-charge profiles. While this is not so much of a problem with high molecular weight analytes, low molecular weight metabolites that have the same molecular weight may be difficult to discriminate, requiring tandem mass spectrometric investigations or discrimination based on isotope abundance ratios. Nevertheless, the application of appropriate analytical conditions can result in information-rich spectra that can be used for functional genomic investigations (vide supra). Moreover among contemporary techniques the signal-to-noise with respect to the information content is still high enough for it to merit applications in microbial characterizations for functional genomics. A major advantage of the technique is its ability to generate gas-phase ions from the liquid phase, at atmospheric pressure, which enables the preservation of the native state of the analyte to be investigated. It also enables coupling to sample cleanup or brief separation stages where required to increase the information content. The development of automated strategies such as the robot-scientist[92] will enable reproducible and high-throughput operations involving DIESMS of crude cell extracts practical. The combination of LC or GC separations with ESMS will improve resolution, where DIESMS approaches do not yield sufficient information.

DIESMS complements other approaches, such as (matrix-assisted) laser desorption ionization ((MA)LDI) MS of whole cells, NMR and FT-IR of whole cells and cell-free supernatants within current rapid strategies for proteomic and metabolomic investigations. It is ideally suited for the development of high-throughput strategies for screening samples prior to GC(GC)-MS, LC(LC)-MS, and CE-MS for more elaborate analyses. Although crude cell extracts can provide information of discriminatory value for functional genomics, where possible it will merit analysing different biochemical fractions independently and integrating the information obtained from the different classes in order to obtain a holistic picture. In a more global sense, integration of data from different "omic" levels and from different strategies can provide not only more accurate estimations of the function of novel genes but also a holistic picture of the system under investigation from a "systems" biology perspective.

ACKNOWLEDGMENT

The authors thank BBSRC, UK for financial assistance, Dr. Nial Logan and Prof. Gus Priest for the supply of bacterial strains, and Mr. Jim Heald and Mr. Russel Morphew for technical assistance.

REFERENCES

1. Oliver, S. G. Functional genomics: Lessons from yeast. *Philos. Trans. R. Soc. Lond. B Biol. Sci.* 2002, **357**, 17–23.
2. Fenn, J. B.; Mann, M.; Meng, C. K.; Wong, S. F.; Whitehouse, C. M. Electrospray ionization for mass spectrometry of large biomolecules. *Science* 1989, **246**, 64–71.

3. Fleischmann, R. D.; Adams, M. D.; White, O.; Clayton, R. A.; Kirkness, E. F.; Kerlavage, A. R.; Bult, C. J.; Tomb, J. F.; Dougherty, B. A.; Merrick, J. M.; McKenney, K.; Sutton, G.; Fitzhugh, W.; Fields, C.; Gocayne, J. D.; Scott, J.; Shirley, R.; Liu, L.; Glodek, A.; Kelley, J. M.; Weidman, J. F.; Phillips, C. A.; Spriggs, T.; Hedblom, E.; Cotton, M. D.; Utterback, T. R.; Hanna, M. C.; Nguyen, D. T.; Saudek, D. M.; Brandon, R. C.; Fine, L. D.; Fritschman, J. L.; Fuhrmann, J. L.; Geoghagen, N. S. M.; Gnehm, C. L.; McDonald, L. A.; Small, K. V.; Fraser, C. M.; Smith, H. O.; Venter, J. C. Whole-genome random sequencing and assembly of *Haemophilus-Influenzae* Rd. *Science* 1995, **269**, 496–512.
4. Fraser, C. M.; Eisen, J.; Flischmann, R. D.; Ketchum, K. A.; Peterson, S. Comparative genomics and understanding of microbial biology. *Emerg. Infect. Dis.* 2000, **6**, 505–512.
5. Lockhart, D. J.; Winzeler, E. A. Genomics, gene expression and DNA arrays. *Nature* 2000, **405**, 827–836.
6. Melton, L. Protein arrays: proteomics in multiplex. *Nature* 2004, **429**, 101–107.
7. Gygi, S. P.; Rochon, Y.; Franza, B. R.; Aebersold, R. Correlation between protein and mRNA abundance in yeast. *Mol. Cell Biol.* 1999, **19**, 1720–1730.
8. Cole, R. B. Some tenets pertaining to electrospray ionization mass spectrometry. *J. Mass Spectrom.* 2000, **35**, 763–772.
9. Dole, M.; Cox, H. L.; Gieniec, J. Electrospray mass spectroscopy. *Abstr. Papers Am. Chem. Soc.* 1971, 44–47.
10. Smith, R. D.; Loo, J. A.; Edmonds, C. G.; Barinaga, C. J.; Udseth, H. R. New Developments in biochemical mass spectrometry: Electrospray ionization. *Anal. Chem.* 1990, **62**, 882–899.
11. Mora, J. F.; Van Berkel, G. J.; Enke, C. G.; Cole, R. B.; Martinez-Sanchez, M.; Fenn, J. B. Electrochemical processes in electrospray ionization mass spectrometry. *J. Mass Spectrom.* 2000, **35**, 939–952.
12. Cole, R. B. *Electrospray Ionization Mass Spectrometry: Fundamentals, Instrumentation and Applications*. New York: Wiley, 1997.
13. Cech, N. B.; Krone, J. R.; Enke, C. G. Predicting electrospray response from chromatographic retention time. *Anal. Chem.* 2001, **73**, 208–213.
14. Kebarle, P. A brief overview of the present status of the mechanisms involved in electrospray mass spectrometry. *J. Mass Spectrom.* 2000, **35**, 804–817.
15. Dole, M.; Mack, L. L.; Hines, R. L. Molecular beams of macroions. *J. Chem. Phys.* 1968, **49**, 2240–2249.
16. Irbarne, J. V.; Thomson, B. A. Evaporation of small ions from charge droplets. *J. Chem. Phys.* 1976, **64**, 2287–2294.
17. Thomson, B. A.; Iribarne, J. V. Field-induced ion evaporation from liquid surfaces at atmospheric-pressure. *J. Chem. Phys.* 1979, **71**, 4451–4463.
18. Vaidyanathan, S.; Kell, D. B.; Goodacre, R. Flow-injection electrospray ionization mass spectrometry of crude cell extracts for high-throughput bacterial identification. *J. Am. Soc. Mass Spectrom.* 2002, **13**, 118–128.
19. Wilm, M.; Mann, M. Analytical properties of the nanoelectrospray ion source. *Anal. Chem.* 1996, **68**, 1–8.
20. Juraschek, R.; Dulcks, T.; Karas, M. Nanoelectrospray—More than just a minimized flow electrospray ionization source. *J. Am. Soc. Mass Spectrom.* 1999, **10**, 300–308.

21. Wang, G.; Cole, R. B. In *Electrospray Ionization Mass Spectrometry: Fundamentals Instrumentation and Applications*. Cole, R. B. (Ed.). New York: Wiley, 1997, pp. 137–174.
22. Loo, J. A.; Edmonds, C. G.; Udseth, H. R.; Smith, R. D. Effect of reducing disulfide-containing proteins on electrospray ionization mass-spectra. *Anal. Chem.* 1990, **62**, 693–698.
23. Cech, N. B.; Enke, C. G. Relating electrospray ionization response to nonpolar character of small peptides. *Anal. Chem.* 2000, **72**, 2717–2723.
24. Cech, N. B.; Enke, C. G. Practical implication of some recent studies in electrospray ionization fundamentals. *Mass Spectrom. Rev*. 2001, **20**, 362–387.
25. Chowdhury, S. K.; Katta, V.; Chait, B. T. Probing conformational-changes in proteins by mass spectrometry. *J. Amer. Chem. Soc.* 1990, **112**, 9012–9013.
26. Konermann, L.; Douglas, D. J. Acid-induced unfolding of cytochrome c at different methanol concentrations: Electrospray ionization mass spectrometry specifically monitors changes in the tertiary structure. *Biochemistry* 1997, **36**, 12296–12302.
27. Loo, J. A.; Loo, R. R. O.; Udseth, H. R.; Edmonds, C. G.; Smith, R. D. Solvent-induced conformational-changes of polypeptides probed by electrospray-ionization mass-spectrometry. *Rapid Comm. Mass Spectrom*. 1991, **5**, 101–105.
28. Grandori, R. Origin of the conformation dependence of protein charge-state distributions in electrospray ionization mass spectrometry. *J. Mass Spectrom*. 2003, **38**, 11–15.
29. Amad, M. H.; Cech, N. B.; Jackson, G. S.; Enke, C. G. Importance of gas-phase proton affinities in determining the electrospray ionization response for analytes and solvents. *J. Mass Spectrom*. 2000, **35**, 784–789.
30. Anhalt, J. P.; Fenselau, C. Identification of bacteria using mass-spectrometry. *Anal. Chem*. 1975, **47**, 219–225.
31. Fenselau, C. *Mass Spectrometry for the Characterization of Microorganisms*. Washington, DC: American Chemical Society, 1994.
32. Goodacre, R.; Kell, D. B. Pyrolysis mass spectrometry and its applications in biotechnology. *Curr. Opin. Biotechnol.* 1996, **7**, 20–28.
33. Goodacre, R.; Shann, B.; Gilbert, R. J.; Timmis, E. M.; McGovern, A. C.; Alsberg, B. K.; Kell, D. B.; Logan, N. A. Detection of the dipicolinic acid biomarker in *Bacillus* spores using Curie-point pyrolysis mass spectrometry and fourier transform infrared spectroscopy. *Anal. Chem*. 2000, **72**, 119–127.
34. Claydon, M. A.; Davey, S. N.; Edwards-Jones, V.; Gordon, D. M. The rapid identification of intact microorganisms using mass spectrometry. *Nature Biotech*. 1996, **14**, 1584–1586.
35. Holland, R. D.; Duffy, C. R.; Rafi, F.; Sutherland, J. B.; Heinze, T. M.; Holder, C. L.; Voorhees, K. J.; Lay, J. O. Identification of bacterial proteins observed in MALDITOF mass spectra from whole cells. *Anal. Chem*. 1999, **71**, 3226–3230.
36. Krishnamurthy, T.; Ross, P. L.; Goode, M. T.; Menking, D. L.; Rajamani, U.; Heroux, K. S. Biomolecules and mass spectrometry. *J. Nat. Toxins* 1997, **6**, 121–162.
37. Smedsgaard, J.; Frisvad, J. C. Using direct electrospray mass spectrometry in taxonomy and secondary metabolite profiling of crude fungal extracs. *J. Microbiol. Meth*. 1996, **25**, 5–17.

38. Snyder, A. P. Electrospray: A popular ionization technique for mass spectrometry. *ACS Symp. Ser.* 1996, **619**, 1–20.
39. Black, G. E.; Snyder, A. P.; Heroux, K. S. Chemotaxonomic differentiation between the *Bacillus cereus* group and *Bacillus subtilis* by phospholipid extracts analyzed with electrospray ionization tandem mass spectrometry. *J. Microbiol. Meth.* 1997, **28**, 187–199.
40. Fang, J.; Barcelona, M. J. Structural determination and quantitative analysis of bacterial phospholipids using liquid chromatography electrospray ionization mass spectrometry. *J. Microbiol. Meth.* 1998, **33**, 23–35.
41. Wang, W.; Liu, Z.; Ma, L.; Hao, C.; Liu, S.; Voinov, V. G.; Kalinosvskaya, N. I. Electrospray ionization multiple-stage tandem mass spectrometric analysis of digycosyldiacylglycerol glycolipids from the bacteria *Bacillus pumilus. Rapid Comm. Mass Spectrom.* 1999, **13**, 1189–1196.
42. Li, J.; Thibault, P.; Martin, A.; Richards, J. C.; Wakarchuk, W. W.; der Wilp, W. Development of an on-line preconcentration method for the analysis of pathogenic lipopolysaccharides using capillary electrophoresis-electrospray mass spectrometry—Application to small colony isolates. *J. Chromatogr. A* 1998, **817**, 325–336.
43. Gibson, B. W.; Phillips, N. J.; John, C. M.; Melaugh, W. Lipooligosaccharides in pathogenic *Haemophilus and Neisseria* species—Mass spectrometric techniques for identification and characterization. *ACS Sympo. Ser.* 1994, **541**, 185–202.
44. Black, G. E.; Fox, A.; Fox, K.; Snyder, A. P.; Smith, P. B. Electrospray tandem mass spectrometry for analysis of native muramic acid in whole bacterial cell hydroslysates. *Anal. Chem.* 1994, **66**, 4171–4176.
45. Wunschel, D.; Fox, K. F.; Fox, A.; Nagpal, M. L.; Kim, K.; Steward, G. C.; Shahgholi, M. Quantitative analysis of neutral and acidic sugars in whole bacterial cell hydrolysates using high-performance anion-exchanges liquid chromatography-electrospray ionization tandem mass spectrometry. *J. Chromatogr. A* 1997, **776**, 205–519.
46. Krishnamurthy, T.; Davis, M. T.; Stahl, D. C.; Lee, T. D. Liquid chromatography/microspray mass spectrometry for bacterial investigations. *Rapid Comm. Mass Spectromr.* 1999, **13**, 39–49.
47. Liu, C. L.; Hofstadler, S. A.; Bresson, J. A.; Udseth, H. R.; Tsukuda, T.; Smith, R. D.; Snyder, A. P. On line dual microdialysis with ESI-MS for direct analysis of complex biological samples and microorganism lysates. *Anal. Chem.* 1998, **70**, 1797–1801.
48. Fang, J. S.; Barcelona, M. J.; Semrau, J. D. Characterization of methanotrophic bacteria on the basis of intact phospholipid profiles. *FEMS Microbiol. Lett.* 2000, **189**, 67–72.
49. Dalluge, J. J. Mass spectrometry for direct determination of proteins in cells: Applications in bitechnology and microbiology Fresenius. *J. Anal. Chem.* 2000, **366**, 701–711.
50. Dunlop, K. Y.; Li, L. Automated Mass Analysis of low-molecular-mass bacterial proteome by liquid chromatography-electrospray ionization mass spectrometry. *J. Chromatogr. A* 2001, **925**, 123–132.
51. Zhou, X. H.; Gonnet, G.; Hallett, M.; Munchbach, M.; Folkers, G.; James, P. Cell fingerprinting: An approach to classifying cells according to mass profiles of digests of protein extracts. *Proteomics* 2001, **1**, 683–690.

52. Smith, P. B. W.; Snyder, A. P.; Harden, C. S. Characterization of bacterial phospholipids by electrospray ionization tandem mass spectrometry. *Anal. Chem.* 1995, **67**, 1824–1830.

53. Rostom, A. A.; Fucine, P.; Benjamin, D. R.; Juenemann, R.; Nierhaus, K. H.; Hartl, F. U.; Dodson, C. M.; Robinson, C. V. Detection and selective dissociation of intact ribosomes in a mass spectrometer. *Proc. Nat. Acad. Sci. USA* 2000, **97**, 5185–5190.

54. Despeyroux, D.; Phillpotts, R.; Watts, P. Electrospray mass spectrometry for detection and characterization of purified cricket paralysis virus (CrPV). *Rapid Comm. Mass Spectrom.* 1996, **10**, 937–941.

55. Siuzdak, G. Probing viruses with mass spectrometry. *J. Mass Spectrom.* 1998, **33**, 203–211.

56. Goodacre, R.; Heald, J. K.; Kell, D. B. Characterization of intact microorganisms using electrospray ionization mass spectrometry. *FEMS Microbiol. Lett.* 1999, **176**, 17–24.

57. Vaidyanathan, S.; Rowland, J. J.; Kell, D. B.; Goodacre, R. Discrimination of aerobic endospore-forming bacteria via electrospray-ionization mass spectrometry of whole cell suspensions. *Anal. Chem.* 2001, **73**, 4134–4144.

58. Logan, N. A. *Bacillus* species of medical and veterinary importance. *J. Med. Microbiol.* 1988, **25**, 157–165.

59. Logan, N. A.; Berkeley, R. C. In *The Aeorobic Endospore-Forming Bacteria.* Berkley, R. C. W., Goodfellow, M. (Eds.). San Diego: Academic Press for the Society of General Microbiology, London, 1981, pp. 105–140.

60. Logan, N. A.; Berkeley, R. C. W. Identification of *Bacillus* strains using the API system. *J. Gen. Microbiol.* 1984, **130**, 1871–1882.

61. Lopez-Diez, E. C.; Goodacre, R. Characterization of microorganisms using UV resonance Raman spectroscopy and chemometrics. *Anal. Chem.* 2004, **76**, 585–591.

62. Plattner, R. D.; Weisleder, D.; Poling, S. M. Analytical determination of fumonisims and other metabolites produced by *Fusarium moniliforme* and related species on corn. *Adv. Exp. Med. Biol.* 1996, **392**, 57–64.

63. Prasain, J. K.; Ueki, M.; Stefanowicz, P.; Osada, H. Rapid screening and identification of cytochalasins by electrospray tandem mass spectrometry. *J. Mass Spectrom.* 2002, **37**, 283–291.

64. Schneiter, R.; Brugger, B.; Sandhoff, R.; Zellnig, G.; Leber, A.; Lampl, M.; Athenstaedt, K.; Hrastnik, C.; Eder, S.; Daum, G.; Paltauf, F.; Wieland, F. T.; Kohlwein, S. D. Electrospray ionization tandem mass spectrometry (ESI-MS/MS) analysis of the lipid molecular species composition of yeast subcellular membranes reveals acyl chain-based sorting/remodeling of distinct molecular species en route to the plasma membrane. *J. Cell Biol.* 1999, **146**, 741–754.

65. Figeys, D.; Aebersold, R. High sensitivity identification of proteins by electrospray ionization tandem mass spectrometry: inital comparison between an ion trap mass spectrometer and a triple quadrupole mass spectrometer. *Electrophoresis* 1997, **18**, 360–368.

66. Enke, C. G. A predictive model for matrix and analyte in electrospray ionization of singly-charged ionic analytes. *Anal. Chem.* 1997, **69**, 4885–4893.

67. Ilk, N.; Kosma, P.; Puchberger, M.; Egelseer, E. M.; Mayer, H. F.; Sleytr, U. B.; Sara, M. Structural and functional analyses of the secondary cell wall polymer of *Bacillus sphaericus* CCM 2177 that serves as an S-layer-specific anchor. *J. Bacteriol.* 1999, **181**, 7643–7646.

68. Priest, F. G.; Ebdrup, L.; Zahner, V.; Carter, P. E. Distribution and characterization of mosquitocidal toxin genes in some strains of *Bacillus sphaericus*. *Appl. Environ. Microbiol.* 1997, **63**, 1195–1198.

69. Zahner, V.; Momen, H.; Priest, F. G. Serotype H5a5b is a major clone with mosquito-pathogenic strains of *Bacillus sphaericus*. *Syst. Appl. Microbiol.* 1998, **21**, 162–170.

70. Zahner, V.; Priest, F. G. Distribution of restriction endonucleases among some entomopathogenic strains of *Bacillus sphaericus*. *Letts. Appl. Microbiol.* 1997, **24**, 483–487.

71. Krishnamurthy, T.; Rajamani, U.; Ross, P. L.; Jabhour, R.; Nair, H.; Eng, J.; Yates, J.; Davis, M. T.; Stahl, D. C.; Lee, T. D. Mass spectral investigations on microorganisms. *J. Toxicol. Toxin Rev.* 2000, **19**, 95–117.

72. Ryzhov, V.; Fenselau, C. Characterization of the portein subset desorbed by MALDI from whole bacterial cells. *Anal. Chem.* 2001, **73**, 746–750.

73. Domin, M. A.; Welham, K. J.; Ashton, D. S. The effect of solvent and matrix combinations on the analysis of bacteria by MALDI-TOF MS. *Rapid Comm. Mass Spectrom.* 1999, **13**, 222–226.

74. Welham, K. J.; Domin, M. A.; Scannell, D. E.; Cohen, E.; Ashton, D. S. The characterization of micro-organisms by MALDI-TOFMS. *Rapid Comm. Mass Spectrom.* 1998, **12**, 176–180.

75. Dai, Y.; Li, L.; Roser, D. C.; Long, S. R. Detection and identification of low-mass peptides and proteins from solvent suspensions of *Escherichia coli* by high performance liquid chromatography fractionation and matrix-assisted laser desorption/ionization mass spectrometry. *Rapid Comm. Mass Spectrom.* 1999, **13**, 73–78.

76. Vaidyanathan, S.; Winder, C. L.; Wade, S. C.; Kell, D. B.; Goodacre, R. Sample preparation in matrix-assisted laser desorption ionization mass spectrometry of whole bacterial cells and the detection of high mass (>kDa) proteins. *Rapid Comm. Mass Spectrom.* 2002, **16**, 1276–1286.

77. Hunt, S. M.; Sheil, M. M.; Belov, M.; Derrick, P. J. Probing the effects of cone potential in the electrospray ion source: Consequences for the determination of molecular weight distributions of synthetic polymers. *Anal. Chem.* 1998, **70**, 1812–1822.

78. Voyskner, R. D.; Pack, T. Investigation of collisional-activation decomposition process and spectra in the transport oregion of an electrospray single-quadrupole mass-spectrometer. *Rapid Comm. Mass Spectrom.* 1991, **5**, 263–268.

79. Vaidyanathan, S.; Broadhurst, D. I.; Goodacre, R.; Kell, D. B. Explanatory optimisation of protein mass pectrometry via genetic search. *Anal. Chem.* 2003, **75**, 6679–6686.

80. Delneri, D.; Brancia, F. L.; Oliver, S. G. Towards a truly integrative biology through the functional genomics of yeast. *Curr. Opin. Biotechnol.* 2001, **12**, 87–91.

81. Oliver, S. G.; Winson, M. K.; Kell, D. B.; Baganz, F. Systematic functional analysis of the yeast genome. *Trends Biotechnol.* 1998, **16**, 373–378.

82. El-Moghazy, A. N.; Zhang, N.; Ismail, T.; Wu, J.; Butt, A.; Khan, S. A.; Merlotti, C.; Woodwark, K. C.; Gardner, D. C.; Gaskell, S. J.; Oliver, S. G. Functional analysis of six novel ORGs on the left arm of chromosome XII in *Saccharomyces cerevisiae* reveals two essential genes, one of which is under cell cycle control. *Yeast* 2000, **16**, 277–288.

83. Planta, R. J.; Brown, A. J.; Cadahia, J. L.; Cerdan, M. E.; de Jonge, M.; Gent, M. E.; Hayes, A.; Kolen, C. P.; Lombardia, L. J.; Sefton, M.; Oliver, S. G.; Thevelein, J.; Tournu, H.; van Delft, Y. J.; Verbart, D. J.; Winderickx, J. Transcript analysis of 250 novel yeast genes from chromosome XIV. *Yeast* 1999, **15**, 329–350.

84. Zhang, N.; Ismail, T.; Wu, J.; Woodwark, K. C.; Gardner, D. C.; Walmsley, R. M.; Oliver, S. G. Disruption of six novel ORGs on the left arm of chromosome XII reveals one gene essential for vegetative growth of *Saccharomyces cerevisiae*. *Yeast* 1999, **15**, 1287–1296.

85. Zhang, N.; Merlotti, C.; Wu, J.; Ismail, T.; El-Moghazy, A. N.; Khan, S. A.; Butt, A.; Gardner, D. C.; Sims, P. F.; Oliver, S. G. Functional Analysis of six novel ORFs on the left arm of Chromosome XII of *Saccharomyces cerevisiae* reveals three of them responding to S-starvation. *Yeast* 2001, **18**, 325–334.

86. Pratt, J. M.; Petty, J.; Riba-Garcia, I.; Robertson, D. H.; Gaskell, S. J.; Oliver, S. G.; Beynon, R. J. Dynamics of protein turnover, a missing dimension in proteomics. *Mol. Cell Proteomics* 2002, **1**, 579–591.

87. Raamsdonk, L. M.; Teusink, B.; Broadhurst, D. I.; Zhang, N.; Hayes, A.; Walsh, M. C.; Berden, J. A.; Brindle, K. M.; Kell, D. B.; Rowland, J. J.; Westerhoff, H. V.; van Dam, K.; Oliver, S. G. A functional genomics strategy that uses metabolome data to reveal the phenotype of silent mutations. *Nature Biotechnol.* 2001, **19**, 45–50.

88. Maharjan, R. P.; Ferenci, T. Global metabolite analysis: the influence of extraction methodology on metabolome profiles of *Escherichia coli*. *Anal. Biochem*. 2003, **313**, 154–154.

89. Castrillo, J. L.; Hayes, A.; Mohammed, S.; Gaskell, S. J.; Oliver, S. G. An optimized protocol for metabolome analysis in yeast using direct infusion electrospray mass spectrometry. *Phytochemistry* 2003, **62**, 929–937.

90. Allen, J.; Broadhurst, H. M.; Heald, J. K.; Rowland, J. J.; Oliver, S. G.; Kell, D. B. High-throughput classification of yeast mutants for functional genomics using metabolic footprinting. *Nat. Biotechnol.* 2003, **21**, 692–696.

91. Kaderbhai, N. N.; Broadhurst, D. I.; Ellis, D. I.; Goodacre, R.; Kell, D. B. Functional genomics via metabolic footprinting: monitoring metabolite secretion by *Escherichia coli* tryptophan metabolism mutants using FT-IR and direct injection electrospray mass spectrometry. *Compar. Func. Genomics* 2003, **4**, 376–391.

92. King, R. D.; Whelan, K. E.; Jones, F. M.; Reiser, P. G.; Bryant, C. H.; Muggleton, S. H.; Kell, D. B.; Oliver, S. G. Functional genomic hypothesis generation and experimentation by a robot scientist. *Nature* 2004, **427**, 247–252.

12

BIOINFORMATICS FOR FLEXIBILITY, RELIABILITY, AND MIXTURE ANALYSIS OF INTACT MICROORGANISMS

Catherine Fenselau and Patrick Pribil

Department of Chemistry and Biochemistry, University of Maryland, College Park, MD 20742

12.1 INTRODUCTION

This chapter will address the applications of protein-based bioinformatics to analysis of microorganisms introduced intact into the instrumental system for rapid processing and analysis. Strategies that require offline extraction and fractionation of proteins will not be discussed. Although the amplification of nucleic acids is a powerful approach, especially coupled with mass spectrometry,[1–5] it requires extraction and processing, and thus is not included.

Intact bacteria were first introduced into a mass spectrometer for analysis of molecular biomarkers without processing and fractionation around 1975.[6] The ionization techniques available at the time limited analysis to secondary metabolites that could be volatilized, such as quinines and diglycerides, and vigorous pyrolysis of bacteria was explored as an alternative.[7] Although biomarkers were destroyed in pyrolysis strategies, computer-supported cluster analysis was developed to characterize pure samples.

The introduction of fast-atom bombardment (FAB) half a decade later extended the analyst's biomarker repertoire to intact polar lipids, desorbed

Identification of Microorganisms by Mass Spectrometry, Edited by Charles L. Wilkins and Jackson O. Lay, Jr.

from bacteria added intact, without processing, to the FAB matrix.[8] Biomarkers accessible by FAB included phospholipids, sulfonolipids, and glycolipids,[9–11] which in themselves served to distinguish bacteria from algae and yeast.[12] In support of this effort, computer algorithms based on simultaneous equations were developed[13] to provide, for the first time, baseline subtractions for FAB spectra, library matching for identification, and semiquantitative mixture deconvolution. Potential roles for tandem mass spectrometry (MSMS) were illustrated with FAB spectra,[11,14] and the use of information from the abscissa, but not the ordinate, *m/z* values but not relative abundances, was proposed for analysis, in conjunction with structural interpretation and biochemical information. For example, fatty acid components of phospholipids could be identified based on molecular masses and fragmentation, and correlated with prokaryote biosynthesis.[11]

The introduction and eventual commercialization of matrix-assisted laser desorption/ionization (MALDI) and electrospray (ESI) allowed biomarker status to be extended to proteins in 1996.[15–17] With a few exceptions, ESI has been used in conjunction with extractions and high-pressure liquid chromatography (HPLC) interfaced with mass spectrometry. MALDI, on the other hand, has been widely adapted for rapid analysis of intact organisms, supported by bioinformatics.[18,19]

It is now widely accepted that ions detected with masses above 4000 Da in MALDI and ESI spectra obtained from unfractionated microorganisms are translated proteins. These ionization techniques and the methods used for sample preparation are highly favorable for protein desorption, in contrast, for example, to nucleic acids. A number of laboratories have provided experimental evidence for the protein character of these ions, mostly involving sequencing after extraction and fractionation[20–24] or genetic modifications that change the molecular masses.[25] When the entire microorganism is introduced into a MALDI ionization source, most of the proteins observed have molecular masses below 20,000 Da. Fortunately the distribution of prokaryote proteins as a function of mass maximizes around 15,000 Da.[26] Thus the mass range most readily achievable in MALDI sources provides access to the largest set of protein biomarkers. However, even within this favored mass range, only a fraction of the proteins present are observed. This reflects differences in the number of copies of each protein synthesized by the microorganism, the propensity of different proteins for ionization (basicity in the case of positive ion formation), and the features that govern ion suppression,[27] which are yet to be clearly defined. In considering the validity of relating observed ions to the expected molecular masses of proteins, it should also be pointed out that proteins in prokaryotes are post-translationally modified by a much smaller set of reactions than eukaryotic proteins.

A number of polypeptide biomarkers have also been identified in the mass range below 4000,[28–31] which are cyclic secondary metabolites bonded to lipids or sugars. These peptide sequences are not directly translated from DNA,[32] and cannot be characterized by bioinformatics, that is, by reference to the genome. They are, of, course, susceptible to analysis by other approaches.

12.2 LIBRARY MATCHING

The analysis of chemicals by reference to a set of library mass spectra was facilitated by the establishment almost 40 years ago of databases such as the NIST/EPA/NIH reference library of electron impact mass spectra (http://www.nist.gov/srd/nist1a.htm). Experimentally derived mass spectra are compared to spectra in the library, and the matches are graded by various algorithms. This comparison is valid because electron impact ionization requires that the sample be vaporized and thus isolated from its sampling history.

A similar straight-forward approach has been advocated by many workers for the identification of microorganisms by MALDI time-of-flight mass spectrometry.[33–42] The library is constructed of spectra of pure organisms obtained under standard conditions. Matches can be made to these previously identified targets using a variety of algorithms. Most workers agree that the bacteria must be grown under standard conditions and the spectra must be run under standard conditions.[43,44] The requirement to culture the sample under standard conditions prolongs the analysis time. Different proteins are desorbed and ionized from species in the mixture when different matrices or different matrix:sample ratios are used.[38,45–47] The peaks observed and their abundances also depend on the laser power[48] and other instrumental conditions. These stringent experimental requirements would appear to exclude the use of library matching in field-portable applications.

12.3 MACHINE LEARNING

In more recent and powerful variations on spectral matching the computer is programmed to identify common core elements in thousands of spectra run under varying growth and instrumental conditions. At least one group has accomplished this using Bayesian belief networks based on probabilistic analysis.[49] This approach can provide experimental flexibility in the identification of individual microorganisms; however, it is likely to be inadequate for analyzing mixtures of microorganisms where suppression effects cannot be predicted. It also requires that targets are identified beforehand so that the machine can be trained.

12.4 BIOINFORMATICS

For decades microorganisms have served as a source of proteins for structural and mechanistic protein biochemistry. Throughout the last decade, as new proteomics strategies were developed, these have usually been demonstrated first on bacteria or yeast. It is not surprising then that proteomic scientists proposed early on to identify microorganisms on the basis of proteins extracted and separated by HPLC,[50] or of proteins extracted and converted to tryptic peptides for identification by LCMS.[51,52] Bioinformatics methods for identification of extracted proteins are described in textbooks and laboratory manuals on standard proteomic methods[53–55] and depend on searching protein/genome databases. Database entries for proteins include the organisms from which the proteins are derived, and thus identification of a protein also reveals the identity of its microorganism source.

The present chapter does not consider analysis of extracted protein biomarkers but rather focuses on strategies for rapid chemotaxonomic analysis of intact microorganisms with automated sample manipulation. "Rapid" means less than 5 minutes. Advantages of the application of bioinformatics and proteomics strategies for rapid identification of microorganisms include the following:

- Experimental observations are referenced to genomic databases, and are independent of growth conditions, matrices, and instrumental conditions.
- One database supports all ionization techniques and all protein ionization techniques are applicable.
- The approach has proved to be successful with spores, viruses, vegetative bacteria and toxins.
- The approach enables facile identification of specific biomarkers and can establish the uniqueness of biomarkers. The biomarker spectrum is interpreted rather than matched or correlated.
- The use of proteomics exploits one of the fastest-growing analytical technologies, with scientists worldwide contributing to the protein/genome databases, and to improved algorithms to search databases and identify proteins.

For development and testing, scientists have used constantly expanding, publicly accessible protein/genome databases, for example, NCBInr (http://www.ncbi.nlm.nih.gov/entrez/query.fcgi), MSDB (http://csc-fserve.hh.med.ic.ac.uk/msdb.html), and SwissProt (http://us.expasy.org/). Search time is reduced if the database can be limited to prokaryote entries. Several groups are assembling special databases comprising the proteins coded only in selected subsets of organisms (www.tigr.org).[56,57] Ideally proteins are characterized in these databases by molecular weights, sequences, taxonomic sources, and other chemical and biochemical properties, and references are provided to relevant literature. It is true that the organism cannot be identified by bioinformatics unless its genome, or the genome of a closely related species, has been sequenced.

Proteomics algorithms have been developed to search databases of protein sequences for matches to short sequences determined experimentally from proteins or peptides in the laboratory,[56–59] and for theoretical matches to molecular masses of sets of peptides produced from purified proteins with residue-specific cleavage agents.[60] The former approach is often called microsequencing, and the latter, peptide mass mapping. Such search engines are publicly available (e.g., http://us.expasy.org/tools/peptident.html; http://www.matrixscience.com; http://prowl.rockefeller.edu/; http://fields.scripps.edu/sequest/). Also in use are algorithms that seek to identify peptides (and thereby parent proteins) by comparing uninterpreted fragmentation spectra of peptides, produced by low-energy collisions, with theoretical spectra whose m/z values reflect sequences of tryptic peptides from proteins in the database.[61] Although most database searches in use rely only on experimental m/z values (x-axis in Figure 12.1), ongoing research seeks the ability to predict relative abundances as well (y-axis in Figure 12.1).[62–64]

These classical bioinformatics strategies are not all directly applicable to analysis of intact microorganisms, which each contain a mixture of protein biomarkers. However, the principle of comparing experimental observations to protein/genome databases (bioinformatics) does hold the key to flexibility,

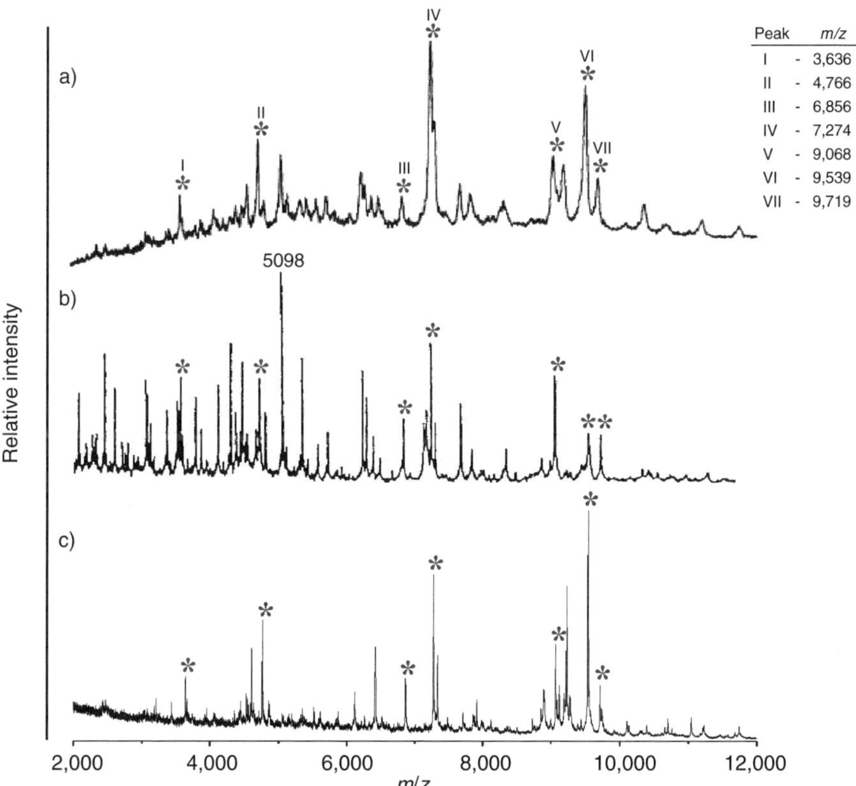

Figure 12.1 MALDI spectra obtained from intact *Escherichia coli* (ATCC# 11775) in three different laboratories.[44,82] Asterisks mark peaks that are common to all three.

reliability, and analysis of mixtures. Three major approaches will be described, based on exploitation of protein molecular masses, peptide maps with special peptide databases, and microsequences.

It should be pointed out that FAB, MALDI, and ESI can be used to provide ions for peptide mass maps or for microsequencing and that any kind of ion analyzer can support searches based only on molecular masses. Fragment or sequence ions are provided by instruments that can both select precursor ions and record their fragmentation. Such mass spectrometers include ion traps, Fourier transform ion cyclotron resonance, tandem quadrupole, tandem magnetic sector, several configurations of time-of-flight (TOF) analyzers, and hybrid systems such as quadrupole-TOF and ion trap-TOF analyzers.

For analyses in the field, the ionization chamber will usually be connected to an automated module for sample collection and preparation.

12.5 PROTEIN MOLECULAR MASSES

In response to the difficulty of reproducing MALDI spectra, a bioinformatics strategy has been proposed in which the molecular masses of proteins observed in mass spectra are matched with the molecular masses of proteins predicted from protein or gene sequences. Spectra of *Escherichia coli* ATCC #11775, for example, published by three different laboratories are shown in Figure 12.1. Although these spectra have some peaks in common (marked by the asterisks), unique proteins are also detected in each of the spectra. Despite this variability a majority of the peaks in each spectrum can be matched to protein masses predicted from *Escherichia coli*'s genome sequence.[26] Thus all three can be identified as spectra of *Escherichia coli* by reference to genes in that genome. Similarly suites of peaks in spectra from bacteria harvested at different growth times, or spectra obtained using different matrices, will not provide fingerprint matches but do correspond well to protein masses predicted from the *Escherichia coli* protein database.[26]

An algorithm has been developed to search databases for suites of protein masses automatically (Pineda et al.).[65] In this case data from only one axis is being used, m/z values for the protonated molecular ions. Relative intensities are not used in the analysis. The algorithm also assesses the probability of accidental matches and false identifications. Significance testing allows a value to be assigned to each microorganism identification.

In analyses where molecular masses are being matched, more accurate mass measurements provide more reliable matches and identifications.[26,65,66] In a reference laboratory the mass accuracy to several decimal points, provided by a Fourier transform ion cyclotron resonance mass analyzer, may be desirable. In field or portable systems there is usually a trade-off in mass accuracy for size and ruggedness. Reliable identifications can be made with moderate mass accuracy, even ±1 Da, if a large enough suite of molecular ions is recorded and used to search the database. If both positive ion and negative ion spectra are recorded, correct integer masses can usually be deduced by the mass difference between ions formed by addition of a proton and molecular ions formed by loss of a proton.

Not all of the protein peaks in the *Escherichia coli* spectra in Figure 12.1 can be matched to protein masses calculated from the genome. These differences reflect posttranslational modifications (PTM), such as acetylation or methylation,[20] or removal of the transcription initiator, *N*-terminal methionine (Met). The latter is the major PTM in bacteria, estimated to occur in 50% of bacterial proteins.[67] Careful study by several laboratories has defined the control of Met removal according to the nature of the amino acid that follows Met in the sequence.[68,69] For mass spectrometry analysis three groups of proteins may be defined, those with *N*-terminal Met in place, those with *N*-terminal Met removed, and those that exhibit both states. These biochemical rules have been incorporated into a search algorithm to provide a check on the second residue in any protein assignment, according to whether the molecular weight is assigned with or without Met.[70] Other PTM could also be considered by computer programs; however, methylation and others are less common.

The significance of each protein mass match, and thus the reliability of microorganism identification, depends on the size of the genome as well as on the accuracy of the mass measurements. Thus a match for a protein from *Helicobacter pylori*, whose genome codes for approximately 1000 proteins (http://us.expasy.org/) (450 in the mass range 4000 to 20,000 Da), is more significant than a match to a protein from *Escherichia coli*, whose genome codes for more than 6000 proteins (about 2000 in the mass range 4000 to 20,000 Da). This realization has led to the proposal that a ribosomal database be established for bacteria and spores[71] (http://www.pinedalab.jhsph.edu/microOrgID/). The entry for each species would comprise approximately 70 ribosomal proteins, a small number, and one that would be nearly equal for all species. In one series of interrogations with a suite of masses from MALDI spectra of *Helicobacter pylori* strain 26695, the overall significance factor, expressed such that a smaller value reflects a more reliable identification, moved from 0.036 for an unedited search of the entire protein/genome database, to 0.002 when removal of Met is incorporated into the calculation of protein masses, to 0.00001 when only ribosomal proteins are considered (see Table 12.1). As Table 12.1 indicates, the sample was clearly distinguished from another strain, *Helicobacter pylori* J99.

In another approach to improving the significance of mass matches, Demirev et al. have constructed a truncated database comprising proteins predicted to be abundant by estimating codon bias.[72,73] The size of the entries for each organism can be selected, for example, as the 10 proteins predicted to be most abundant.

It is also important to have multiple biomarker masses to use in the search. In samples of protein toxins, and of many viruses (e.g., see Figure 12.2), only one of two proteins may be ionized and analyzed for database searching.[15,74–76]

TABLE 12.1 Web-Based Search Results for 35 Biomarker Peaks from *M. pylori* 26995[70]

Rank	Organism	Partial (4–20 kDa) Proteome Size	Number of Matches	Significance Level
		No search restrictions		
1	*H. pylori* 26995	443	17	0.036
2	*H. pylori* J99	291	10	0.065
3	*M. leprae*	656	15	0.198
		After considering N-terminal Met loss		
1	*H. pylori* 26995	443	17	0.002
2	*R. prowazekii*	207	6	0.268
3	*T. pallidum*	251	6	0.427
		After considering N-terminal Met loss, searching ribosomal protein database[a]		
1	*H. pylori* 26995	37	6	0.00001
2	*P. aeruginosa*	39	3	0.018
3	*T. maritima*	34	2	0.085

[a] P. Demirev, private communication.

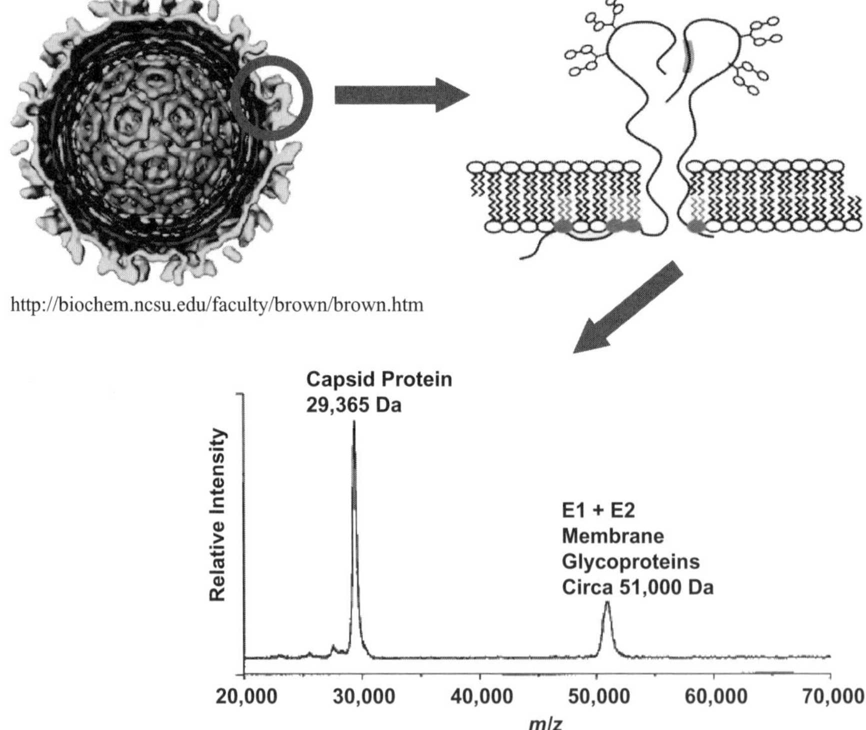

Figure 12.2 MALDI analysis of Sindbis virus.[74]

The reliability of the identification can be improved if that protein can be cleaved to two or more polypeptides.[77–79]

12.6 PEPTIDE MAPS

In part to address the problem described in the preceding paragraph, a strategy has been proposed[77–79] in which peptides are generated in situ to identify microorganisms directly, omitting the identification of individual proteins. Peptide databases are constructed to replace protein databases, comprising the predicted peptide products of, for example, tryptic digestion of the complete set of viral proteins, of protein toxins, of the ribosomal proteins, or of protein families that can be selectively solublized, such as the small acid-soluble protein (SASP) family from *Bacillus* spores. Although proof of concept has been carried out using immobilized trypsin, other selective cleavage agents could be invoked to create the database and used in the experimental protocol. These peptide databases can be interrogated to provide identifications of microorganisms by algorithms analogous to those developed to search protein databases for suites of protein mass matches.[80,81]

This strategy requires a more complex system for sample processing; however, it engages the advantages of improved mass accuracy and better sensitivity that derive from working with lower mass ions. Peptide maps from protein mixtures have been demonstrated to provide reliable identifications of microorganisms that are reasonably pure.[79–81]

As is the case with identifications based on protein molecular masses, it appears that the use of tryptic or other peptide masses as the basis for identification is extended with difficulty to mixtures of microorganisms. This reflects unpredictable suppression. Another limitation is redundancy of peptide masses across several microorganisms. For example, the most abundant proteins (SASPs), and thus the most abundant peptides, in spores of *Bacillus anthracis* and the closely related pesticide *Bacillus thuringiensis* have extensive sequence homology.[25,82]

12.7 MICROSEQUENCES FROM PEPTIDES AND PROTEINS

Peptide microsequencing has proved to be a powerful approach to identify protein components of mixtures,[53–55,61] and has been applied to extracted proteins as a method to identify biological threats.[56,83–85] Commercial and internet accessible search engines and databases provide adequate bioinformatics support to use sequence information produced by the mass spectrometer (e.g., http://www.matrixscience.com; http://prospector.ucsf.edu/ucsfhtml4.0/mstagfd.htm; http://prowl.rockefeller.edu/).

Peptide microsequencing also provides an alternative approach to characterizing intact microorganisms; but, the application where its potential appears to be unique and unrivaled is the identification of individual species in mixtures of microorganisms. Such mixtures might be encountered in environmental collections from air or water, for example, or in clinical samples. The requirement to carry out the analysis in situ and in less than 5 minutes has been realized by solublization and analysis of only a limited number of proteins and by using trypsin immobilized on beads.[56,83–88] If a limited number of proteins is solublized for cleavage, then a limited number of peptide molecular ions will be ionized, and more of these will carry sufficient ion current to produce sequence ion spectra in subsequent tandem mass spectrometry experiments. In addition it is more likely that protein (and thus species) identifications can be based on sequences from more than one peptide, and thus be more reliable.

Larger proteins can be sampled in this way, as well as small ones. For example, flagellin, molecular weight 32.6kDa, has been selectively solublized from Gram-positive vegetative *Bacillus subtilis* sp. 168, cleaved and identified by microsequencing, as illustrated in Figure 12.3. Several peptides derived from flagellin are detected in the top spectrum (Figure 12.3*a*), marked with solid circles. The lower spectrum (Figure 12.3*b*) presents ions produced by unimolecular decomposition from the precursor peptide detected at m/z 2606 (in a TOF analyzer with a curved field reflectron). Protein identification was based on microsequences from this and other peptide ions, with 39% coverage.[87] Flagellins are abundant in many species, and have species-characteristic sequences in their central domains. Flagellin genes are already in use as biomarkers for studies of population genetics and epidemiological analysis.[89] Four other proteins that contributed abundant peptides to the MALDI spectrum (Figure 12.3*a*) are also identified in the figure legend.

Time can be saved and sensitivity increased if a targeted analysis strategy is used that skips the survey scan and looks directly for the peptide(s) that is considered to be characteristic of each analyte of interest. Species-specific peptides for *Bacillus anthracis* spores, for example, have been predicted by genomic analysis of many *Bacillus* species, coupled with experimental screening.[82] In Figure 12.4 a spectrum of peptide ions from SASPs released in situ from a 10:1 mixture of *Bacillus cereus* and *Bacillus anthracis* sp. Sterne is shown. These peaks are categorized in the figure as characteristic of the former bacterium, the latter, or belonging to a set of peaks that are characteristic of two or more of the members of the *Bacillus cereus* group. *Bacillus anthracis* sp. Sterne can be identified in this mixture through selection and sequencing (in an ion trap) of its peptide at m/z 1528.

Both collisional activation (in ion traps) and post source decay (in curved field reflectron TOF analyzers) have been used successfully to obtain sequence ions from peptides prepared in situ on the sample holder. Single Dalton mass windows are advantageous for precursor selection, as are realized in ion-trap and trap-TOF configurations. Publicly available search algorithms can be used

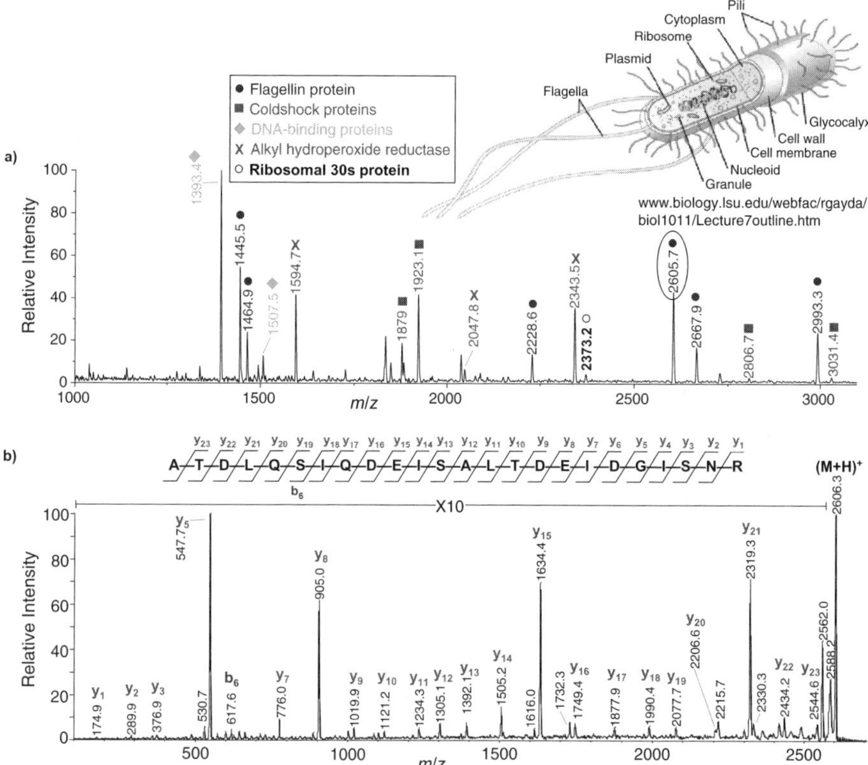

Figure 12.3 MALDI analysis of peptides from *Bacillus subtilis* sp. 168 vegetative cells prepared in situ.[87] (*a*) Survey spectrum of peptide products. Protein assignments are listed in the figure. (*b*) Spectrum of product ions of unimolecular decomposition of the peptide with *m*/*z* 2606.

or adapted. As search programs are improved in the broader field of proteomics, these will also be brought to this application.

Multiply charged proteins can also be partially sequenced, and microsequences of proteins isolated from several microorganisms have been reported, accomplished with electrospray ionization and FTMS.[23,90] Nonadjacent fragment ions may be used to identify bacterial proteins in these top-down strategies.[91] In all cases these sequences could be related by bioinformatics to the parent species. An obvious extension would be to characterize proteins from intact microorganisms in this way. In at least one instance a microsequence has been obtained from a protein released from a contaminated intact bacteriophage sample (MS2) to provide a chemotaxonomic identification.[77] This work was carried out in an ion trap mass spectrometer.

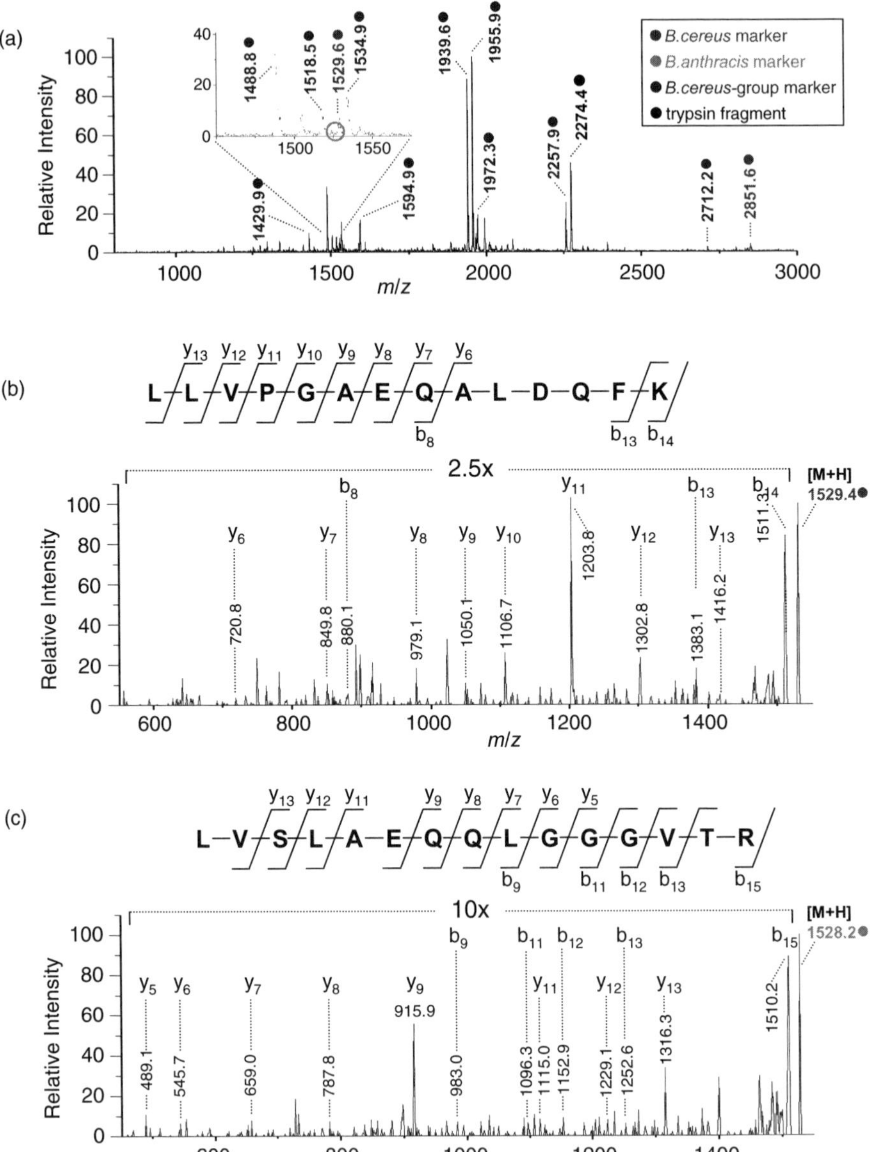

Figure 12.4 MALDI analysis of peptides prepared *in situ* from a 10:1 mixture of *Bacillus cereus* and *Bacillus anthracis* sp. Sterne.[82] (*a*) Survey spectrum of peptide products. Species assignments are indicated on the figure. (*b*) Spectrum of fragment ions produced by low-energy collisions of the *Bacillus cereus*–specific peptide of mass 1529. (*c*) Spectrum of fragment ions produced by low-energy collisions of ions of the *Bacillus anthracis* peptide of mass 1528.

12.8 REMAINING CHALLENGES

Although the feasibility of the proteomics or bioinformatics approach has been demonstrated, considerable room remains for improved methods for selective solublization of protein biomarkers and for rapid cleavage to produce peptides. There is also demand for advanced instrumentation to collect, process, and analyze microorganisms.

Both absolute quantitation and relative quantitation of species in mixtures is of interest in some circumstances. Quantitation in a 5-minute analysis can be achieved by addition of an internal standard, ideally the target microorganism grown in special media to incorporate heavy isotopes[92–95] and determination of the relative peak heights of pairs of proteins from the analyte and the standard. Isotope-labeled proteins or peptides, selected to match proteins or peptides characteristic of target microorganisms, can also serve as internal standards for isotope ratio measurement. The addition of unmatched proteins or peptides is less reliable for either ESI or MALDI measurements because of unpredictable suppression in the variable mixture.

The analysis of microorganisms in mixtures provides a continuing analytical challenge, for which a number of solutions have been proposed:

- One proven approach is the analysis of individual colonies from the mixture grown on a culture plate.
- The use of computer algorithms to deconvolute MALDI or ESI spectra of mixtures of microorganisms obtained under uncontrolled conditions is less likely to be successful, as discussed above.
- A third possibility is that partial separation will occur on the MALDI target, such as during matrix crystallization,[96] which can be distinguished by rastering the laser beam.
- Single particle analysis offers improved resolution of species in mixtures.
- Selective biocapture, such as with immobilized antibodies, adhesions or carbohydrates, can be used to concentrate targeted analysis online or in situ.[97–99]
- Microsequencing many peptides in a digested mixture may be coupled with bioinformatics to reveal information on all the component species.
- Alternatively, species-specific peptides can be targeted and partially sequenced to confirm the presence or absence of microorganisms of high interest.

A final problem for bioinformatics and bioanalytical scientists is the characterization of engineered microorganisms. Whole-cell analysis by mass spectrometry has been used to confirm the introduction of therapeutic genes into adenovirus vectors,[100] to confirm the expression of recombinant proteins in bacteria,[101,102] and also in vaccinology.[103] In the broader case, identification of both the host species and recognition of the engineered protein would be desired.

12.9 CONCLUSIONS

A major advantage of mass spectrometry is that everything has a mass spectrum. Other chemical and biological detectors ask "Is it there?" The broadband mass spectrometer asks "What is there?" At the same time mass spectrometry is a very specific technique that measures an intrinsic property: mass. Both molecular and fragment masses provide detailed fingerprints that can also be interpreted and related to the universal database, the genome database. Mass spectrometric analyses can be made quickly, with competitive sensitivity, and can be automated. The use of bioinformatics to reference various kinds of mass spectra to genomic information provides critical experimental flexibility, while providing chemotaxonomic characterization and a reliable approach to the identification of components of mixtures.

ACKNOWLEDGMENTS

The authors thank the Defense Advanced Research Projects Agency for support.

REFERENCES

1. Hurst, G. B.; Doktycz, M. J.; Vass, A. A.; Buchanan, M. V. Detection of bacterial DNA polymerase chain reaction products by matrix-asssisted laser desorption/ionization mass spectrometry. *Rapid Comm. Mass Spectrom.* 1996, **10**, 377–382.
2. Jiang, Y.; Hofstadler, S. A. A highly efficient and automated method for purifying and desalting PCR products for analysis by electrospray ionization mass spectrometry. *Anal. Biochem.* 2003, **316**, 50–57.
3. Muddiman, D. C.; Wunschel, D. S.; Liu, C.; Pasa-Tolic, L.; Fox, K. F.; Fox, A.; Anderson, G. A.; Smith, R. D. Characterization of PCR products from *Bacilli* using electrospray ionization FTICR mass spectrometry. *Anal. Chem.* 1996, **68**, 3705–3712.
4. Wunschel, D. S.; Fox, K. F.; Bruce, J. E.; Muddiman, D. C.; Smith, R. D. Analysis of double-stranded polymerase chain reaction products from the *Bacillus cereus* group by electrospray ionization Fourier transform ion cyclotron resonance mass spectrometry. *Rapid Comm. Mass Spectrom* 1996, **10**, 29–35.
5. von Wintzingerode, F.; Bocker, S.; Schlotelburg, C.; Chiu, N. H.; Storm, N.; Jurinke, C.; Cantor, C. R.; Gobel, U. B.; van den Boom, D. Base-Specific fragmentation of amplified 16S rRNA genes analyzed by mass spectrometry: A tool for rapid bacterial identification. *Proc. Nat. Acad. Sci. USA* 2002, **99**, 7039–7044.

6. Anhalt, J. P.; Fenselau, C. Identification of bacteria using mass spectrometry. *Anal. Chem.* 1975, **47**, 219–225.
7. Snyder, A. P.; Smith, P. B. W.; Dworzanski, J. P.; Meuzelaar, H. L. In *Mass Spectrometry for the Characterization of Microorganisms.* Washington, DC: American Chemical Society, 1994, pp. 62–84.
8. Heller, D. N.; Cotter, R. J.; Fenselau, C.; Om, U. Profiling of bacteria by fast atom bombardment mass spectrometry. *Anal. Chem.* 1987, **59**, 2806–2809.
9. Cole, M. J.; Enke, C. G. In *Mass Spectrometry for the Characterization of Microorganisms.* Washington, DC: American Chemical Society, 1994, pp. 36–61.
10. Heller, D. N.; Fenselau, C.; Cotter, R. J.; Demirev, P. A.; Olthoff, J. K.; Honovich, J.; Uy, M.; Tanaka, T.; Kishimoto, Y. Mass spectral analysis of complex lipids desorbed directly from lyophilized membranes and cells. *Biochem. Biophys. Res. Comm.* 1987, **142**, 194–199.
11. Heller, D. N.; Murphy, C. M.; Cotter, R. J.; Fenselau, C.; Uy, O. M. Constant neutral loss scanning for the characterization of bacterial phospholipids desorbed by fast atom bombardment. *Anal. Chem.* 1988, **60**, 2787–2791.
12. Fenselau, C. In *Mass Spectrometry for the Characterization of Microorganisms.* Washington, DC: American Chemical Society, 1994, pp. 1–7.
13. Platt, J. A.; Uy, O. M.; Heller, D. N.; Cotter, R. J.; Fenselau, C. Computer-based linear regression analysis of desortion mass spectra of microorganisms. *Anal. Chem.* 1998, **60**, 1415–1419.
14. Cole, M. J.; Enke, C. G. Direct determination of phospholipid structures in microorganisms by fast atom bombardment triple quadrupole mass spectrometry. *Anal. Chem.* 1991, **63**, 1032–1038.
15. Despeyroux, D.; Phillpotts, R.; Watts, P. Electrospray mass spectrometry for detection and characterization of purified cricket paralysis virus (CrPV). *Rapid Comm. Mass Spectrom.* 1996, **10**, 937–941.
16. Holland, R. D.; Wilkes, J. G.; Rafi, F.; Sutherland, J. B.; Persons, C. C.; Voorhees, K. J.; Lay, J. O. Rapid identification of whole bacteria based on spectral patternas using matrix-assisted laser desorption/ionization with time-of-flight mass spectrometry. *Rapid Comm. Mass Spectrom.* 1996, **10**, 1227–1232.
17. Krishnamurthy, T.; Ross, P. L. Rapid identification of bacteria by direct matrix-assisted laser desorption/ionization mass spectrometric analysis of whole cells. *Rapid Comm. Mass Spectrom.* 1996, **10**, 1992–1996.
18. Fenselau, C.; Demirev, P. A. Characterization of intact microorganisms by MALDI mass spectrometry. *Mass Spectrom. Rev.* 2001, **20**, 157–171.
19. Lay, J. O. MALDI-TOF mass spectrometry of bacteria. *Mass Spectrom. Rev.* 2001, **20**, 172–194.
20. Arnold, R. J.; Reilly, J. P. Observation of *Escherichia coli* ribosomal proteins and their possttranslational modifications by mass spectrometry. *Anal. Biochem.* 1999, **269**, 105–112.
21. Cain, T. C.; Lubman, D. M.; Weber, W. J. Differentiation of bacteria using protein profiles from matrix-assisted laser desorption time-of-flight mass spectrometry. *Rapid Comm. Mass Spectrom.* 1994, **8**, 1026–1030.
22. Dai, Y.; Li, L.; Roser, D. C.; Long, S. R. Detection and Identification of low-mass peptides and proteins from solvent suspensions of *Escherichia coli* by high

performance liquid chromatography fractionation and matrix-assisted laser desorption/ionization mass spectrometry. *Rapid Comm. Mass Spectrom.* 1999, **13**, 73–78.

23. Demirev, P. A.; Ramirez, J.; Fenselau, C. Tandem mass spectrometry of intact proteins for characterization of biomarkers from *Bacillus cereus* T spores. *Anal. Chem.* 2001, **73**, 5725–5731.
24. Holland, R. D.; Duffy, C. R.; Rafi, F.; Sutherland, J. B.; Heinze, T. M.; Holder, C. L.; Voorhees, K. J.; Lay, J. O. Identification of bacterial proteins observed in MALDI TOF mass spectra from whole cells. *Anal. Chem.* 1999, **71**, 3226–3230.
25. Hathout, Y.; Setlow, B.; Cabrera-Martinez, R. M.; Fenselau, C. Small, acid-soluble proteins as biomarkers in mass spectrometry analysis of *Bacillus* spores. *Appl. Environ. Microbiol.* 2003, **69**, 1100–1107.
26. Demirev, P. A.; Ho, Y. P.; Ryzhov, V.; Fenselau, C. Microorganism identification by mass spectrometry and protein database searches. *Anal. Chem.* 1999, **71**, 2732–2738.
27. Fenselau, C. MALDI MS and strategies for protein analysis. *Anal. Chem.* 1997, **69**, 661A–665A.
28. Hathout, Y.; Ho, Y. P.; Ryzhov, V.; Demirev, P. A.; Fenselau, C. Kurstakins: A new class of lipopeptides isolated from *Bacillus thuringiensis*. *J. Nat. Prod.* 2000, **63**, 1492–1496.
29. Leenders, F.; Stein, T. H.; Kablitz, B.; Franke, P.; Vater, J. Rapid typing of *Bacillus subtilis* strains by their secondary metabolites using matrix-assisted laser desorption/ionization mass spectrometry of intact cells. *Rapid Comm. Mass Spectrom.* 1999, **13**, 943–949.
30. Welker, M.; Brunke, M.; Preussel, K.; Lippert, I.; von Dohren, H. Diversity and distribution of *microcystis* (Cyanobacteria) oligopeptide chemotypes from natural communities studied by single-colony mass spectrometry. *Microbiology* 2004, **150**, 1785–1796.
31. Williams, B. H.; Hathout, Y.; Fenselau, C. Structural characterization of lipopeptide biomarkers isolated from *Bacillus globigii*. *J. Mass Spectrom.* 2002, **37**, 259–264.
32. Moffitt, M. C.; Neilan, B. A. The expansion of mechanistic and organismic diversity associated with non-ribosomal peptides. *FEMS Microbiol Lett.* 2000, **191**, 159–167.
33. Arnold, R. J.; Reilly, J. P. Fingerprint matching of *E. coli* strains with matrix-assisted laser desorption/ionization time-of-flight spectrometry of whole cells using a modified correlation approach. *Rapid Comm. Mass Spectrom.* 1998, **12**, 630–636.
34. Bright, J. J.; Claydon, M. A.; Soufian, M.; Grodon, D. B. Rapid typing of bacteria using matrix-assisted laser desorption ionisation time-of-flight mass spectrometry and pattern recognition software. *J. Microbiol. Meth.* 2002, **48**, 127–138.
35. Conway, G. C.; Smole, S. C.; Sarracino, D. A.; Arbeit, R. D.; Leopold, P. E. Phyloproteomics: Species identification of *Enterobacteriaceae* using matrix-assisted laser desorption/ionization time-of-flight mass spectrometry. *J. Microbiol. Biotechnol.* 2001, **3**, 103–112.

36. Haag, A. M.; Taylor, S. N.; Johnston, K. H.; Cole, R. B. Rapid identification and speciation of *Haemophilus* bacteria by matrix-assisted laser desorption/ionization time-of-flight mass spectrometry. *J. Mass Spectrom.* 1998, **33**, 750–756.

37. Hayek, C. S.; Pineda, F. J.; Doss, O. W.; Lin, J. S. Computer-assisted interpretation of mass spectra. *JHU-APL Tech. Digest* 1999, **20**, 363–370.

38. Jarman, K. H.; Cebula, S. T.; Saenz, A. J.; Peterson, C. E.; Valentine, N. B.; Kingsley, M. T.; Wahl, K. L. An algorithm for automated bacterial identification using matrix-assisted laser desorption/ionization mass spectrometry. *Anal. Chem.* 2000, **72**, 1217–1223.

39. Madonna, A. J.; Basile, F.; Ferrer, I.; Meetani, M. A.; Rees, J. C.; Voorhees, K. J. On-probe sample pretreatment for the detection of proteins above 15 KDa from whole cell bacteria by matirx-assisted laser desorption/ionization time-of-flight mass spectrometry. *Rapid Comm. Mass Spectrom.* 2000, **14**, 2220–2229.

40. Nilsson, C. L. Fingerprinting of *Helicobacter pylori* strains by matrix-assisted laser desorption/ionization mass spectrometric analysis. *Rapid Comm. Mass Spectrom.* 1999, **13**, 1067–1071.

41. Wahl, K. L.; Wunschel, S. C.; Jarman, K. H.; Valentine, N. B.; Petersen, C. E.; Kingsley, M. T.; Zartolas, K. A.; Saenz, A. J. Analysis of microbial mixtures by matrix-assisted laser desorption/ionization time-of-flight mass spectrometry. *Anal. Chem.* 2002, **74**, 6191–6199.

42. Winkler, M. A.; Uher, J.; Cepa, S. Direct analysis and identification of *Helicobacter* and *Campylobacter* species by MALDI-TOF mass spectrometry. *Anal. Chem.* 1999, **71**, 3416–3419.

43. Saenz, A. J.; Petersen, C. E.; Valentine, N. B.; Gantt, S. L.; Jarman, K. H.; Kingsley, M. T.; Wahl, K. L. Reproducibility of matrix-assisted laser desorption time-of-flight mass spectrometry for replicate bacterial culture analysis. *Rapid Comm. Mass Spectrom.* 1999, **13**, 1580–1585.

44. Wang, Z.; Russon, L.; Li, L.; Roser, D. C.; Long, S. R. Investigation of spectral reproducibility in direct analysis of bacteria proteins by matrix-assisted laser desorption/ionization time-of-flight mass spectrometry. *Rapid Comm. Mass Spectrom.* 1998, **12**, 456–464.

45. Domin, M. A.; Welham, K. J.; Ashton, D. S. The effect of solvent and matrix combinations on the analysis of bacteria by matrix-assisted laser desorption/ionisation time-of-flight mass spectrometry. *Rapid Comm. Mass Spectrom.* 1999, **13**, 222–226.

46. Elhanany, E.; Barak, R.; Fisher, M.; Kobiler, D.; Altboum, Z. Detection of specific *Bacillus anthracis* spore biomakers by matrix-assisted desorption/ionization time-of-flight mass spectrometry. *Rapid Comm. Mass Spectrom.* 2001, **15**, 2110–2116.

47. Vaidyanathan, S.; Winder, C. L.; Wade, S. C.; Kell, D. B.; Goodacre, R. Sample preparation in matrix-assisted laser desorption/ionization mass spectrometry of whole bacterial cells and detection of high mass (>20 kDa) proteins. *Rapid Comm. Mass Spectrom.* 2002, **16**, 1276–1286.

48. Ramirez, J.; Fenselau, C. Factors contributing to peak broadening and mass accuracy in the characterization of intact spores using matrix-assisted laser desorption/ionization coupled with time-of-flight mass spectrometry. *J. Mass Spectrom.* 2001, **36**, 929–936.

49. Feldman, A. B.; Scholl, P.; Antoine, M.; Benson, R.; Carlson, M.; Cebula, S.; Collins, B.; Cornish, T.; Cotter, R. J.; Demirev, P. A.; Ecelberger, S.; Fenselau, C.; Jarman, K.; Joseph, R.; Lewis, D.; Lin, J. S.; McLaughlin, M.; Miragliotta, J.; Phillips, T.; Pineda, F. J.; Pitt, L.; Prieto, M.; Ramirez, J.; Reilly, J. P.; Velky, J.; Wahl, K.; Wall, J.; Bryden, W. *Proc. 49th Conf. on Mass Spectrom. and Allied Topics*, Santa Fe, NM; Am. Soc. for Mass Spectrom., 2001.

50. Liang, X.; Zheng, K.; Qian, M. G.; Lubman, D. M. Determination of bacterial protein profiles by matrix-assisted laser desorption/ionization mass spectrometry with high-performance liquid chromatography. *Rapid Comm. Mass Spectrom.* 1996, **10**, 1219–1226.

51. Krishnamurthy, T.; Davis, M. T.; Stahl, D. C.; Lee, T. D. Liquid chromatography/microspray mass spectrometry for bacterial investigations. *Rapid Comm. Mass Spectrom.* 1999, **13**, 39–49.

52. Krishnamurthy, T.; Rajamani, U.; Ross, P. L.; Eng, J. K.; Davis, M.; Lee, T. D.; Stahl, D. S.; Yates, J. R. In *Natural and Selected Synthetic Toxins: Biological Implications*. Washington, DC: American Chemical Society, 2000, pp. 67–97.

53. James, P. *Proteome Research: Mass Spectrometry.* New York: Springer, 2001.

54. Liebler, D. C. *Introduction to Proteomics: Tools for the New Biology.* Totowa, NJ: Humana Press, 2002.

55. Simpson, R. J. *Proteins and Proteomics.* Cold Spring Harbor, NY: Cold Spring Harbor Laboratory Press, 2003.

56. Dworzanski, J. P.; Snyder, A. P.; Chen, R.; Zhang, H.; Wishart, D.; Li, L. Identification of bacteria using tandem mass spectrometry combined with a proteome database and statistical scoring. *Anal. Chem.* 2004, **76**, 2355–2366.

57. Wang, Z.; Dunlop, K.; Long, S. R.; Li, L. Mass spectrometric methods for generation of protein mass database used for bacterial identification. *Anal. Chem.* 2002, **74**, 3174–3182.

58. Mann, M.; Wilm, M. Error-tolerant identification of peptides in sequence databases by peptide sequence tags. *Anal. Chem.* 1994, **66**, 4390–4399.

59. Mortz, E.; O'Connor, P. B.; Roepstorff, P.; Kelleher, N. L.; Wood, T. D.; McLafferty, F. W.; Mann, M. Sequence tag identification of intact proteins by matching tandem mass spectral data against sequence data bases. *Proc Nat. Acad. Sci. USA* 1996, **93**, 8264–8267.

60. Henzel, W. J.; Billeci, T. M.; Stults, J. T.; Wong, S. C.; Grimley, C.; Watanabe, C. Identifying proteins from two-dimensional gels by molecular mass searching of peptide fragments in protein sequence databases. *Proc. Nat. Acad. Sci. USA* 1993, **90**, 5011–5015.

61. Eng, J. K.; McCormack, A. L.; Yates, J. R. An approach to correlate tandem mass spectra data of peptides with amino acid sequences in a protein database. *J. Am. Soc. Mass Spectrom.* 1994, **5**, 976–989.

62. Breci, L. A.; Tabb, D. L.; Yates, J. R.; Wysocki, V. H. Cleavage *N*-terminal to proline: Analysis of a database of peptide tandem mass spectra. *Anal. Chem.* 2003, **75**, 1963–1971.

63. Tabb, D. L.; Huang, Y.; Wysocki, V. H.; Yates, J. R. Influence of basic residue content on fragment ion peak intensities in low-energy collision-induced dissociation spectra of peptides. *Anal. Chem.* 2004, **76**, 1243–1248.

64. Tsaprailis, G.; Nair, H.; Somogyi, A.; Wysocki, V. H.; Zhong, W.; Futrell, J. H.; Summerfield, S. G.; Gaskell, S. J. Influence of secondary structure on the fragmentation of protonated peptides. *J. Am. Chem. Soc.* 1999, **121**, 5142–5154.

65. Pineda, F. J.; Lin, J. S.; Fenselau, C.; Demirev, P. A. Testing the significance of microorganism identification by mass spectrometry and proteome database search. *Anal. Chem.* 2000, **72**, 3739–3744.

66. Jones, J. J.; Stump, M. J.; Fleming, R. C.; Lay, J. O.; Wilkins, C. L. Investigation of MALDI-TOF and FT-MS techniques for analysis of *Escherichia ecoli* whole cells. *Anal. Chem.* 2003, **75**, 1340–1347.

67. Gonzales, T.; Robert-Baudouy, J. Bacterial aminopeptidases: Properties and functions. *FEMS Microbiol Rev.* 1996, **18**, 319–344.

68. Dalboge, H.; Bayne, S.; Pedersen, J. In vivo processing of *N*-terminal methionine in *E. coli. FEBS Lett.* 1990, **266**, 1–3.

69. Miller, C. G.; Strauch, K. L.; Kukral, A. M.; Miller, J. L.; Wingfield, P. T.; Mazzei, G. J.; Werlen, R. C.; Graber, P.; Movva, N. R. *N*-terminal methionine-specific peptidase in *Salamonella typhimurium. Proc. Nat. Acad. Sci. USA* 1987, **84**, 2718–2722.

70. Demirev, P. A.; Lin, J. S.; Pineda, F. J.; Fenselau, C. Bioinformatics and mass spectrometry for microorganism identification: Proteome-wide post-translational modifications and database search algorithms for characterization of intact *H. Pylori. Anal. Chem.* 2001, **73**, 4566–4573.

71. Pineda, F. J.; Antoine, M. D.; Demirev, P. A.; Feldman, A. B.; Jackman, J.; Longenecker, M.; Lin, J. S. Microorganism identification by matrix-assisted laser/desorption ionization mass spectrometry and model-derived ribosomal protein biomarkers. *Anal. Chem.* 2003, **75**, 3817–3822.

72. Demirev, P. A.; Feldman, A. B.; Lin, J. S. Bioinformatics-based strategies for rapid microorganism identification by mass spectrometry. *JHU-APL Tech Dig.* 2004, **25**, 27–37.

73. Demirev, P. A.; Feldman, A. B.; Lin, J. S.; Pineda, F. J.; Resch, C. L., *Proc. 51st Conf. on Mass Spectrom. and Allied Topics*, Santa Fe, NM; Am. Soc. for Mass Spectrom., 2003

74. Kim, Y. J.; Freas, A.; Fenselau, C. Analysis of viral glycoproteins by MALDI-TOF mass spectrometry. *Anal. Chem.* 2001, **73**, 1544–1548.

75. She, Y. M.; Haber, S.; Siefers, D. L.; Loboda, A.; Chernushevich, I.; Perreault, H.; Ens, W.; Standing, K. G. Determination of the complete amino acid sequence for the coat protein of brome mosaic virus by time-of-flight mass spectrometry: Evidence for mutations associated with change of propagation host. *J. Biol. Chem.* 2001, **246**, 20039–20047.

76. Thomas, J. J.; Falk, B.; Fenselau, C.; Jackman, J.; Ezzell, J. Viral characterization by direct analysis of capsoid proteins. *Anal. Chem.* 1998, **70**, 3863–3867.

77. Cargile, B. J.; McLuckey, S. A.; Stephenson, J. L. Identification of bacteriophage MS2 coat protein from *E. coli* lysates via ion trap collisional activation of intact protein ions. *Anal. Chem.* 2001, **73**, 1277–1285.

78. Fredriksson, S.; Hulst, A. G.; Artursson, E.; de Jong, A. L.; Nilsson, C.; van Baar, B. L. Forensic identification of neat ricin and of ricin from crude castor bean extracts by mass spectrometry. *Anal. Chem.* 2005, **77**, 1545–1555.

79. Yao, Z. P.; Demirev, P. A.; Fenselau, C. Mass spectrometry-based proteolytic mapping for rapid virus identification. *Anal. Chem.* 2002, **74**, 2529–2534.

80. English, R. D.; Warscheid, B.; Fenselau, C.; Cotter, R. J. *Bacillus* spore identification via proteolytic peptide mapping with a miniaturized MALDI TOF mass spectrometer. *Anal. Chem.* 2003, **75**, 6886–6893.

81. Yao, Z. P.; Afonso, C.; Fenselau, C. Rapid microorganism identification with on-slide proteolytic digestion followed by matrix-assisted laser desorption/ionization tandem mass spectrometry and database searching. *Rapid Comm. Mass Spectrom.* 2002, **16**, 1953–1956.

82. Pribil, P.; Patton, E.; Black, G.; Doroshenko, V. M.; Fenselau, C. Rapid characterization of *Bacillus* spores targeting species-unique peptides produced with an atmospheric-pressure MALDI source. *J. Mass Spectrom.* 2005, **40**, 464–474.

83. Cooper, B.; Eckert, D.; Andon, N. L.; Yates, J. R.; Haynes, P. A. Investigative proteomics: Identification of an unknown plant virus from infected plants using mass spectrometry. *J. Am. Soc. Mass. Spectrom.* 2003, **14**, 736–741.

84. Harris, W. A.; Reilly, J. P. On-probe digestion of bacterial proteins for MALDI-MS. *Anal. Chem.* 2002, **74**, 4410–4416.

85. VerBerkmoes, N. C.; Hervey, W. J.; Shah, M.; Land, M.; Hauser, L.; Larimer, F. W.; Van Berkel, G. J.; Goeringer, D. E. Evaluation of "shotgun" proteomics for identification of biological threat agents in complex enviromental matrices: Experimental simulations. *Anal. Chem.* 2005, **27**, 923–932.

86. Warscheid, B.; Fenselau, C. Characterization of *Bacillus* spores species and their mixtures using possource decay with a curved-field reflectron. *Anal. Chem.* 2003, **75**, 5618–5627.

87. Warscheid, B.; Fenselau, C. A targeted proteomics approach to the rapid identification of bacterial cell mixtures by MALDI mass spectrometry. *Proteomics* 2004, **4**, 2877–2892.

88. Warscheid, B.; Jackson, K.; Sutton, C.; Fenselau, C. MALDI analysis of *Bacilli* in spore mixtures by applying a quadrupole ion trap time-of-flight tandem mass spectrometer. *Anal. Chem.* 2003, **75**, 5608–5617.

89. Winstanley, C.; Morgan, J. A. The bacterial flafellin gene as a biomarker for detection, population genetics and epidemiological analysis. *Microbiology* 1997, **143**, 3071–3084.

90. Xiang, F.; Anderson, G. A.; Veenstra, T. D.; Lipton, M. S.; Smith, R. D. Characterization of microorganisms and biomarker development from global ESI-MS/MS analyses of cell lysates. *Anal. Chem.* 2000, **72**, 2475–2481.

91. Meng, F.; Cargile, B. J.; Miller, L. M.; Forbes, A. J.; Johnson, J. R.; Kelleher, N. L. Informatics and multiplexing of intact protein identification in bacteria and the archaea. *Nat. Biotechnol.* 2001, **19**, 952–957.

92. Czerwieniec, G. A.; Russell, S. C.; Tobias, H. J.; Fergenson, D. P.; Steele, P.; Pitesky, M. E.; Horn, J. M.; Frank, M.; Gard, E. E.; Lebrilla, C. B. Stable isotope labeling of entire *Bacillus atrophaeus* spores and vegetative cells using bio-aerosol mass spectrometry. *Anal. Chem.* 2005, **77**, 1081–1087.

93. Oda, Y.; Huang, K.; Cross, F. R.; Cowburn, D.; Chait, B. T. Accurate quantitation of protein expression and site-specific phsphorylation. *Proc. Nat. Acad. Sci. USA* 1999, **96**, 6591–6596.

94. Stump, M. J.; Jones, J. J.; Fleming, R. C.; Lay, J. O.; Wilkins, C. L. Use of double-depleted 13C and 15N culture media for analysis of whole cell bacteria by MALDI time-of-flight and Fourier transform mass spectrometry. *J. Am. Soc. Mass Spectrom.* 2003, **14**, 1306–1314.
95. Veenstra, T. D.; Martinovic, S.; Anderson, G. A.; Pasa-Tolic, L.; Smith, R. D. Proteome analysis using selective incorporation of isotopically labeled amino acids. *J. Am. Soc. Mass Spectrom.* 2000, **11**, 78–82.
96. Feldman, A. B.; Antoine, M.; Lin, J. S.; Demirev, P. A. Covariance mapping in matrix-assisted laser desorption/ionization time-of-flight mass spectrometry. *Rapid Comm. Mass Spectrom* 2003, **17**, 991–995.
97. Bundy, J.; Fenselau, C. Lectin-based affinity capture for MALDI-MS analysis of bacteria. *Anal. Chem.* 1999, **71**, 1460–1463.
98. Bundy, J. L.; Fenselau, C. Lectin and carbohydrate affinity capture surfaces for mass spectrometric analysis of microorganisms. *Anal. Chem.* 2001, **73**, 751–757.
99. Madonna, A. J.; Basile, F.; Furlong, E.; Voorhees, K. J. Detection of bacteria from biological mixtures using immunomagnetic separation combined with matrix-assisted laser desorptio/ioniaztion time-of-flight mass spectrometry. *Rapid Comm. Mass Spectrom.* 2001, **15**, 1068–1074.
100. Carrion, M.; Smith, J.; Harris, B.; McVey, D. *Proc. 49th Conf. on Mass Spectrom. and Allied Topics*, Santa Fe, NM; Am. Soc. for Mass Spectrom., 2001.
101. Winkler, M. A.; Hickman, R. K.; Golden, A.; Aboleneen, H. Analysis of recombinant protein expression by MALDI-TOF mass spectrometry of bacterial colonies. *Biotechniques* 2000, **28**, 890–892, 894–895.
102. Winkler, M. A.; Xu, N.; Wu, H.; Aboleneen, H. MALDI-TOFMS of chemically modified recombinant hepatitis B surface antigen. *Anal. Chem.* 1999, **71**, 664A–667A.
103. Nilsson, C. L. Bacterial proteomics and vaccine development. *Am. J. Pharmacogenomics* 2002, **2**, 59–65.

13

MALDI-FTMS OF WHOLE-CELL BACTERIA

JEFFREY J. JONES, MICHAEL J. STUMP, AND CHARLES L. WILKINS
Department of Chemistry, University of Arkansas, Fayetteville, AR 72701

13.1 INTRODUCTION

For years mass spectrometrists have sought to develop rapid and accurate means of detecting and characterizing microorganisms. Recently taxonomic identification of pure bacteria has become possible through the use of matrix-assisted laser desorption/ionization (MALDI) mass spectrometry, directly detecting abundant proteins and other components that comprise cells during rapid cellular growth.[1–3] The resultant mass measurement of whole cell bacteria derives primarily from desorbed, intact proteins and lipid components, specific to cells in culture. MALDI mass spectra appear as characteristic profiles that are comprised of biomarkers, unique within particular classes of microorganisms that are sometimes referred to as "fingerprint patterns."[4–8]

These biomarkers, mass to charge observations unique to an organism, are not always assigned as specific proteins or lipids. Rather, low-resolution MALDI spectra are treated as "fingerprints." [9–12] Subtle changes in either growth of the cell colony or MALDI sample preparations can cause fluctuations in recorded mass fingerprints, distorting biomarkers and complicating microorganism identification.[13–15] However, reproducibility is sufficiently good that differences in mass fingerprints allows researchers to characterize and identify species of microorganisms.[1,3] Recently MALDI Fourier transform mass spectrometry (FTMS) of environmental samples has proved its advan-

Identification of Microorganisms by Mass Spectrometry, Edited by Charles L. Wilkins and Jackson O. Lay, Jr.

tage, showing that high resolution is the answer to making accurate assignments in complex mixtures.[16–18] As will be discussed in this chapter, FTMS takes whole-cell identification one step further by allowing assignment of molecular identities to observed masses. For proteins, accurate mass determination coupled with high resolution, eliminates the ambiguity that can arise from other components interfering with the assignment to known protein or lipid constituents. With its improved mass accuracy and much higher resolving power than MALDI-TOF mass spectrometry, FTMS allows use of specifically identified biomarkers for bacterial taxonomic purposes, a significant improvement over "fingerprint" approaches based on use of patterns of unidentified substances.

13.2 FUNDAMENTALS OF MALDI-FTMS

13.2.1 Whole-Cell MALDI

MALDI whole-cell bacterial samples often employ 2,5-dihydroxybenzoic acid (DHB) as the matrix compound because it has strong absorbance at 337 nm, has a low thermal desorption energy, and crystallizes well with intact proteins.[19] Whole-cell biological analysis must promote cell lysis and diffusion of cellular materials and proteins into the matrix solution. Consequently there are a variety of reported methods for the optimization of cellular lysis, including both chemical and physical methods.[20] This process ensures that cellular lysis takes place. The type of laser used for desorption in MALDI-FTMS has a significant effect on the quality of spectra. Nitrogen lasers operating at 337 nm and Nd-YAG lasers operating at 355 nm are the two most commonly used lasers in MALDI mass spectrometry. Both lasers routinely produce acceptable spectra, provided that the optimal power and beam spot size are carefully determined and controlled.

13.2.2 Fourier Transform Mass Spectrometry

Fourier transform mass spectrometry is made possible by the measurement of an AC current produced from the movement of ions within a magnetic field under ultra-high vacuum, commonly referred to as ion cyclotron motion.[21] Ion motion, or the frequency of each ion, is recorded to the precision of one thousandth of a Hertz and may last for several seconds, depending on the vacuum conditions. Waveform motion recorded by the mass analyzer is subjected to a Fourier transform to extract ion frequencies that yield the corresponding mass to charge ratios. To a first approximation, motion of a single ion in a magnetic field can be defined by the equation

$$f = \frac{z \cdot B}{m},$$

where m is the mass of the ion, z is the charge of the ion, B is the magnetic field in Tesla, and f is the frequency of the ion as it is cycling in the magnetic field.

Some typical sizes for superconducting solenoid magnets are 3.0, 4.7, 7.0, 9.4, and 12.0 Tesla.[22,23] The equation above shows that an increase in magnetic field results in an increase in ion frequency. It is not the increase in ion frequency that improves the mass resolution but the difference in frequencies between masses. Increased frequency separation results in increased resolution, afforded by higher magnetic fields. Additionally higher magnetic fields restrict larger mass ions to smaller magnetron motions and thus contain them longer. Therefore, when studying high molecular weight compounds, such as proteins, a larger magnetic field is desirable.

This equation is illustrated in Figure 13.1, detailing the effect magnetic field has on cyclotron frequencies of given mass/charge ions. For convenience, this plot can be divided into three regions, two of which display a nearly linear relation between frequency and mass/charge, the low and high mass regions, and the third that shows a more exponential relationship, the middle mass region. For this reason it is best to calibrate each of the three ranges separately. The same division of mass ranges, using a 7.0 Tesla or higher field magnet, is convenient for grouping biologically related compounds. Although there are exceptions, in general, lipids and phospholipids can be grouped into the low-mass region, peptides and protein fragments can be grouped into the middle-mass region, and whole proteins are observed almost exclusively in

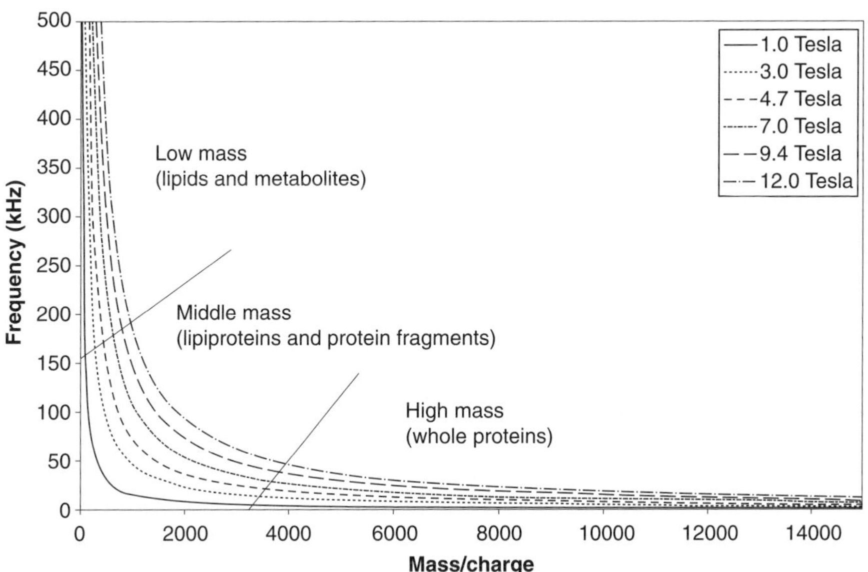

Figure 13.1 Plot of equation for various magnetic fields.

the high-mass region. Typically this plot is displayed on a logarithmic scale; however, for the present discussion purposes, it is displayed in this form to illustrate the point that large mass/charge ratios are difficult to measure with high accuracy and resolution, mainly because frequency differences between heavy ions are smaller. This fact is one reason for the current popularity of electrospray ionization FTMS for protein analysis, where the multiple charged ions typically produced result in much lower m/z ratios.

In almost all FTMS measurements, a set of standards are used to calibrate the instrument, wherein the magnetic field is calculated along with calibration constants. In order to minimize space-charge effects[24–26] a common experimental strategy is to eliminate by an ejection procedure all ions outside of those of the specific mass range of interest. This way errors can be reduced significantly over those for a corresponding broadband spectrum. With IonSpec instruments isolated and broadband spectra are commonly calibrated with the function designed by Li and coworkers,[24] with insulin, ubiquitin, and polyethylene glycol standards. Calibrations are built using the standards, starting out at low mass and working up to calibrants in the m/z region of interest while taking careful precaution not to mistake the A_1 (the one carbon-13 isotopic peak) as the A_0 (the all carbon-12 isotopic peak) to propagate a one Dalton error in mass measurement.[27]

13.3 FUNDAMENTALS OF COMPLEX BIOLOGICAL ANALYSIS

13.3.1 Whole-Cell MALDI-FTMS Analysis

FTMS accurately identifies and characterizes protein constituents and cellular components measured in experiments involving whole-cell bacteria. Although it has not been reported whether manipulation of the growth environment changes the spectra produced by MALDI-FTMS, it is presumed that spectra from altered bacteria will reflect environmental manipulation.[28–30] Such differences in spectra complicate pattern recognition techniques. Therefore, for MALDI-TOF spectra, considerable effort has been devoted to building mass fingerprint libraries that can be interrogated with elaborate search algorithms matching unknown spectra with archived species-specific spectra.[14,31–33] Additionally it is well known that small changes in growth conditions, matrix selection, or instrumental parameters affect biomarker abundance and presence, complicating identifications based on mass fingerprints.[13,33] FTMS, by nature, avoids these problems by measuring the molecular mass of desorbed bacterial components to ppm accuracy.[29]

Approximately 90% to 95% of whole-cell MALDI MS profiles are representative of the ribosomal proteins abundant in rapid growth whole cells. Although the identification of proteins, from whole bacterial cells, by MALDI-TOF MS analysis is ambiguous, at best, due to the low-mass accuracy and resolving power, several researchers realized that many of the observed masses potentially corresponded to ribosomal proteins. Arnold and Reilly[13] isolated ribosomal proteins from a strain of *E. coli* and analyzed them by MALDI-TOF analysis. In the course of their research, anomalies to database sequences were found, including posttranslational modifications previously unreported. They developed a set of rules to determine *N*-terminus methionine retention, based on the identity of the second amino acid in the protein sequence. Until this study was performed, it was not known why certain proteins retained the *N*-terminus methionine. It has also been documented[34,35] that several *E. coli* ribosomal proteins undergo posttranslation modifications, atypical for *E. coli*. These results from the work of Arnold and Reilly provided valuable background information, necessary for further development of MALDI-FTMS analysis of whole-cell bacteria.

13.3.2 Library Searching

Identification of individual compounds within a mass fingerprint from a whole cell bacterium is impossible with low-resolution mass data. High-resolution mass data afforded from FTMS analysis facilitates proteomics and genomics database searching. An overwhelming amount of data is easily accessible, evident by the number of characterized and sequenced proteins that exist in any reasonable online database. Several online databases exist, populated with amino acid sequences to proteins generated from open-reading-frame predictions to genomics data, Edmond degradation sequences and NMR and crystallographic solved structures and sequences.[36,37] Sorting through existing databases, in order to use them with high-resolution FTMS data, is an overwhelming task. Extensive database augmentation is necessary in order accommodate accurate mass and posttranslational modifications, and there is no currently available tool for searching accurate parent mass databases. Software, whose development was driven by the utility of ESI, exists only for accurate fragment mass data from pure parent mass precursors. Furthermore there are no such databases for accurate mass data of lipid and phospholipid parent ions. The entirety of published work on MALDI-FTMS whole-cell analysis was accomplished by tedious hand calculations for each of the presumed parent ions matching suspected, greatest abundance ribosomal proteins.

13.4 WHOLE-CELL CHARACTERIZATION THROUGH MALDI-FTMS

13.4.1 Whole-Cell FTMS Glyceride, Phospholipid, and Glycolipid Elucidation

The low molecular weight *m*/*z* region of spectra derived from microorganism whole cells is generally comprised of mass spectral peaks derived from glycerides, phospholipids, glycolipids, and small protein fragments. This mass spectral information should be examined as a compliment to protein biomarker detection. To date, taxonomic identification by mass spectrometry has focused on the identification of protein biomarkers to characterize microorganisms.[1–3] Identification of glycerides, phospholipids, and glycolipids is the basis of numerous bacterial pyrolysis[38–40] and the FAME[41–43] (fatty acid methyl ester) tests generally the accepted methods of classification by microbiologists. Only recently has mass spectrometry combined both protein and lipid analysis into one mass spectrometric technique.[44] FTMS is the only rapid analysis technique that can achieve from the same sample such a dynamic range of observations, including lipids, phospholipids, peptides, and proteins. A theoretical calculation of all possible lipid structures, degrees of unsaturation, and cation attachments reveals that for this application, mass accuracy of approximately 0.05 Da is needed, corresponding to a resolving power of 20,000 at an m/z of 1000 Da. When analyzing complicated mixtures, such as whole bacterial cells, this resolution and mass measurement is only achievable by FTMS analysis.

The identification of bacteria using high-resolution MALDI-FTMS is in part based on the accurate mass and isotopic abundance values obtained from each experiment. Ho and Fenselau[30] were able to demonstrate that the low molecular weight compounds contained in bacterial cell cultures were readily detectable by MALDI-FTMS and could be done with sufficient resolution and mass accuracy, such that mass assignments could be made regarding the chemical composition of peak groups. Jones et al. were able to reproduce similar results by detecting the lipid components of *Escherichia coli* JM109 in a higher field magnet, using a nitrogen laser.[29] Stump et al. elaborated on this work and achieved improved accuracy by using isotopically depleted growth media during the culture phase.[28]

It is plausible that several hundred peaks could be observed over a very narrow mass range, greatly increasing the difficulty of data interpretation. One approach developed at the University of Arkansas is to employ a rapid visual inspection system that eases the identification process. Jones et al. were able to show that using a set of logical rules based on lipid chemistry, mathematical formulas, and data collected with a high field 9.4 Tesla FTMS instrument, all peaks in a spectrum can be assigned with relatively high confidence.[44] This confidence was based on a relatively simple data treatment first employed by Kendrick[45–47] to characterize accurate mass data derived from the analysis of crude oil. This approach is based on the observation that any accurate mass can be separated into two distinct numbers, the integer (nominal mass) and the fractional mass. For example, the mass 12.34 is separated into 12 and 0.34 and plotted using Cartesian coordinates with nominal mass as abscissa and the fractional mass as the ordinate. Figure 13.2 shows all of the theoretical plotted points for six common phospholipid classes with potassium attached as the cationizing agent. Figure 13.3 shows the general molecular formulas of these phospholipids.

Mass defect plots utilized by Jones and coworkers[44] highlight impossible mass defects for the compounds considered and suggest chemical com-

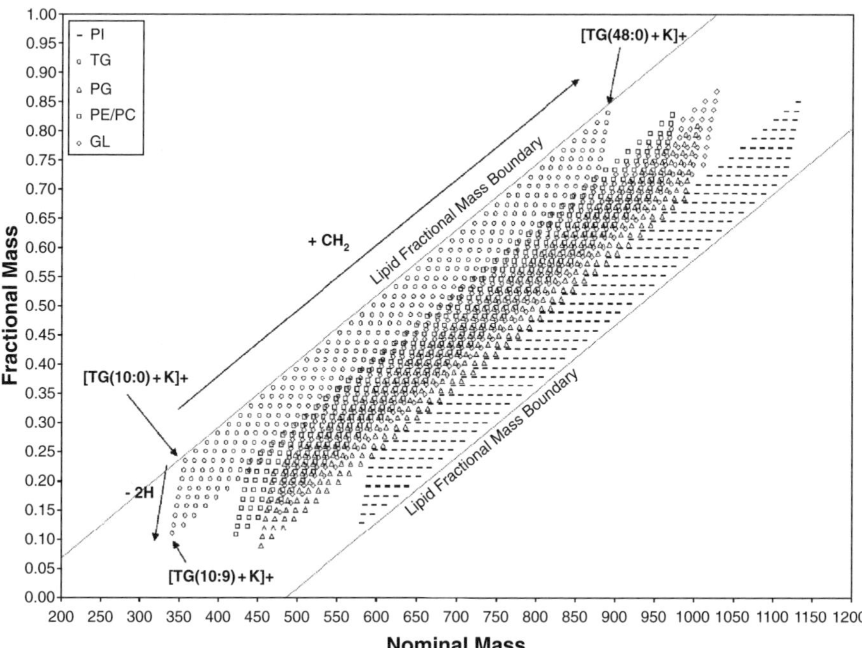

Figure 13.2 Theoretical mass defect plot for six common classes of phospholipids.

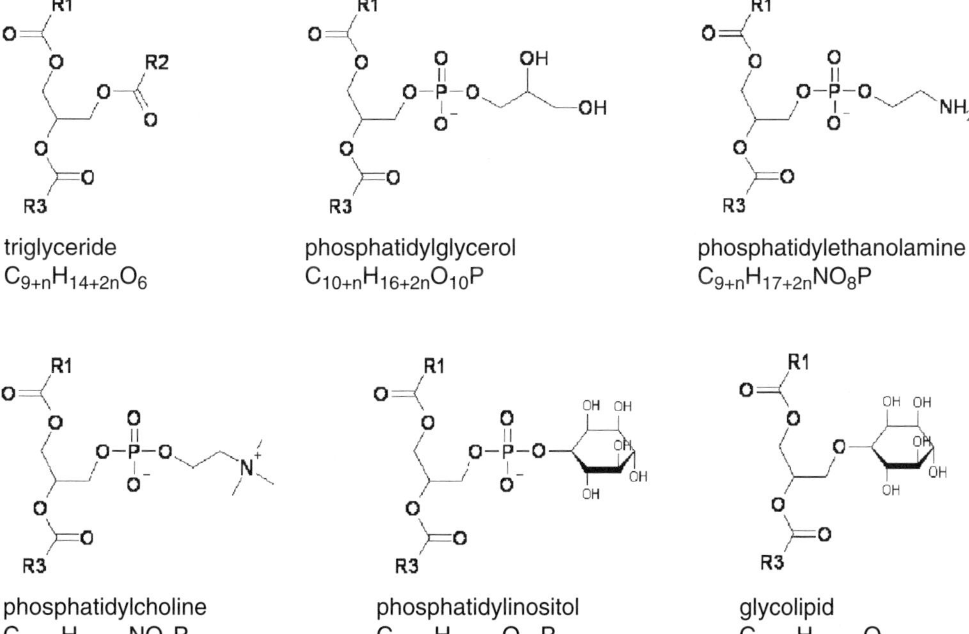

Figure 13.3 General structures for six common phospholipids.

positions for unknown or unpredicted ions. For example, ions corresponding to phospholipid fragments missing the polar head group ($C_3H_5O_4(CH_2)_n(CH_3)_2$–$PO_4H_2(CH_2)_2N(CH_3)_3$) were at first classified as unknown, until the fractional mass of each was revealed to contain four oxygen atoms inclusive of carbon and hydrogen hydrophobic chains. Knowing what atoms contribute to the mass defect for each unknown ion aided in the realization that these ions are possibly fragments originating from phospholipids whose observed *m/z* values have been identified in literature reporting mass spectra of isolated phospholipids.[48] Above the line in Figure 13.2 labeled "lipid fractional mass boundary," the fractional mass of any peak is inconsistent with expected values from lipid species and is unknown; these peaks may arise from either the presence of unknown chemical constituents or electronic noise. Some metabolites rich in oxygen do have small mass defects. These can be distinguished from common electronic noise by peak width, which is too narrow to be considered as arising from ions. Below the line bounding the lower limit of the calculations, the fractional mass of any peak plotted is too small to be consistent with assignment as a lipid or phospholipid. Peptides and protein fragments can be calculated and plotted according to their nominal and fractional masses, similar to lipids and phospholipids. Possible peptide masses are described in a similar manner and follow similar analysis.[49]

Misidentification of a mass as arising from a lipid and not a peptide is also unlikely because lipids follow distinct patterns of homologous series (i.e., unsaturation and carbon chain length) similar to a polymer series, whereas peptides generally do not. The elemental composition that governs the chemical makeup of proteins is richer in oxygen than lipids, and therefore, when plotted in the same manner as the lipids and phospholipids, protein appear in a separate region that is readily visualized on a mass defect plot, overlapping lipid compositions only in the low fractional mass region, typically below 0.3. The majority of low mass peaks from *Saccharomyces cerevisiae* (Fig. 13.4) have fractional masses that fall within the limits of the calculations for lipids and plot accordingly in the mass defect plot of Figure 13.5. This representation enhances visualization of obvious homologous series not evident in a conventional mass spectral display. Outlined are groupings of lipids and phospholipids as assigned from data analysis and mass searching a table of lipid compositions and masses. Distinct clustering of fragments is easily visualized as are species of cationized phosphatidyl choline lipids. Grouping of ions into distinct regions is expected for mass defect plots because each of the lipid head groups and contributing cation are shared for each grouping, a feature not evident in conventional mass spectral display.

13.4.2 Uncharacterized Proteins and Other Compounds

The middle-mass region, which includes almost entirely compounds that have eluded characterization and assignment, provides a readily observable spectrum. Mass assignments in this region are difficult due to ambiguity in complex

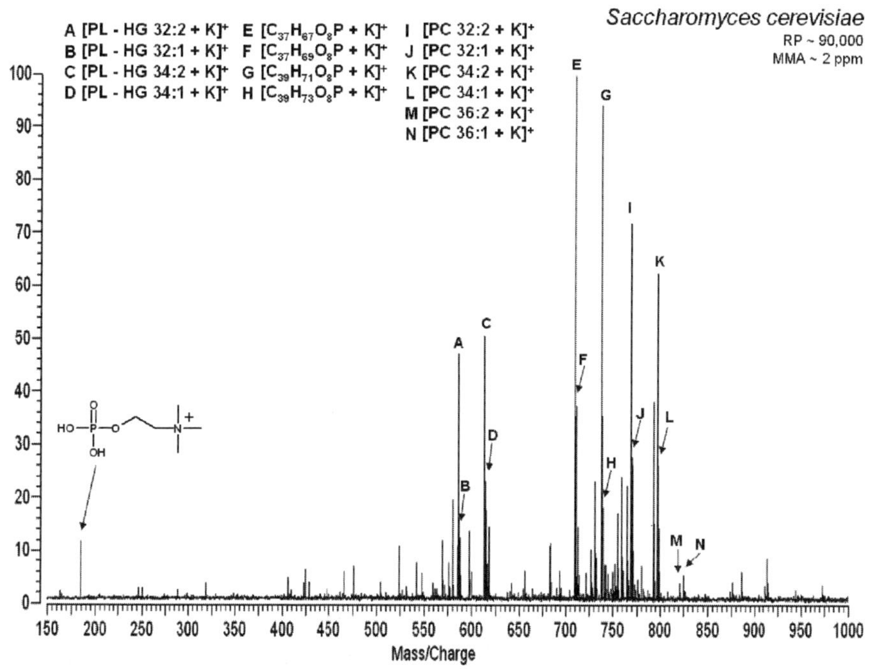

Figure 13.4 Low-mass lipid region of the 9.4 T MALDI-FTMS spectrum of *Saccharomyces cerevisiae.*

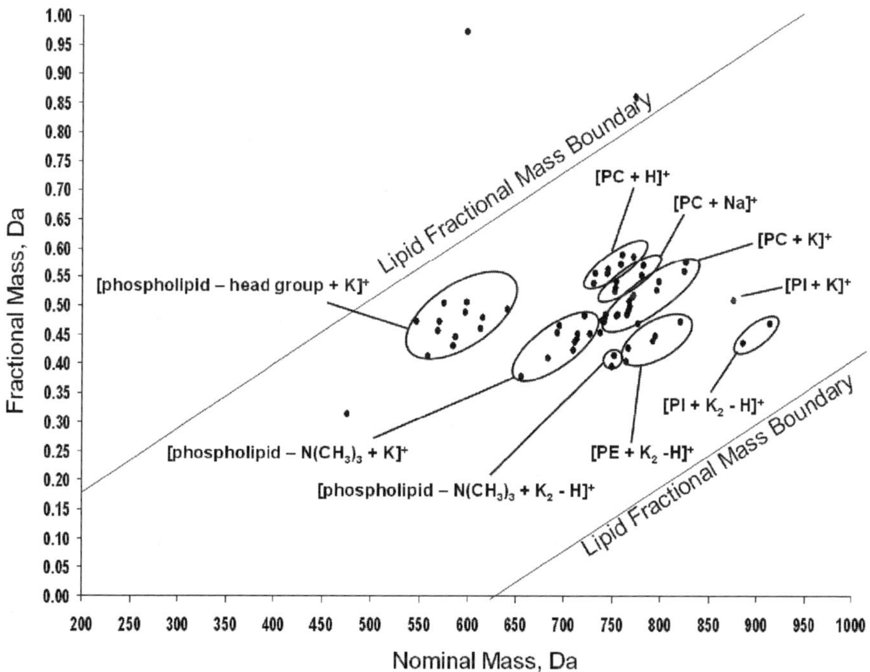

Figure 13.5 Mass defect plot of a 9.4 T MALDI-FTMS spectrum of *Saccharomyces cerevisiae.*

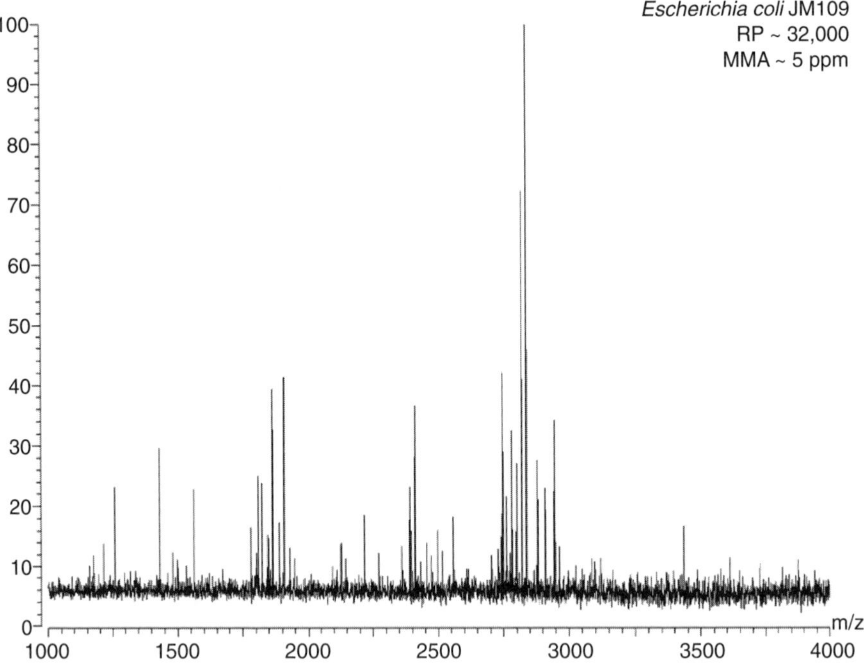

Figure 13.6 9.4 T MALDI-FTMS spectrum of *E. coli*, of the middle-mass region.

whole cell analysis, because none of the observable *m/z* peaks can be matched to a reliable database entry.[36,37] Each of the peaks in Figure 13.6 can be identified only as protein in origin, having a water-loss series, characteristic of MALDI-FTMS spectra of proteins measured with an external source instrument. There is reason to suspect that these compounds are fragments of lipoproteins observable by FTMS under acidic ionization conditions. However, the greatest majority of peaks relevant to the characterization of the organism, as understood to date, appear exclusively in the low- and high-mass regions.

13.4.3 Whole-Cell FTMS Protein Elucidation

The *m/z* 4000 to 12,000 range is defined as the bacterial mass fingerprint region for TOF data, and the same region is considered to be the high-mass range in FTMS analysis. Signals from high-abundance proteins, predominantly ribosomal proteins, most commonly found in whole cell bacteria appear in this region, as detailed in Figure 13.7. In order to accurately identify each of these biomarkers, it is necessary to have a database of known proteins or a complete DNA map of the organism. Using the tools available at EMBOSS,[50] Demirev and coworkers[51] predicted the relative abundances and probabilities

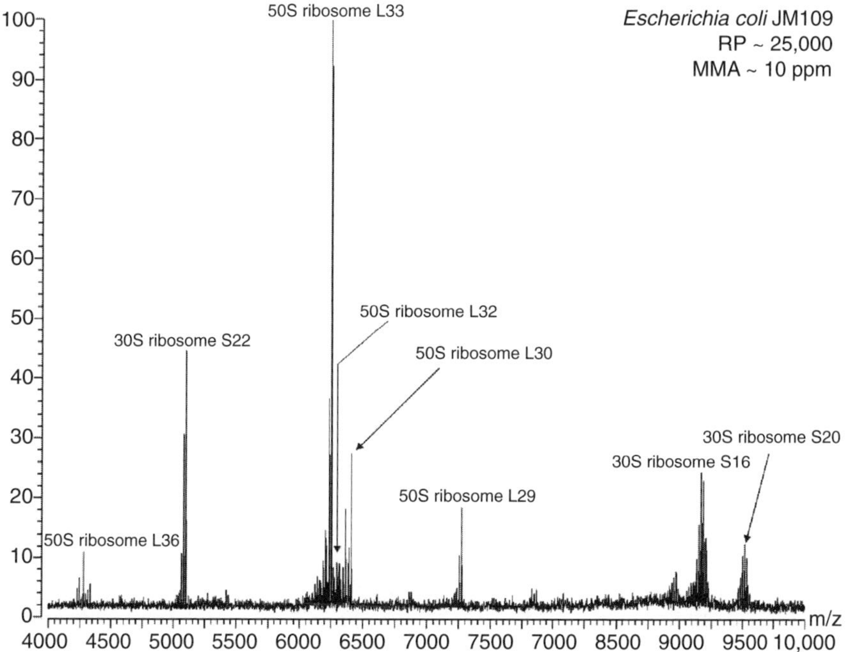

Figure 13.7 Broadband 9.4 T MALDI-FTMS spectrum of *E. coli* JM109, for the high-mass protein region.

of observation of cellular proteins by mass spectrometry. Their theory is based on the relative usage of codons, also considering gene regulation, and assumes that the more frequent a codon sequence is employed, without genetic repression, the higher the probability that a given gene sequence will express a protein. Their conclusions were later confirmed by the results of Stump and coworkers,[28] who observed nearly all the *E. coli* proteins predicted by Demirev and coworkers,[52] mostly ribosomal proteins. Even assuming the majority of the proteins being detected are ribosomal and the structure and mass are known, posttranslational modifications (PTMs) are also seen. Thus a percentage of the proteins found in any MALDI-MS analysis are unexpected, and in the absence of high-precision measurements, a mass fingerprint cannot be used to deduce the correct protein identities. Arnold and Reilly[35] tested this theory by isolating all known ribosomal proteins from *Escherichia coli* and then performing MALDI-TOF MS analysis on the individual isolates. The results allowed for the determination of *N*-terminus methionine retention.[34] These findings were later confirmed by MALDI-FTMS analysis of whole cells by Jones et al. and in greater detail by Stump et al.[28,29] Almost all the proteins observed in the MALDI-FTMS analysis of *Escherichia coli* whole cells are identified as ribosomal proteins.

Bacterial proteins from MALDI FTMS experiments are identifiable using the Swiss-Prot database.[53] In order to identify any microorganism, it is neces-

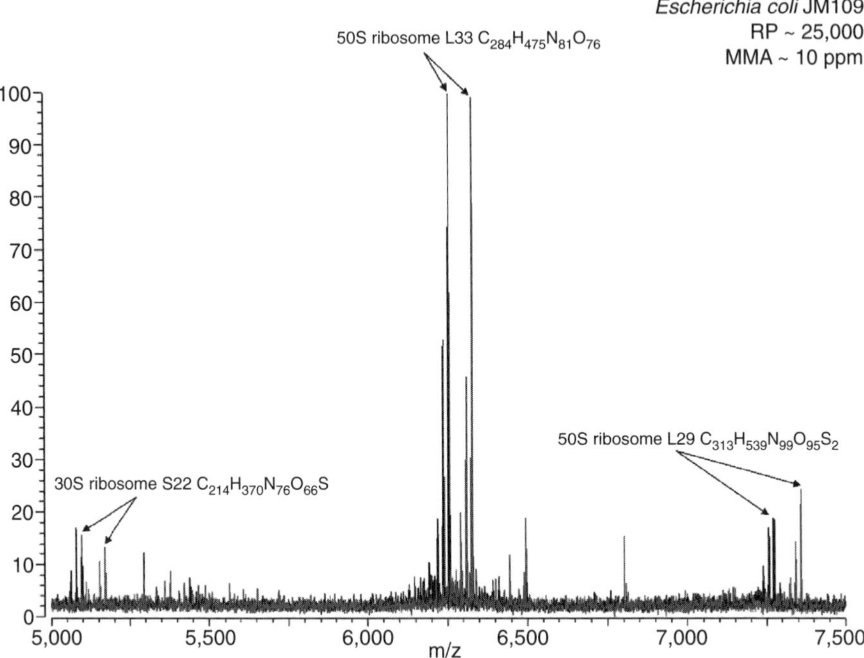

Figure 13.8 MALDI-FTMS spectrum of *E. coli* in natural abundance media (*black*) and >98.00% ^{15}N and >99.95% ^{12}C media (*blue*).

sary that the genome be known or at least a majority of the organism's proteome cataloged and solved by amino acid sequencing.[54] However, there are additions to Swiss-Prot database entries where TOF mass spectrometry is referenced for characterization of a posttranslational modification.[35] For example, the *E. coli* ribosomal protein L33, Swiss-Prot accession number P02436, is noted to exist in only the methylated form, as observed by mass spectrometry.

Assignment to each of the observed ribosomal proteins is confirmed by high-resolution FTMS data,[28] and through the use of ^{15}N enhancement as demonstrated in Figure 13.8. By growing bacteria in growth media that is >98.00% ^{15}N and >99.95% ^{12}C the number of nitrogen atoms can be counted simply by calculating the difference between masses measured in natural isotopic media and the latter. This type of analysis provides a measured characteristic other than the accurate mass. Any mass measured for a protein will usually have more than one match in a protein database, such as the entire Swiss-Prot and NCBI databases[36,37] that are similar in exact mass, within the mass measurement error. Therefore, by quantifying the number of nitrogen atoms for an observable protein mass, the certainty of the assignment increases. The average minimum required mass accuracy needed to make a

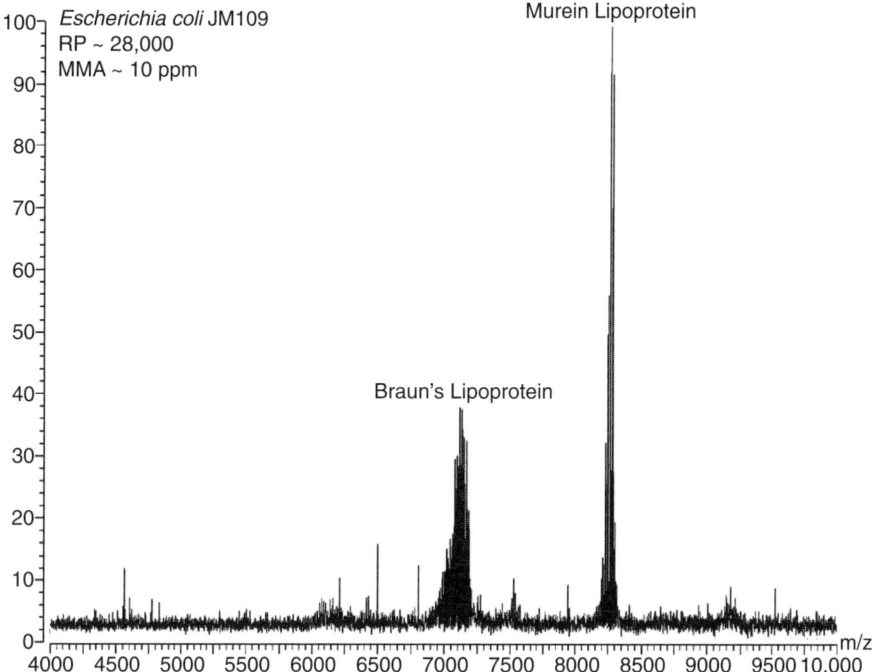

Figure 13.9 9.4 T MALDI-FTMS spectrum of *E. coli* JM109 whole cells showing both Braun's and Murein lipoproteins.

positive identification for unmodified proteins in the database with knowledge of the number of nitrogen atoms is 125 ppm, with a standard deviation of 23 ppm. If common posttranslational modifications are included, the average minimum required mass accuracy for protein identification becomes 50 ppm, with a standard deviation of 24 ppm. The same technique has been applied to lipid and phospholipids to confirm head group assignment, especially between phosphatidyl ethanolamine and phosphatidyl choline.[55]

Although ribosomal proteins are readily observed as in Figures 13.7 and 13.8 altered matrix conditions can alter the relative ionization of bacterial whole-cell compounds. A systematic analysis involving laser power/fluence and sample preparation conditions reveals that if the concentrated trifluoroacetic acid is added and the laser power increased above optimal conditions, ionization of bacterial surface compounds can be enhanced. Figure 13.9 is the resulting 9.4 T MALDI-FTMS, seen are both the Braun's lipoprotein[56,57] and the Murein lipoprotein. Both of these compounds are complex combinations of hydrocarbon lipids attached to a protein base. This is the first MALDI-FTMS observation of surface proteins desorbed directly from whole cells by influencing ionization conditions.

13.4.4 Isotopically Depleted Growth Media

In order to obtain the most accurate FTMS mass measurement for ions of interest, the best strategy is to eject other ions to maintain the ion population as constant as possible and consistent with that used for calibration. Space-charge and trapping effects on cyclotron resonance frequencies have been well studied,[24–26,58–60] and current calibration equations are designed to correct for the resulting frequency shifts.[24,26] The monoisotopic mass, which is lowest mass in an isotopic multiplet, is less detectable at high masses for ions with high carbon and nitrogen content.[27] Enhancing this mass by depletion of the ^{13}C and ^{15}N isotopes results in spectra of high-mass proteins that are dominated by a prominent single C^{12} peak instead of the more complicated normal isotopic distribution. Spectra observed without the use of the depleted growth media have large distributions of isotopic multiplets, sometimes containing up to 20 peaks with the monoisotopic peak in such low abundance it is not observable. The advantage to measuring isotopically depleted samples is a reduction in chemical noise from isotopic contributions. Enhancement of the monoisotopic mass leads to increases in the number of detectable ions at that mass with a resulting increase in mass accuracy.[61] Isotopically depleted bacterial whole-cell MALDI-FTMS spectra result in observations of proteins of predominately the monoisotopic mass, and reveals previously unobserved masses that are obscured by an isotopic multiplet.

Stump and coworkers analyzed bacteria grown in media doubly depleted in ^{13}C and ^{15}N isotopes (>99.95% ^{12}C and >99.97% ^{14}N) and observed the enhancements shown in Figure 13.10. When Figure 13.10*a* is compared with Figure 13.10*b*, it is immediately obvious that the MALDI-FTMS spectrum of the sample grown on isotopically depleted media reveals much more chemical detail than that for the natural isotopic abundance media sample. Figure 13.10*a* is an expanded scale plot that includes the ribosomal protein with Swiss-Prot accession number P02436 grown in natural isotopic abundance media. The ribosomal protein P02436 is posttranslationally modified with the addition of a methyl group. From MALDI-TOF spectral data using natural abundance growth media it was concluded that only the methylated form is present.[35] However, the MALDI-FTMS spectrum from the sample grown on depleted media clearly shows the presence of a significant amount of unmodified protein (labeled P02436 parent in the figure).

Along with improved mass resolution and enhancement of the monoisotopic mass for unambiguous assignment, isotopically depleted samples improve mass accuracy. Figure 13.10 is a partial mass spectrum of *Escherichia coli* grown in isotopically depleted media revealing the ribosomal protein P02429 and the cold shock protein P36996.[62] Because these proteins have masses above 7000 Da and because there is only a 3 Da difference between the masses of the proteins, it was difficult to assign the MALDI-TOF spectra to one of these two proteins. Understandingly, literature references disagree on the assignment.[63,64] Although the cold shock protein is not

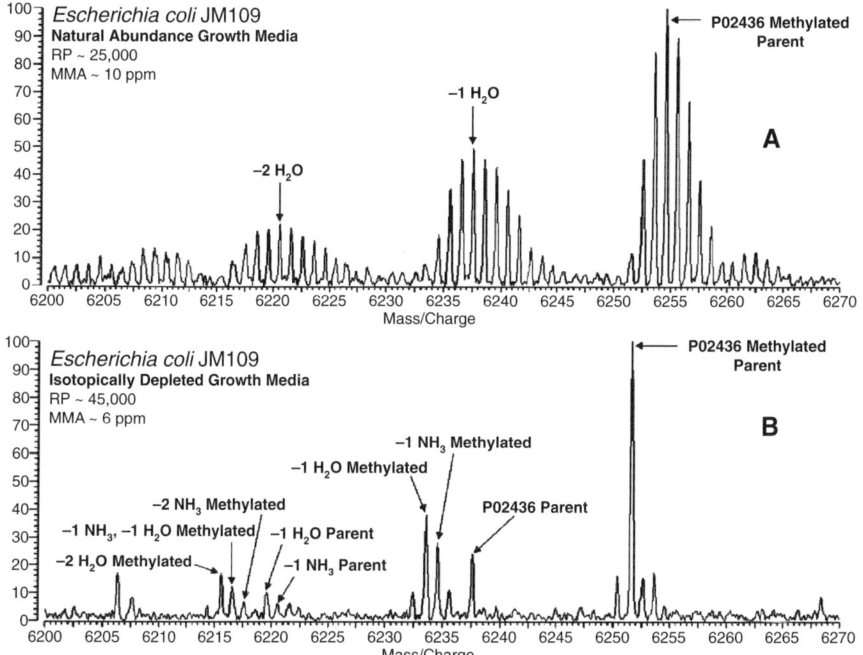

Figure 13.10 Partial 9.4 T MALDI-FTMS spectrum of *E. coli* JM109, showing differences in observable desorbed proteins between natural abundance and isotopically depleted growth media.

observed in its native state by MALDI-FTMS, Figure 13.11 is a partial spectrum showing both proteins accurately assigned with errors of less than 5 ppm in one spectrum.

The advantages of isotopically depleted media are not exclusive to proteins, spectra of lower molecular weight compounds, such as lipids and phospholipids, also benefit from improved mass accuracy in those cases where the isotopic-depletion stragegy can be employed.

13.5 RECOMBINANT OVEREXPRESSED PROTEINS DESORBED DIRECTLY FROM WHOLE CELLS

The utility of MALDI-FTMS analysis for use in chemotaxonomic applications has been established, but this method can be applied to other areas of interest, such as biomedical and environmental analyses. A common method used by biochemists and biologists today is recombinant overexpression of proteins using bacterial whole cells in cases where large quantity of a protein is desired. The main method presently used to determine if the overexpression was successful is the use of SDS-PAGE (sodium dodecylsulfate-poly acrylamide gel

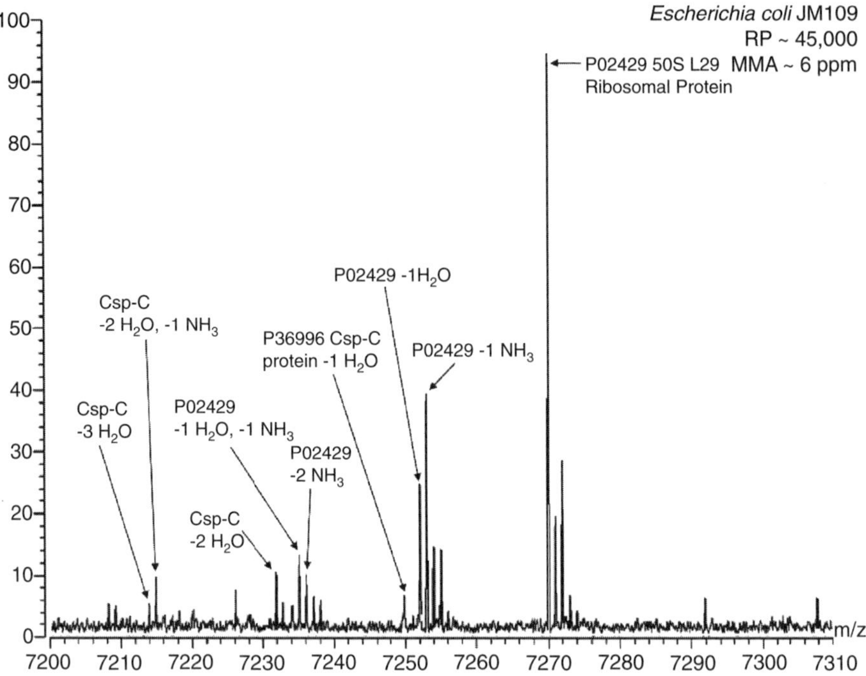

Figure 13.11 Partial 9.4 T MALDI-FTMS spectrum of *E. coli* JM109, showing expression of cold shock protein.

electrophoresis) method.[65] This method is known to be highly inaccurate for molecular weight determinations, and often results in incomplete separation of proteins in a sample. MALDI-MS can be used directly monitor the overexpression of proteins, which can be conveniently monitored by sampling at various times during the culture process. Easterling and coworkers demonstrated that monitoring protein overexpression could be accomplished using MALDI-TOF-MS analysis.[66] Additionally Parker et al. noted similar findings with the analysis of recombinant HIV-1IIIB p26 protein.[67]

As noted with the chemotaxonomic studies, the limited resolving power and mass accuracy of MALDI-TOF complicates identification of unknown proteins. If the greatly improved resolving power and accuracy of MALDI-FTMS can be used to monitor overexpressed proteins, it could have significant advantages. Figure 13.12 is a MALDI-FTMS spectrum of *E. coli* whole cells that have been genetically altered to produce the soluble core domain mammalian cytochrome b5 protein, which consists of 98 amino acids.

Note that the dominant peak in the observed mass spectrum is from the overexpressed protein. The observation of several ribosomal proteins allows for an internal calibration to be performed on the mass spectrum, greatly improving the mass measurement accuracy.

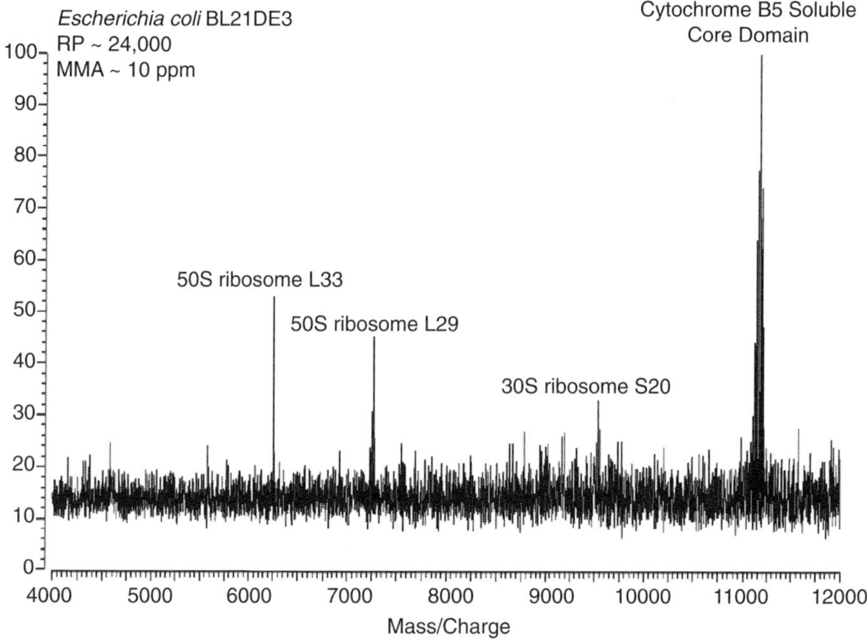

Figure 13.12 9.4 T MALDI-FTMS spectrum of *E. coli* BL21DE3 overexpressing the soluble core domain cytochrome b5 protein.

13.6 CONCLUSIONS

The results for bacterial whole-cell analysis described here establish the utility of MALDI-FTMS for mass spectral analysis of whole-cell bacteria and (potentially) more complex single-celled organisms. The use of MALDI-measured accurate mass values combined with mass defect plots is rapid, accurate, and simpler in sample preparation then conventional liquid chromatographic methods for bacterial lipid analysis. Intact cell MALDI-FTMS bacterial lipid characterization complements the use of proteomics profiling by mass spectrometry because it relies on accurate mass measurements of chemical species that are not subject to posttranslational modification or proteolytic degradation.

With its proved high-mass resolving power and accuracy MALDI-FTMS is quite suitable for application to bacterial taxonomy. As discussed above, use of isotopically depleted growth media as applied to analysis of bacterial whole cells results in MALDI-FTMS spectra showing increases in resolving power as a result of reduction of chemical noise. There are also significant gains in sensitivity over spectra from bacteria grown on natural abundance growth media. Thus in the future it is expected that MALDI-FTMS will continue to be used for such demanding applications.

REFERENCES

1. Lay, J. O., Jr. MALDI-TOF mass spectrometry of bacteria. *Mass Spectrom. Rev.* 2001, **20**, 172–194.
2. Fenselau, C.; Demirev, P. A. Characterization of intact microorganisms by MALDI mass spectrometry. *Mass Spectrom. Rev.* 2001, **20**, 157–171.
3. van Baar, B. L. M. Characterization of bacteria by matrix-assisted laser desorption/ionisation and electrospray mass spectrometry. *FEMS Microbiol. Rev.* 2000, **24**, 193–219.
4. Chong, B. E.; Kim, J.; Lubman, D. M.; Tiedje, J. M.; Kathariou, S. Use of nonporous reversed-phase high-performance liquid chromatography for protein profiling and isolation of proteins induced by temperature variations for Siberiam permafrost bacteria with identification by matrix-assisted laser desorption/ionization time-of-flight mass spectrometry and capillary electrophoresis-electrospray ionization mass spectrometry. *J. Chromatogr. B* 2000, **748**, 167–177.
5. Jacobs, A.; Dahlman, O. Enhancement of the quality of MALDI mass spectra of highly acidic oligosaccharides by using a Nafion-coated probe. *Anal. Chem.* 2001, **73**, 405–410.
6. Li, Y.; Tang, K.; Little, D. P.; Köster, H.; Hunter, R. L.; McIver, R. T. High-resolution MALDI Fourier transform mass spectrometry of oligonucleotides. *Anal. Chem.* 1996, **68**, 2090–2096.
7. Nelson, R. W.; Jarvik, J. W.; Taillon, B. E.; Tubbs, K. A. BIA/MS of epitope-tagged peptides directly from *E. coli* lysate: Multiplex detection and protein identificatin at low-femtomole to subfemtomole levels. *Anal. Chem.* 1999, **71**, 2858–2865.
8. Wall, D. B.; Lubman, D. M.; Flynn, S. J. Rapid profiling of induced proteins in bacteria using MALDI-TOF mass spectrometric detection of nonporous RP HPLC-separated whole cell lysates. *Anal. Chem.* 1999, **71**, 3894–3900.
9. Amiri-Eliasi, B.; Fenselau, C. Characterization of protein biomarkers desorbed by MALDI from whole fungal cells. *Anal. Chem.* 2001, **73**, 5228–5231.
10. Anhalt, J. P.; Fenselau, C. Identification of bacteria using mass spectrometry. *Anal. Chem.* 1975, **47**, 219–225.
11. Arnold, R. J.; Reilly, J. P. Fingerprint matching of *E. coli* strains with matrix-assisted laser desorption/ionization time-of-flight mass spectrometry of whole cells using a modified correlation approach. *Rapid Comm. Mass Spectrom.* 1998, **12**, 630–636.
12. Holland, R. D.; Duffy, C. R.; Rafii, F.; Sutherland, J. B.; Heinze, T. M.; Holder, C. L.; Voorhees, K. J.; Lay, J. O. Identification of bacterial proteins observed in MALDI TOF mass spectra from whole cells. *Anal. Chem.* 1999, **71**, 3226–3230.
13. Arnold, R. J.; Karty, J. A.; Ellington, A. D.; Reilly, J. P. Monitoring the growth of a bacteria culture by MALDI-MS of whole cells. *Anal. Chem.* 1999, **71**, 1990–1996.
14. Jarman, K. H.; Cebula, S. T.; Saenz, A. J.; Peterson, C. E.; Valentine, N. B.; Kingsley, M. T.; Wahl, K. L. An algorithm for automated bacterial identification using matrix-assisted laser desorption/ionization TOF mass spectrometry. *Anal. Chem.* 2000, **72**, 1217–1223.
15. Jarman, K. H.; Daly, D. S.; Peterson, C. E.; Saenz, A. J.; Valentine, N. B.; Wahl, K. L. Extracting and visualizing matrix-assisted laser desorption/ionization time-

of-flight mass spectral fingerprints. *Rapid Comm. Mass Spectrom.* 1999, **13**, 1586–1594.

16. Stenson, A. C.; Landing, W. M.; Marshall, A. G.; Cooper, W. T. Ionization and fragmentation of humic substances in electrospray ionization Fourier transform-ion cyclotron resonance mass spectrometry. *Anal. Chem.* 2002, **74**, 4397–4409.
17. Wu, Z.; Jernstroem, S.; Hughey, C. A.; Rodgers, R. P.; Marshall, A. G. Resolution of 10,000 compositionally distinct components in polar coal extracts by negative-ion electrospray ionization Fourier transform ion cyclotron resonance. *Mass Spectrom. Ener. Fuels* 2003, **17**, 946–953.
18. Hughey, C. A.; Rodgers, R. P.; Marshall, A. G. Resolution of 11,000 compositionally distinct components in a single electrospray ionization Fourier transform ion cyclotron resonance. *Mass Spectrum Crude Oil Anal. Chem.* 2002, **74**, 4145–4149.
19. Dreisewerd, K. The desorption process in MALDI. *Chem. Rev.* 2003, **103**, 395–425.
20. Williams, T. L.; Andrzejewski, D.; Lay, J. O.; Musser, S. M. Experimental factors affecting the quality and reproducibility of MALDI TOF mass spectra obtained from whole bacteria cells. *J. Am. Soc. Mass Spectrom.* 2003, **14**, 342–351.
21. Marshall, A. G.; Hendrickson, C. L.; Jackson, G. S. Fourier transform cyclotron resonance mass spectrometry: A primer. *Mass Spectrom. Rev.* 1998, **17**, 1–35.
22. Bruker Daltonics Inc., 40 Manning Road, Manning Park, Billerica, MA 01821, USA (Billerica), 2004.
23. Ion Spec, World Headquarters, 20503 Crescent Bay Drive, Lake Forest, California 92630 US, 2004.
24. Li, Y.; McIver, R. T., Jr.; Hunter, R. L. High-accuracy molecular mass determination for peptides and proteins by Fourier transform mass spectrometry. *Anal. Chem.* 1994, **66**, 2077–2083.
25. Francl, T. J.; Sherman, M. G.; Hunter, R. L.; Locke, M. J.; Bowers, W. D.; McIver, R. T. Experimental determination of the effects of space charge on ion cyclotron resonance frequencies. *Int. J. Mass Spectrom. Ion Phys.* 1983, **54**, 189–199.
26. Ledford, E. B., Jr.; Rempel, D. L.; Gross, M. L. Space charge effects in Fourier transform mass spectrometry: Mass calibration. *Anal. Chem.* 1984, **56**, 2744–2748.
27. Senko, M. W.; Beu, S. C.; McLafferty, F. W. Determination of monoisotopic masses and ion populations for large biomolecules from resolved isotopic distributions. *J. Am. Soc. Mass Spectrom.* 1995, **6**, 229–233.
28. Stump, M. J.; Jones, J. J.; Fleming, R. C.; Lay, J. O., Jr.; Wilkins, C. L. Use of double-depleted ^{13}C and ^{15}N culture media for analysis of whole cell bacteria by MALDI time-of-flight and Fourier transform mass spectrometry. *J. Am. Soc. Mass Spectrom.* 2003, **14**, 1306–1314.
29. Jones, J. J.; Stump, M. J.; Fleming, R. C.; Lay, J. O., Jr.; Wilkins, C. L. Investigation of MALDI-TOF and FT-MS techniques for analysis of *Escherichia coli* whole cells. *Anal. Chem.* 2003, **75**, 1340–1347.
30. Ho, Y. P.; Fenselau, C. Applications of 1.06-μm IR laser desorption on a Fourier transform mass spectrometer. *Anal. Chem.* 1998, **70**, 4890–4895.
31. Bright, J. J.; Claydon, M. A.; Soufian, M.; Gordon, D. B. Rapid typing of bacteria using matrix-assisted laser desorption ionisation time-of-flight mass spectrometry and pattern recognition software. *J. Microbiol. Methods* 2002, **48**, 127–138.

32. Dare, D. J.; Sutton, H. E.; Keys, C. J.; Shah, H. N.; Wells, G.; McDowall, M. A. Optimisation of a database for rapid identification of intact bacterial cells of *Escherichia coli* by matrix-assisted desorption ionization time-of-flight mass spectrometry, *Proc. 51st ASMS Conference*, Montreal, Quebec, Canada, June 8–12, 2003.
33. Wang, Z.; Russon, L.; Li, L.; Roser, D. C.; Long, S. R. Investigation of spectral reproducibility in direct analysis of bacteria poteins by marix-assisted laser desorption/ionization time-of-flight mass spectrometry. *Rapid Comm. Mass Spectrom.* 1998, **12**, 456–464.
34. Sherman, F.; Stewart, J. W.; Tsunasawa, S. Methionine or not methionine at the beginning of a protein. *Bio. Essays* 1985, **3**, 27–31.
35. Arnold, R. J.; Reilly, J. P. Observation of *Escherichia coli* ribosomal proteins and their posttranslational modifications by mass spectrometry. *Anal. Biochem.* 1999, **269**, 105–112.
36. Gasteiger, E.; Gattiker, A.; Hoogland, C.; Ivanyi, I.; Appel, R. D.; Bairoch, A. ExPASy: The proteomics server for in-depth protein knowledge and analysis. *Nucl. Acids Res.* 2003, **31**, 3784–3788.
37. National Center for Biotechnology Information (NCBI), US National Library of Medicine, 8600 Rockville Pike, Bethesda, MD 20894, 2004.
38. Gharaibeh, A. A.; Voorhess, K. J. Characterization of lipid fatty acids in whole-cell microorganisms using in situ supercritical fluid derivatization/extraction and gas chromatography/mass spectrometry. *Anal. Chem.* 1996, **68**, 2805–2810.
39. Kurkiewicz, S.; Dzierzewicz, Z.; Wilczok, T.; Dworzanski, J. P. GC/MS determination of fatty acid picolinyl esters by direct Curie-point pyrolysis of whole bacterial cells. *J. Am. Soc. Mass Spectrom.* 2003, **14**, 58–62.
40. Snyder, A. P.; Maswadeh, W. M.; Parsons, J. A.; Tripathi, A.; Meuzelaar, H. L. C.; Dworzanski, J. P.; Kim, M. G. Field detection of *Bacillus* spore aerosols with stand-alone pyrolysis–gas chromatography–ion mobility spectrometry. *Field Anal. Chem. Technol.* 1999, **3**, 315–326.
41. Xu, M.; Voorhees, K. J.; Hadfield, T. L. Repeatability and pattern recognition of bacterial fatty acid profiles generated by direct mass spectrometric analysis of in situ thermal hydrolysis/methylation of whole cells. *Talanta* 2003, **59**, 577–589.
42. Xu, M.; Basile, F.; Voorhees, K. J. Differentiation and classification of user-specified bacterial groups by in situ thermal hydrolysis and methylation of whole bacterial cells with tert-butyl bromide chemical ionization ion trap mass spectrometry. *Anal. Chim. Acta* 2000, **418**, 119–128.
43. Basile, F.; Beverly, M. B.; Abbas-Hawks, C.; Mowry, C. D.; Voorhees, K. J.; Hadfield, T. L. Direct mass spectrometric analysis of in situ thermally hydrolyzed and methylated lipids from whole bacterial cells. *Anal. Chem.* 1998, **70**, 1555–1562.
44. Jones, J. J.; Stump, M. J.; Lay, J. O., Jr.; Wilkins, C. L. Investigation of basic membrane chemistry and components of bacteria by FTMS using fundamental preparations and simple matrix conditions, *Proc. 5st ASMS Conference*, Orlando, FL, June 1–5, 2002.
45. Hughey, C. A.; Hendrickson, C. L.; Rodgers, R. P.; Marshall, A. G. Kendrick mass defect spectrum: A compact visual analysis for ultrahigh-resolution broadband mass spectra. *Anal. Chem.* 2001, **73**, 4676–4681.

46. Kendrick, E. A mass scale based on CH = 14.0000 for high resolution mass spectrometry of organic compounds. *Anal. Chem.* 1963, **35**, 2146–2154.

47. Wu, Z.; Rodgers, R. P.; Marshall, A. G. Two- and three-dimensional van Krevelen diagrams: A graphical analysis complementary to the Kendrick mass plot for sorting elemental compositions of complex organic mixtures based on ultrahigh-resolution broadband Fourier transform ion cyclotron resonance mass measurements. *Anal. Chem.* 2004, **76**, 2511–2516.

48. Al-Saad, K. A.; Zabrouskov, V.; Siems, W. F.; Knowles, N. R.; Hannan, R. M.; Hill, H. H. Matrix assisted laser desorption/ionization time-of-flight mass spectrometry of lipids ionization and prompt fragmentation patterns. *Rapid Comm. Mass Spectrom.* 2003, **17**, 87–96.

49. Zubarev, R. A.; Hakansson, P.; Sundqvist, B. Accuracy requirements for peptide characterization by monoisptopic molecular mass measurements. *Anal. Chem.* 1996, **68**, 4060–4063.

50. Rice, P.; Longden, I.; Bleasby, A. EMBOSS: The European molecular biology open software suite. *Trends Genet.* 2000, **16**, 276–277.

51. Demirev, P. A.; Lin, J. S.; Pineda, F. J.; Fenselau, C. Bioinformatics and mass spectrometry for microorganism identification: Proteome-wide post-translational modifications and database search algorithms for characterization of intact *H. pylori*. *Anal. Chem.* 2001, **73**, 4566–4573.

52. Demirev, P. A.; Ho, Y. P.; Ryzhov, V.; Fenselau, C. Microorganism identification by mass spectrometry and protein database searches. *Anal. Chem.* 1999, **71**, 2732–2738.

53. SwissProt/TrEMBL.

54. Demirev, P. A.; Feldman, A. B.; Lin, J. S.; Pineda, F. J.; Resch, C. L. Microorganism identification by mass spectrometry and bioinformatics-generated databases, *Proc. 51st ASMS Conference*, Montreal, Quebec, Canada, June 8–12 2003.

55. Jones, J. J.; Stump, M. J.; Fleming, R. C.; Lay, J. O.; Wilkins, C. L. Strategies and data analysis techniques for lipid and phospholipid chemistry elucidation by intact cell MALDI-FTMS. *J. Am. Soc. Mass Spectrom.* 2004, **15**, 1665–1674.

56. Braun, V.; Bosch, V. Repetitive sequences in the murein-lipoprotein of the cell wall of *Escherichia coli. Proc. Nat. Acad. Sci. USA* 1972, **69**, 970–974.

57. Pittenauer, E.; Quintela, J. C.; Schmid, E. R.; Allmaier, G.; Paulus, G.; de Pedro, M. A. Characterization of Braun's lipoprotein and determination of its attachment sites to peptidoglycan by 252-Cf-PD and MALDI time-of-flight mass spectrometry. *J. Am. Soc. Mass Spectrom.* 1995, **6**, 892–905.

58. Masselon, C.; Tolmechev, A. V.; Anderson, G. A.; Harkewicz, R.; Smith, R. D. Mass measurement errors caused by local frequency perturbations in FTICR mass spectrometry. *J. Am. Soc. Mass Spectrom.* 2002, **13**, 99–107.

59. Becker, E. W.; Walcher, W. Image defects in a mass spectrometer due to space charge and magnetic field saturation. *Z. Physik* 1952, **131**, 395–407.

60. Jeffries, J. B.; Barlow, S. E.; Dunn, G. H. Theory of space-charge shift of ion cyclotron resonance frequencies. *Int. J. Mass Spectrom. Ion Phys.* 1983, **54**, 169–187.

61. Marshall, A. G.; Senko, M. W.; Li, W.; Dillon, S.; Guan, S.; Logan, T. M. Protein molecular mass to 1 Da by 13C, 15N double-depletion and FT-ICR mass spectrometry. *J. Am. Chem. Soc.* 1997, **119**, 433–434.

62. Goldstein, S.; Pelitt, S. N.; Inouye, M. Major cold shock protein of *Escherichia coli. Proc. Nat. Acad. Sci. USA* 1990, **87**, 283–287.

63. Dai, Y.; Li, L.; Roser, D. C.; Long, S. R. Detection and identification of low-mass peptides and proteins from solvent suspensions of *Escherichia coli* by high performance liquid chromatography fractionation and matrix-assisted laser desorption/ionization mass spectrometry. *Rapid Comm. Mass Spectrom.* 1999, **13**, 73–78.

64. Ryzhov, V.; Fenselau, C. Characterization of the protein subset desorbed by MALDI from whole bacterial cells. *Anal. Chem.* 2001, **73**, 746–750.

65. Chen, H.; Wang, R.; Cerniglia, C. E. Molecular cloning, overexpression, purification, and characterization of an aerobic FMN-dependent azoreductase from *Enterococcus faecalis. Protein Expression Purif.* 2004, **34**, 302–310.

66. Easterling, M. L.; Colangelo, C. M.; Scott, R. A.; Amster, J. I. Monitoring protein expression in whole bacterial cells with MALDI time-of-flight mass spectrometry. *Anal. Chem.* 1998, **70**, 2704–2709.

67. Parker, C. E.; Papac, D. I.; Tomer, K. B. Monitoring cleavage of fusion protein by matrix-assisted laser desorption ionization/mass spectrometry: Recombinant HIV-1IIIB p26. *Anal. Biochem.* 1996, **239**, 25–34.

14

A REVIEW OF ANTIBODY CAPTURE AND BACTERIOPHAGE AMPLIFICATION IN CONNECTION WITH THE DIRECT ANALYSIS OF WHOLE-CELL BACTERIA BY MALDI-TOF MS

KENT J. VOORHEES AND JON C. REES

Department of Chemistry and Geochemistry, Colorado School of Mines, Golden, CO 80401

14.1 INTRODUCTION

Events involving deliberate or accidental distribution of bacterial pathogens into our everyday environment have clearly defined the need for a sensitive, specific, and rapid method of bacterial detection. Bioterrorism was first introduced in the United States in 1984 with the *Salmonella typhimurium* attack in The Dalles, Oregon, by a cult group attempting to affect a local election.[1] As a result of this act 751 people contracted salmonellosis, which totally overwhelmed the hospitals and medical clinics with patients. Later our society became keenly aware of the potential of bioterrorism during the last four months of 2001 when *Bacillus anthracis* (anthrax) spores were sent through the US mail in an envelope to several locations. These events had

Identification of Microorganisms by Mass Spectrometry, Edited by Charles L. Wilkins and Jackson O. Lay, Jr.

limited physical consequences but were successful in spreading fear to millions of people.

Many diseases, including anthrax, are most effectively treated before actual manifestation of the symptoms is observed. Presently a presumptive identification of *Bacillus anthracis* can be made in about 3 hours; however, if a full laboratory response network (LRN) confirmation procedure is utilized, the theoretical time increases substantially to approximately 48 hours. During the recent anthrax cases 72 to 96 hours were common to complete the entire LRN protocol. In the meantime antibiotics were administered as a precaution based on the presumptive results to individuals thought to be exposed to *B. anthracis* spores or with anthrax symptoms. The mass administering of antibiotics from a cost standpoint, as well as from medical prudence to prevent the rise of antibiotic-resistant strains, is not the optimal answer to the anthrax infection problem. Therefore it is important that early tests be rapid and reliable with a minimum number of false positive and false negative results.

The LD_{50} for the inhalation of various pathogens, as published in open source agent lists, ranges from 8000 to 10,000 spores for *B. anthracis* to one organism for *Coxiella burnetti* (Q-fever). This requires an instrument to have sub-picogram sensitivity to directly detect an organism that would be fatal to 50% of the exposures. The future potential for widespread mayhem on human health caused by the deliberate use of biological weapons can be estimated based on the many cases of unintentional outbreaks of food borne disease. An outbreak of *Salmonella enteritidis* from contaminated liquid ice cream caused 224,000 to become sick in 1994 in the United States.[2] Similarly in the United States a 1985 outbreak of *S. typhimurium* sickened 170,000 people when pasteurized milk from a dairy plant became contaminated.[3] In what could be the largest incident of widespread illness arising from contaminated food, 300,000 people were sickened by an outbreak of hepatitis A associated with the eating of clams in Shanghai, China.[4]

Another important group of pathogens, *Enterohemorrhagic Escherichia coli* (EHEC), are strains of *E. coli* that are emerging etiological agents of food-borne illness. In the United States an estimated 73,000 cases of EHEC occur annually, resulting in 61 deaths.[5] EHEC cause severe bloody diarrhea and intestinal cramps, with little or no fever. In rare instances EHEC can cause nonbloody diarrhea or may present no symptoms at all. In some individuals, particularly elderly persons or children under the age of 5, EHEC infection leads to hemolytic uremic syndrome (HUS) where red blood cells are destroyed, leading to kidney failure, and often, to death. In the United States, HUS is the leading cause of acute kidney failure in children, and the majority of cases of HUS are caused by EHEC. The previous statistics collected for *Salmonella*, hepatitis A, and *E. coli* demonstrate the enormous scale to which disease may spread should food become contaminated with an infectious agent and a rapid and reliable diagnosis of the pathogen is not attained.

14.2 BACTERIAL IDENTIFICATION

Traditional microbiological methods for the identification of bacteria involve culturing the bacteria (enrichment), staining, and examination of the colonies with microscopy. This approach usually requires 24 to 48 hours for results. In an attempt to reduce the time and labor involved with bacterial identification, several new detection methodologies have been developed. These methodologies can be broadly classified into four major areas: DNA based methods,[6] immunoassay based methods,[7] immuno-latex agglutination methods,[8] and biochemical methods.[9] DNA-based detection methods typically take 24 hours to obtain a result, including steps for enrichment, nucleic acid extraction, in vitro amplification, and detection of the amplified material by gel electrophoresis and/or hybridization techniques. While DNA methods are highly specific and sensitive, they are expensive in terms of initial capital cost for equipment, trained labor to run the test, and assay reagent costs. Immunoassay and immuno-latex agglutination assays are numerous and typically take 24 to 48 hours to obtain a result. These assays are not as sensitive or specific as DNA based assays and require an enrichment step, yet remain attractive because of cost and ease of use.

In addition to the previously described methodologies, there are several new detection strategies based on analytical instruments, such as mass spectrometry, which have been primarily funded by the US Department of Defense (DoD). The DoD mass spectrometer systems have been designed to provide results in less than 5 minutes with a sensitivity approaching 25 colony forming units per liter (CFU/liter) of air for aerosols. These instruments fulfill the requirements of analysis speed and sensitivity; however, the major problem with these approaches is their inability to analyze microorganism mixtures or a single microorganism in a complex biological background. In the mid-1990s there was a clear need for newer techniques that would allow for mixture analysis.

A paper by Holland et al. first reported on the use of matrix-assisted laser desorption ionization–time-of-flight mass spectrometry (MALDI-TOF MS) for the direct analysis of proteins from whole-cell bacteria.[10] In the initial study, conducted in early 1995, five bacteria were analyzed by MALDI MS to obtain reference spectra for future studies involving unknown samples. These spectra were surprisingly simple, and usually contained 3 to 10 peaks over a molecular weight range of 5000 to 16,000 Da. Mass spectra for *Enterobacter cloacae* and *Proteus mirabilis* are shown in Figure 14.1. Some of the peaks, as would be expected, were unique to a particular bacterium while others were present in all of the bacteria studied. The MALDI mass spectra of the five bacteria could be differentiated by visual examination without the need for pattern recognition analysis. Later samples analyzed as unknowns were compared to reference spectra which allowed for the correct identification of each sample.

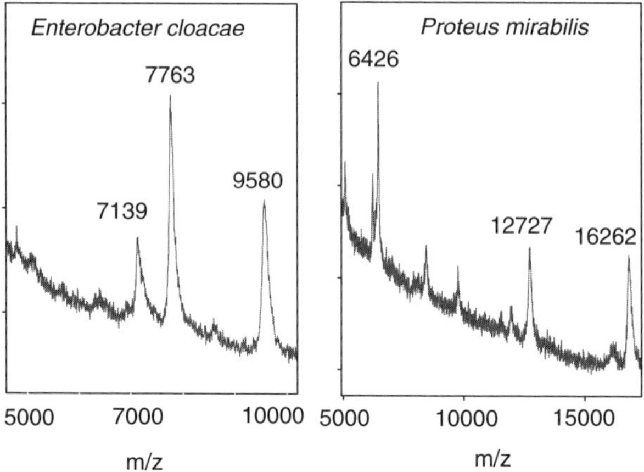

Figure 14.1 MALDI MS of *Enterobacter cloacae* and *Proteus mirabilis.*

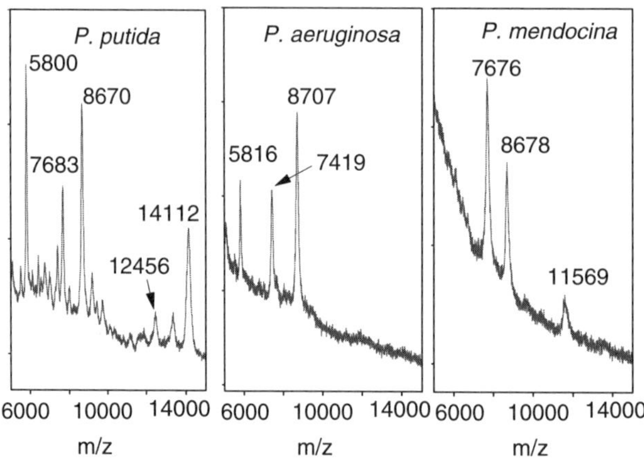

Figure 14.2 MALDI mass spectra of three different species of *Pseudomonas*.

Variations on the spectral peaks from different species of the same genus were also observed. Three species of *Pseudomonas* produced the spectra shown in Figure 14.2. These spectra are clearly unique and were used to correctly identify unknown samples. Because of peak ratio reproducibility issues in bacterial protein profiles obtained by MALDI MS,[11] a fingerprint approach that had been used for other mass spectrometry approaches has not been used. The profile reproducibility problem was first recognized by Reilly et al.[12,13] and later researched by others in the field.[14,15] As a later alternative, a direct comparison of the mass-to-charge ratio (*m*/*z*) of the unknown mass spectral peaks with a database of known protein masses has been used to identify unknown samples.[14]

The achievable mass resolution (peak width = 1.5 Da at 1000 Da) and mass accuracy (±0.02%) of current commercially available MALDI-TOF MS instruments is sufficient to make the MALDI-MS bacterial identification approach practical. For example, in the sodium dodecyl sulfate (SDS) polyacrylamide gel electrophoresis (PAGE) analysis (mass resolution of about 1 kDa) of six isolates of *Encephalitozoon intestinalis* against reference strains of *E. hellem*, and *E. cuniculi*, over 50 bands ranging from 14 to 200 kDa were observed.[16] Despite the complexity and a number of shared bands, characteristic band patterns were observed that could easily differentiate the three species of *Encephalitozoon*. Smaller differences were observed for the six isolates of *E. intestinalis*. Because the mass resolution achieved with SDS-PAGE is relatively low when compared to MALDI-TOF MS, the increased mass resolution of MALDI MS is more than enough to elucidate mass differences as small as 100 Da or lower between two protein signals. In the case of viruses, identification has been accomplished by detection of its capsid protein subunits.[17]

An example of the microorganism differentiation power achievable with the protein profiling approach is illustrated using the MALDI mass spectra of *Escherichia coli* and *Cryptosporidium parvum* (Figure 14.3). Clear

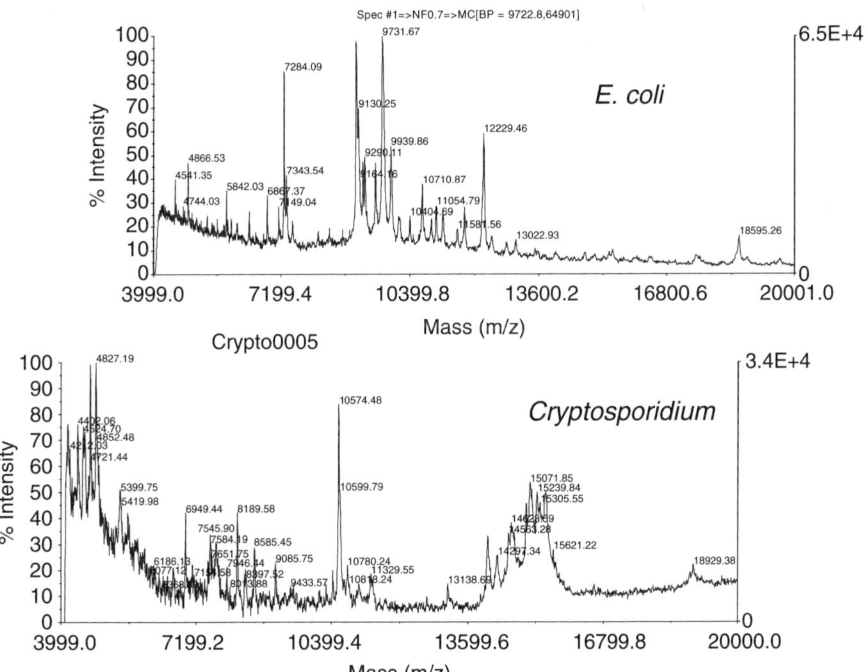

Figure 14.3 MALDI mass spectra of *E. coli* (top) and *Cryptosporidium parvum* (*bottom*).

differentiation between the two microorganisms is achieved by comparing the two spectra. Signals at *m/z* 10,574 and in the 14 to 15kDa are unique to *C. parvum*, and these signals were used to unequivocally identify this pathogen.

MALDI-MS studies conducted on whole-cell bacteria prior to 2000 utilized spectra that rarely contained peaks above 20,000Da. In 2000 a paper describing a methodology was published that extended the mass range for whole-cell bacteria to 100,000Da.[18] The matrix solvent consisted of a mixture of formic acid, acetonitrile, and water in a ratio of 17:33:50. The solvent mixture has been successfully used with most of the common MALDI matrices. The reasons for the success of this solvent are most likely related to the crystal formation of the matrix. When compared to other solvents, smaller and more uniform matrix-analyte crystals were observed with the method using formic acid, acetonitrile, and water as the solution. Figure 14.4 shows an example of *E. coli* obtained utilizing the described methodology.

The two remaining shortfalls with MALDI-MS analysis of whole bacterial cells are sensitivity and mixture analysis. The sensitivity for MALDI-MS analysis of whole-cell bacteria from our experiments and those reported by other laboratories is about 10^7 cells/ml. To realistically utilize MALDI MS as a tool that meets DoD detector sensitivity goals, this should be 10^3 cells/ml or lower.

Many situations exist when MALDI-MS analysis of a sample will involve a bacterial mixture or a single bacterium in a complex matrix. From our experience, MALDI-MS mixture analysis produces spectra that are highly irreproducible. Figure 14.5 represents spectra of two individual bacteria and a 50:50 mixture of the two bacteria. In the spectrum of the mixture the

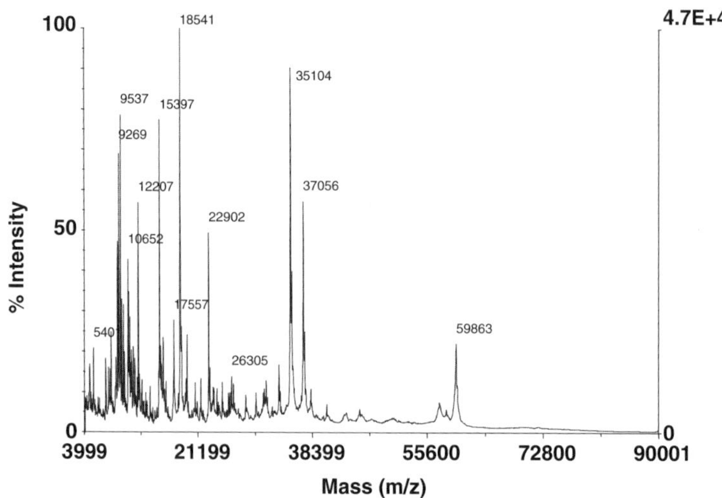

Figure 14.4 MALDI MS of *E. coli* run to 90,000Da.

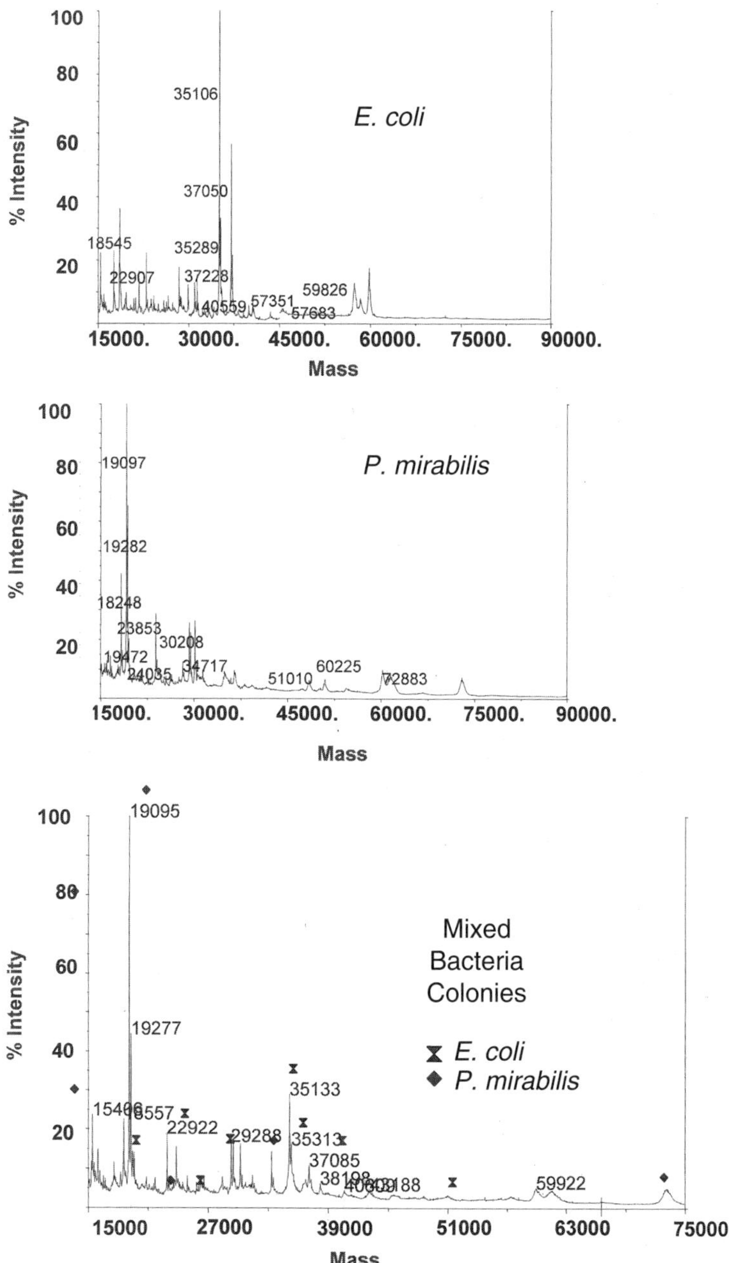

Figure 14.5 Spectra of individual bacteria and a 50:50 mixture of the two bacteria.

peaks from *P. mirabilis* are much more intense than the peaks from *E. coli*. It is quite possible that a re-analysis of the same mixture will produce an entirely different set of peak intensities where the *E. coli* base peaks are most intense.

Wahl et al. have investigated the mixture problem utilizing a statistical approach.[19] In a recent paper 50 simulated mixed bacterial cultures were generated and analyzed by MALDI MS. All of the bacteria used were in their fingerprint library. Using an automated data extraction routine and a PNNL analysis algorithm,[20] all but one sample was correctly identified to the species level. The authors did note that for closely related bacteria some difficulty existed in making the correct assignment.

14.3 IMMUNOCAPTURE OF BACTERIAL MIXTURES

Another approach to mixture analysis is using some type of capture molecule to selectively remove a targeted bacterium. Hutchens and Yip were the first to show that affinity molecules could be directly coupled with MALDI MS to isolate and detect a targeted molecule.[21] In their study single-stranded DNA with a high affinity for lactoferrin was immobilized on agarose beads and mixed with infant urine containing lactoferrin. After removal of the beads from the urine followed by thorough washing, MALDI-MS analysis confirmed the capture and removal of lactoferrin from the sample. In an effort to consolidate affinity capture with MALDI-MS analysis, Brockman and Orlando have covalently attached an antibody directly to a gold-plated MALDI plate, and showed that lysozyme could be extracted from a tear solution and detected by MALDI MS.[22] Other groups have expanded antibody capture of other biomolecules combined with MALDI MS.[23]

The first application of affinity capture of whole-cell bacteria coupled with MALDI analysis was demonstrated by Bundy and Fenselau. This group showed that lectins, protein moieties that bind carbohydrates, can be used to isolate Gram-negative bacteria prior to mass analysis.[24] The lectin Concanavalin A was immobilized to a gold foil taped to a MALDI probe, and used to retrieve *E. coli* from a physiological buffer and a spiked sample of urine. MALDI-MS analysis confirmed that *E. coli* had been captured from the solutions by identifying protein biomarkers in the resultant spectra. A follow-up study by Bundy and Fenselau not only validated the ability of immobilized lectins to isolate bacteria (and viruses) from suspensions but also used immobilized carbohydrates to isolate bacteria by binding to lectins present on the surface of the bacterial cell.[25] A third study conducted by this research group reported on the immobilized lectin motif to a microscope slide, which after immersion in a suspension of bacteria could be directly inserted into the mass spectrometer via a modified MALDI plate.[26]

Madonna et al. evaluated an immunologically based trapping/collection system to isolate bacteria prior to MALDI-MS analysis.[27] Figure 14.6 illus-

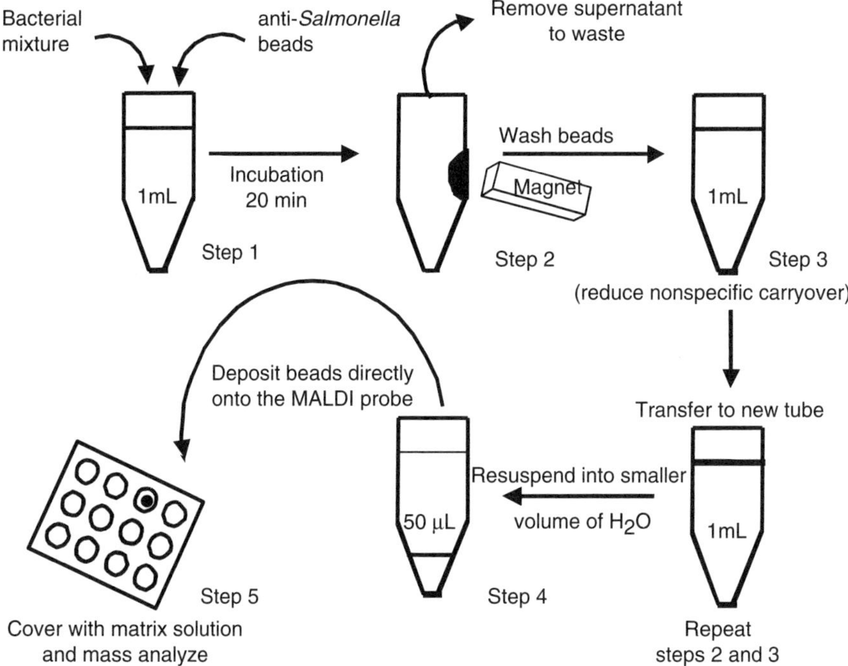

Figure 14.6 Immunomagnetic separation (IMS) MALDI-MS protocol to isolate bacteria from biological mixtures.

trates the Madonna et al. procedure. Immunomagnetic beads with antibodies specific for the target biological agent are mixed with the sample for a short incubation period (20 min), then isolated by a magnet and washed free of background constituents. Following the washing step, the beads are then re-suspended into a smaller volume of double de-ionized water, and applied dropwise to the MALDI sample probe. The target bead complex is then overlaid with microliter amounts of matrix solution and dried at room temperature, forming a crystalline mixture of sample and matrix. Irradiation of the resulting crystalline mass with a high-intensity laser promotes the liberation and ionization of intact cellular proteins that are subsequently detected by a time-of-flight mass spectrometer (TOF MS). The resulting mass spectrum is then evaluated for definitive mass peaks that signify the presence of the suspected organism.

Results from experiments using the protocol above[27] have shown that anti-*Salmonella* immunomagnetic beads could be used to unambiguously determine the presence of *Salmonella choleraesuis* from suspensions of bacterial mixtures. This target organism was also positively identified from spiked samples of river water, human urine, chicken blood, and 1% milk. For the river water and urine samples, no cross-reactivity was observed and only protein

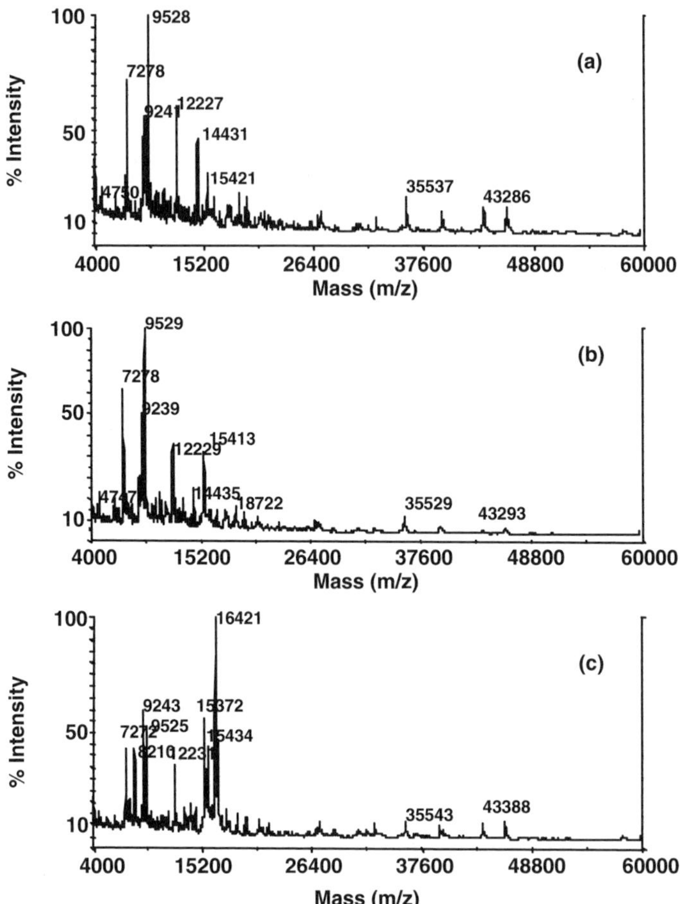

Figure 14.7 Immuno-separated *Salmonella choleraesuis* from (*a*) river water, (*b*) human urine, and (*c*) chicken blood.

signals indicative of *S. choleraesuis* were produced in the resulting mass spectra. High-intensity peaks at 7.3 kDa, 9.2 kDa, 9.5 kDa, 12.2 kDa, 14.4 kDa, 15.4 kDa, 35.5 kDa, and 43.4 kDa are known biomarkers for *S. choleraesuis*. However, the sample of chicken blood showed signals due to cross-reactions between the anti-*Salmonella* antibodies and a protein with a signal at *m/z* 16.4 kDa. This interfering signal is probably due to myoglobin in blood, which has a peak with a mass coinciding with the contaminating peak. The mass spectra from the river water, urine, and chicken blood samples are shown in Figure 14.7. These example mass spectra illustrate the powerful capability of combining immunoassays with mass spectrometry. Although the antibodies used in this assay cross-reacted in the case of the blood, the target could still be determined by its signature protein peaks. More important, had the target not been present, the cross-reaction would not have been mistaken for a positive result. Under similar circumstances methods that rely on binary (yes/no) outputs like fluorescence emission would most likely return a false positive result. Since the nature of the interference is known, steps can be implemented to eliminate or cleanup the sample further prior to immunomagnetic separation (IMS). Although not published in Madonna et al.,[27] later studies have shown that both Gram-positive and Gram-negative bacteria are both captured with antibodies.

14.4 BACTERIOPHAGE AMPLIFICATION OF BACTERIA

As discussed previously, direct MALDI-MS analysis of *E. coli* standard suspensions resulted in detection limits of 10^7 cells/ml. A major goal of the research conducted at the Colorado School of Mines has been to improve detection limits of the bacteria to a level consistent with infectious doses. Improvements investigated for increasing sensitivity include: incorporation of a pre-concentration step (filtration or centrifugation), the evaluation of MALDI probes with microscale wells or posts designed to concentrate the sample to a point-source, thus enhancing the analyte-to-laser-shot ratio; improve the target capture efficiencies of different manufactured beads and antibodies (currently 5–10%). Only minor improvements in the detection limit were realized by optimizing these parameters. The use of virus inoculate below the virus detection limits that would only become detectable in the event of infection and release from the microorganisms became the topic of major focus.

Bacteriophages are viruses that infect bacteria; for example, MS2 coliphage infects *Enterobacteria* including *E. coli*. Bacteriophages are also known to specifically infect at the strain level and have been used in the past as a standard characterization test for bacteria.[28–31] MS2 is composed of a protein shell in the form of an icosahedron containing 180 copies of a single capsid protein. When the bacteriophage attacks the bacterial cell, the virus is able to take over the cell biochemistry such that the virus can replicate itself. In the case of MS2 there are various reports on the number of replicates per cell that range from 10,000 to 20,000.[32] Theoretically a MALDI-MS signal from a single bacterial cell could be minimally amplified to produce a capsid protein signal by a factor of 1.8×10^6 (one bacterial cell times the number of proteins in capsid times the burst volume).

Bacteriophage amplification with MALDI-MS detection has been successfully tested as illustrated in Figure 14.7 using MS2 bacteriophage to infect *E. coli* following IMS.[33] In the initial step bacteriophage is added to a bead–*E. coli* complex at a titer below the MALDI-MS detection level. Following room temperature incubation, which can be as short as 30 minutes, a one microliter aliquot is removed and spotted onto the MALDI plate. The sandwich sample preparation method using ferulic acid as a matrix was used for all analyses. Preliminary results show that *E. coli* could be detected at a concentration of

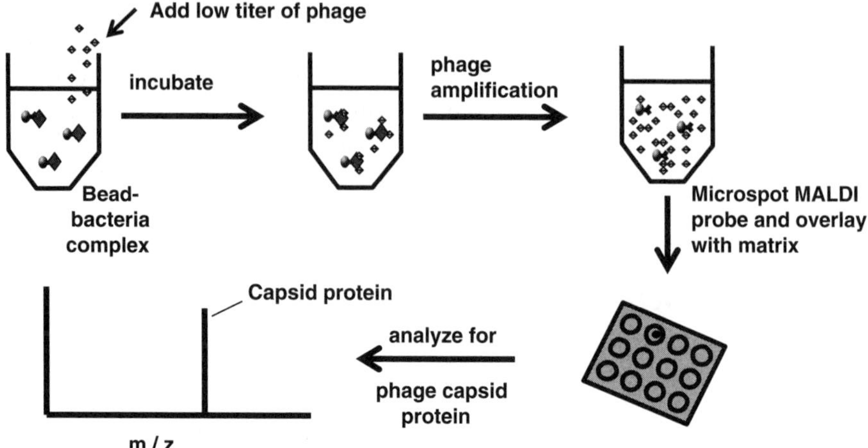

Figure 14.8 Outline of the phage amplification process.

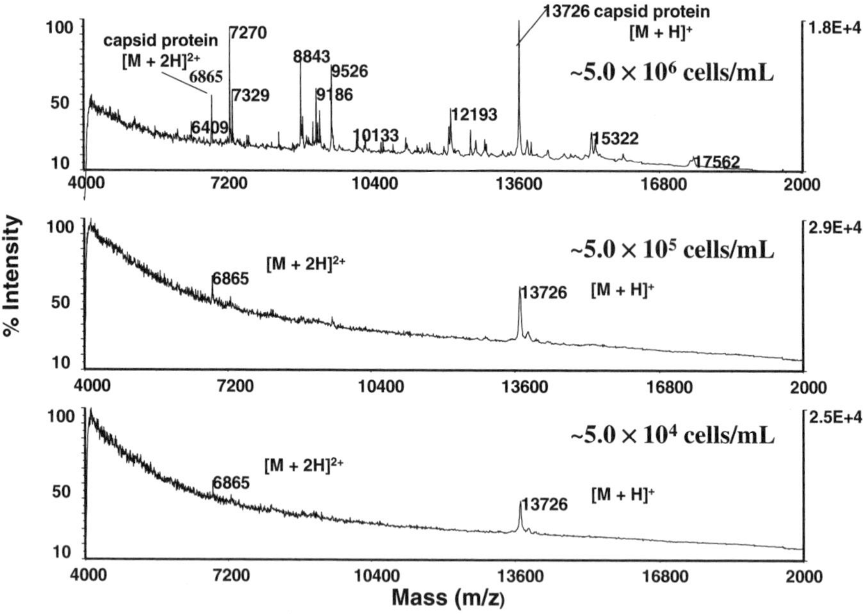

Figure 14.9 Detection of decreasing levels of *E. coli* with bacteriophage (MS-2) amplification.

10^4 cells/ml from aqueous suspensions. Confirmation of the presence of the bacterium was based on the detection of the MS2 capsid protein at 13.7 kDa. Figure 14.9 shows mass spectra of a progressively lower concentration of bacterial suspensions inoculated with MS2 phage. Signals corresponding to

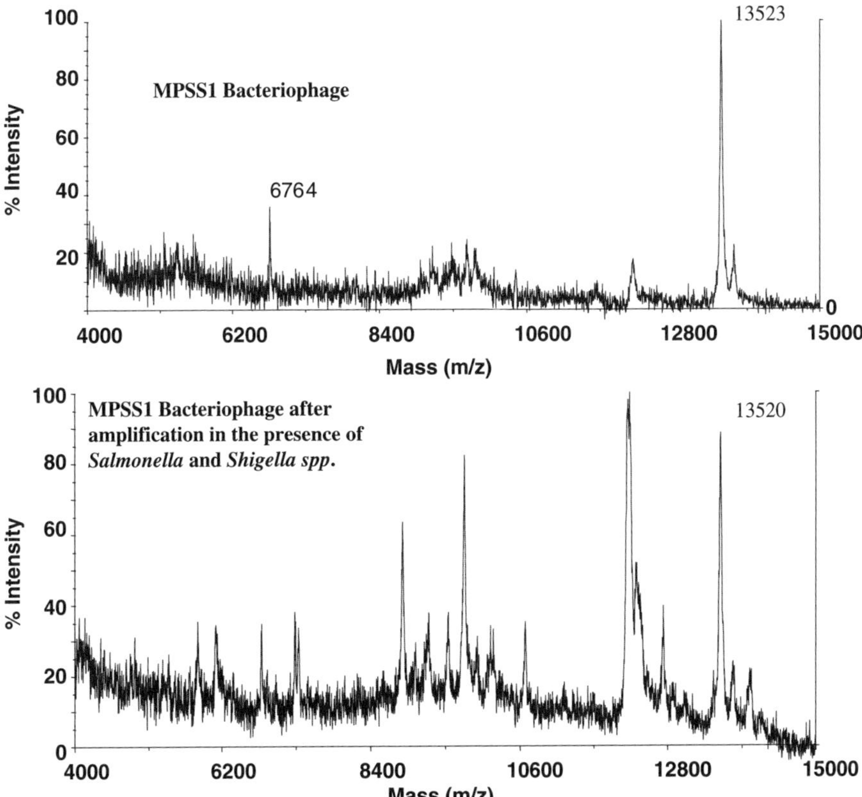

Figure 14.10 Spectra showing the selective bacteriophage amplification of Salmonella in the presence of Shigella.

bacterial proteins are not observed at levels below 10^6 cells/ml. However, the phage capsid protein signal is still observed even at a bacterial cell count of 10^4 cells/ml. This represents at least a second-order magnitude improvement in the detection limit of *E. coli* by MALDI MS. More recent tests have established a detection limit between 10^3 to 10^4 bacterial cells/ml.

Preliminary work on bacterial mixture bacteriophage amplification has been conducted without the antibody (Ab) capture of the bacteria. The specificity of infection mechanism of the bacteriophage allows mixtures of bacteria to be successfully amplified for detection of a targeted organism without the need for Ab collection.[34] As an example, Figure 14.10 shows spectra for MP-SS1 (*Salmonella*) bacteriophage and a mixture of *Shigella spp.* and *Salmonella* incubated with MP-SS1 for four hours. The concentrations of the bacteriophage and the two bacteria were all below the detection limits of MALDI at time = 0. The major peak in the MP-SS1 bacteriophage spectrum is approximately 13,523 Da with minor peaks at approximately 12,210, 10,146, 9766, and

Lawn of bacteria with no virus added

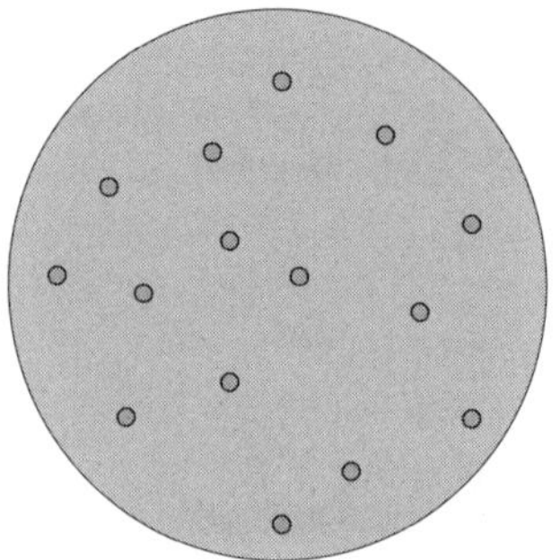

Lawn of bacteria with a low titer of bacteriophage added to the sample. The 'plaques' in the lawn are points where a single virus has undergone many rounds of replication and killed the bacteria within the plaque zone.

Figure 14.11 An illustration of a plaque assay.

8856 Da. The spectrum of the MP-SS1 amplified mixture shows these same peaks indicating the presence of the *Salmonella* bacterium. In other experiments MP-SS1 provided no amplification of the *Shigella spp*., thus demonstrating the specificity of the MP-SS1 toward a *Salmonella* host cell.

The utilization of bacteriophages in bacterial identification is not a new concept. Bacteriophages have been utilized in phage typing schemes for many years. This procedure from classical microbiology involves developing a bacterial lawn, introducing a layer of phage, and observing plaques (areas where bacterial growth has been inhibited by the phage). Figure 14.11 illustrates a bacterial lawn containing plaques. The specificity of the bacteriophage varies from a cross-genera to strain level. For example, the lipid containing bacteriophage PRD1 can infect a broad range of Gram-negative bacterial hosts containing an IncN, IncP, or IncW conjugative plasmid, including *E. coli* and *Salmonella* strains.[35] In contrast, the gamma phage of *Bacillus anthracis* is known to exclusively infect *B. anthracis* strains, and is refractory to essentially all non-anthracis bacilli.[28,36] Even a greater level of specificity exists for bacteriophages designated AR1 and LG1, which have been shown to be highly specific for verotoxigenic *E. coli*, which includes the eminently pathogenic *E. coli* 0157:H7.[37,38,39] The varying degree of natural host specificity among bacteriophages allows for greater flexibility when designing a phage amplification–MALDI-MS assay.

A majority of the work done in our laboratory has been conducted on *E. coli* and *B. anthracis* using MS2 and gamma-phage, respectively. Amplification results have been obtained from several other host bacteria and bacteriophages. Table 14.1 summarizes the materials studied. In all cases studied, no two bacteriophages contained the same proteins.

TABLE 14.1 Summary of Host Bacteria and Phages Studied at CSM

Bacteria	Bacteriophage
Escherichia coli	MS-2
	PRD-1
	T-4
	T-1
Bacillus anthracis	gamma phage
Group B *Streptococcus*	B-30
Salmonella choleraesuis	PRD-1
	P-22
	SL-1
Staphylococcus aureus	187
	47
	53
Yersinia pestis	φA1122

14.5 CONCLUSION

The use of antibodies to capture targeted bacteria has been successfully applied to the analysis of bacterial mixtures and bacteria in complex biological backgrounds. MALDI-TOF MS overcomes the nonselectivity of the antibody and results in acceptable false negatives and positives. The major problem associated with the technique is the overall sensitivity (cells/ml). First, it is recognized that the capture efficiency of antibodies is a few percent. This coupled with the MALDI-TOF MS detection limit of approximately 10^7cells/ml results in an approach that is severely limited in its application to direct analysis of samples without the culturing step. Bacteriophage amplification has been incorporated with Ab capture which lowered the detection limit to 10^3 to 10^4 cells/ml.

Bacteriophage amplification without Ab capture has also been evaluated for bacterial mixtures. Results have been obtained that demonstrate, by using a selective bacteriophage, that a targeted bacterium in a bacterial mixture or complex biological background can be identified. The detection limit for direct bacteriophage amplification is species dependent but is usually an order of magnitude better than is observed with Ab capture/bacteriophage amplification.

One additional point that has not been previously discussed is the elimination for the need of exact growing conditions. It is well established that bacterial protein expression is dependent on media composition and time of growth. For successful use of the fingerprint method for the identification of bacteria, these parameters must be carefully controlled. When phage amplification of bacteria at different growth stages or from different growing conditions is analyzed, the results always show the protein distribution for the phage and are independent of the bacterial sample. It should be noted, however, that the bacteria must be in a viable state to be detected.

REFERENCES

1. Torok, T.; Tauxe, R. V.; Wise, R. P.; Livengood, J.; Sokolow, R.; Mauvais, S.; Birkness, K. A.; Skeels, M. R.; Horan, J. M.; Foster, L. R. A large community outbreak of salmonellosis caused by intentional contamination of restaurant salad bars. *J. Am. Med. Assoc.* 1997, **278**, 389–395.
2. Hennesy, T. W.; Hedberg, C. W.; Slutsker, L.; White, K. E.; Besser-Wiek, J. M.; Moen, M. E.; Feldman, J.; Coleman, W. W.; Edmonson, L. M.; McDonald, K. L.; Osterholm, M. T. A national outbreak of *Salmonella enteritidis* infections from ice cream. *New Eng. J. Med.* 1996, **33**, 1281–1286.
3. Ryan, C. A.; Nickels, M. K.; Hargrett-Bean, N. T.; Potter, M. E.; Endo, T.; Mayer, L.; Langkop, C. W.; Gibson, C.; McDonald, R. C.; Kenny, R. T. Massive outbreak of antimicrobial-resistant *Salmonellosis* traced to pasteurized milk. *J. Am. Med. Assoc.* 1987, **258**, 3269–3274.
4. Halliday, M. L.; et al. An epidemic of hepatitis-a attributable to the ingestion of raw clams in Shanghai, China. *J. Infect. Dis.* 1991, **164**, 852–859.
5. *Escherichia* coli O157:H7. Atlanta, GA: Centers for Disease Control.
6. Alcano, I. E. *DNA Technology: The Awesome Skill.* San Diego: Academic Press, 2000.
7. Van Emon, J. M. *Immunochemical Methods of Environmental Analysis.* Washington, DC: American Chemical Society, 1989.
8. Kusunoki, H.; Latiful Bari, M.; Kita, T.; Sugii, S.; Uemura, T. Flow cytometry for the detection of enterohaemorrhagic *Escherichia coli* O157: H7 with latex beads sensitized with specific antibody. *J. Vet. Med.* 2000, **47**, 551–559.
9. Genta, M.; Heuland, H. *Methods in Biotechnology, 14.* Food Microbiology Protocols, 2000, pp. 11–24.
10. Holland, R. D.; Wilkes, J. G.; Rafii, F.; Sutherland, J. B.; Persons, C. G.; Voorhees, K. J.; Lay, J. O. Rapid identification of intact whole bacteria based on spectral patterns using matrix-assisted laser desorption/ionization with time-of-flight mass spectrometry. *Rapid Comm. Mass Spectrom.* 1996, **10**, 1232–1241.
11. Wang, Z.; Russon, L.; Li, L.; Roser, D. C.; Long, S. R. Investigation of spectral reproducibility in direct analysis of bacteria proteins by matrix-assisted laser desorption/ionization time-of-flight mass spectroscopy. *Rapid Comm. Mass Spectrom.* 1998, **12**, 456–464.
12. Arnold, R. J.; Reilly, J. P., *Proc. 46th American Society for Mass Spectrometry and Allied Topics*, Orlando, FL, 1998, MP 125.
13. Arnold, R. J.; Reilly, J. P., *Proc. 46th American Society for Mass Spectrometry and Allied Topics*; Orlando, FL, 1998, MP 065.
14. Domin, M. A.; Welham, K. J.; Ashton, D. S. The effect of solvent and matrix combinations on the analysis of bacteria by matrix-assisted laser desorption/ionisation time-of-flight mass spectrometry. *Rapid Comm. Mass Spectrom.* 1999, **13**, 222–226.

15. Lay, J. O. MALDI-TOF mass spectrometry and bacterial taxonomy. *Trends Anal. Chem.* 2000, **19**, 507–516.

16. Aguila, C. D.; P., C. G.; Moura, H.; Silva, A. J. D.; Leitch, G. J.; Moss, D. M.; Wallace, S.; Slemenda, S. B.; Peiniazek, N. J.; Wisvesvara, G. S. Ultrastructure, immunofluorescence, Western blot, and PCR analysis of eight isolates of *Encephalitozoon (Septata) intestinalis* established in culture from sputum and urine samples and duodenal aspirates of five patients with AIDS. *J. Clin. Microbiol.* 1998, **36**, 1201–1208.

17. Thomas, J. J.; Falk, B.; Fenselau, C.; Jackman, J.; J., E. Viral characterization by direct analysis of capsid proteins. *Anal. Chem.* 1998, **70**, 3863–3867.

18. Madonna, A. J.; Basile, F.; Ferrer, I.; Metanni, M. A.; Rees, J. C; Voorhees, K. J. On-probe sample pretreatment for the detection of proteins above 15 KDa from whole cell bacteria by matrix-assisted laser desorption/ionization time-of-flight mass spectrometry. *Rapid Comm. Mass Spectrom.* 2000, **14**, 2220–2229.

19. Wahl, K. L.; Wunschel, S. C.; Jarman, K. H.; Valentine, N. B.; Peterson, C. E.; Kingsley, M. T.; Zartolas, K. A.; Saenz, A. J. Analysis of microbial mixtures by matrix-assisted laser desorption/ionization time-of-flight mass spectrometry. *Anal. Chem.* 2002, **74**, 6191–6199.

20. Jarman, K. H.; Daly, D. S.; Petersen, C. E.; Saenz, A. J.; Valentine, N. B.; Wahl, K. L. Extracting and visualizing matrix-assisted laser desorption/ionization time-of-flight mass spectral fingerprints. *Rapid Comm. Mass Spectrom.* 1999, **13**, 1586–1594.

21. Hutchens, T. W.; Yip, T. T. New desorption strategies for the mass-spectrometric analysis of macromolecules. *Rapid Comm. Mass Spectrom.* 1993, **7**, 576–580.

22. Brockman, A. H.; Orlando, R. Probe immobilized affinity-chromatography mass-spectrometry. *Anal. Chem.* 1995, **67**, 4581–4585.

23. Pineda, F. J.; Lin, J. S.; Fenselau, C.; Demirev, P. A. Testing the significance of microorganism identification by mass spectrometry and proteome database search. *Anal. Chem.* 2000, **72**, 3739–3744.

24. Bundy, J. L.; Fenselau, C. Lectin-based affinity capture for MALDI-MS analysis of bacteria. *Anal. Chem.* 1999, **71**, 1460–1463.

25. Bundy, J. L.; Fenselau, C. Lectin and carbohydrate affinity capture surfaces for mass spectrometric analysis of microorganisms. *Anal. Chem.* 2001, **73**, 751–757.

26. Alfonso, C.; Fenselau, C. Use of bioactive glass slides for matrix-assisted laser desorption/ionization analysis: Application to microorganisms. *Anal. Chem.* 2003, **75**, 694–697.

27. Madonna, A.; Basile, F.; Voorhees, K. J. Detection of bacteria from biological mixtures using immunomagnetic separation combined with matrix-assisted laser desorption/ionization time-of-flight mass spectrometry. *Rapid Comm. Mass Spectrom.* 2001, **15**, 1068–1074.

28. Brown, E. R.; Cherry, W. B. Specific identification of *Bacillus*-anthracis by means of a variant bacteriophage. *J. Inf. Dis.* 1955, **96**, 34–39.

29. Girard, G. Sensibilite des *bacilles* pesteux et pseudotuberculeux d'une part des germes du groupe coli-dysenterique d'autre part aux bacteriophages homologues. *Ann. Inst. Pasteur* 1943, **69**, 52–54.

30. Stringer, J. Development of a phage-typing system for group-B streptococci. *J. Med. Microbiol.* 1980, **13**, 133–143.
31. Levy, J. A. *Virology.* Upper Saddle River, NJ: Prentice Hall.
32. Hickman-Brenner, F. W.; Stubbs, A. D.; Farmer, J. J. Phage typing of *Salmonella-Enteritidis* in the United States. *J. Clin. Microbiol.* 1991, **29**, 2817–2823.
33. Madonna, A. J.; Van Cuyk, S.; Voorhees, K. J. Detection of *Escherichia coli* using immunomagnetic separation and bacteriophage amplification coupled with matrix-assisted laser desorption/ionization time-of-flight mass spectrometry. *Rapid Comm. Mass Spectrom.* 2003, **17**, 257–263.
34. Rees, J. C.; Voorhees, K. J. Unpublished results.
35. Grahn, M. A.; Daugelavicius, R.; Bamford, D. H. The small viral membrane-associated protein P32 is involved in bacteriophage PRD1 DNA entry. *J. Virol.* 2002, **78**, 4866–4872.
36. Abshire, T. G.; Brown, J. E.; Teska, J. D.; Ezell, J. W. Validation of the use of gamma phage for identifying *bacillus anthracis*. *4th International Conference on Anthrax*, Annapolis, MD 2001.
37. Goodridge, L.; Chen, J.; Griffiths, M. Development and characterization of a fluorescent-bacteriophage assay for detection of *Escherichia coli* O157: H7. *Appl. Environ. Microbiol.* 1999, **65**, 1397–1404.
38. Ronner, A. B.; Cliver, D. O. Isolation and characterization of a coliphage specific for *Escherichia-coli*-O157-H7. *J. Food Prot.* 1990, **53**, 944–947.

15

DISCRIMINATION AND IDENTIFICATION OF MICROORGANISMS BY PYROLYSIS MASS SPECTROMETRY: FROM BURNING AMBITIONS TO COOLING EMBERS—A HISTORICAL PERSPECTIVE

ÉADAOIN TIMMINS

National Centre for Biomedical Engineering Science, National University of Ireland-Galway, Galway, Ireland

ROYSTON GOODACRE

School of Chemistry, University of Manchester, Manchester, M60 1QD

15.1 INTRODUCTION TO MICROBIAL CHARACTERIZATION

During the 1980s to mid 1990s pyrolysis mass spectrometry (PyMS) was extensively researched as a physico-chemical method for classifying and identifying bacteria. However, by the turn of the century few researchers were still active in this field and had cooled to this fingerprinting approach. This chapter presents a historical perspective of PyMS as applied for the rapid characterization of bacteria and fungi in the later part of the 20th century.

Microbial characterization, including detection, differentiation, and identification, are the most regular tasks of microbiology laboratories worldwide. A

Identification of Microorganisms by Mass Spectrometry, Edited by Charles L. Wilkins and Jackson O. Lay, Jr.

critical problem facing the wide variety of microbiology laboratories, in the performance of these tasks, is the length of time between receipt of a specimen and its final microbiological result.

Shortening the time taken to identify a pathogen in medical laboratories would allow the effective management of patients with infectious diseases. It would accelerate targeted prescription and thus help prevent inappropriate use of antibiotics—a serious public health issue of our day.[1] Also, in medicine, speedy microbial analysis could be used to good effect in epidemiological studies. In industry, rapid (but still accurate) microbiological testing is often important for purely economical reasons. In both the food and pharmaceutical sectors, time and money would be saved by faster quality control procedures on raw materials, in-process samples, and finished products. Furthermore, in the biotechnology industry, rapid microbial analysis would be invaluable in the screening of large numbers of isolates for the production of biologically active metabolites.

Traditionally the length of time between receipt of a specimen and its final microbiological result has been considerable (between 24 and 72 h). Both the quantification and characterization of microorganisms involve culturing, which often requires lengthy periods of incubation. Next, microbial characterization usually requires a plethora of morphological, nutritional, and biochemical tests, which may require further periods of incubation before results can be read and interpreted. A major development in traditional microbiology over the past 15 to 20 years has been the miniaturization of these conventional biochemical and nutritional tests into commercially available kits that offer advantages of convenience and strict standardization. However, these systems usually require overnight incubation (and for slow growing organisms like *Mycobacterium tuberculosis* this can be 3 to 6 weeks), and they are expensive to use for large studies as the cost per test is relatively high. Traditional microbiology has also experienced some improvements in analysis time by the development of rapid diagnostic tests that employ either serological or genetic probes.

Serological methods are based on the interaction between antibodies and antigens. To detect rapidly bacterial antigens in body fluids and tissues, monoclonal and polyclonal antibodies are commercially available for use in enzyme-linked immunoassays and direct immunofluorescence antibody assays. Serological tests are typically used in clinical laboratories where a particular organism might be sought or epidemiological tracing is required. They are increasing in popularity because of their rapidity, ease of use, and specificity, and kits for a wide range of microorganisms are available. For example, in cerebrospinal fluid, antigen detection tests for *Haemophilus influenza*, *Neisseria meningitidis*, and *Streptococcus pneumoniae* have been used for some time.[2] Although extremely specific, the sensitivity of serological methods is variable. In addition each kit can only be used to confirm the identity of a *single* species or organism type, and the cost per test is again relatively high.[3]

The development of DNA technologies has revolutionized several aspects of biological sciences, including the diagnosis of infectious diseases. After the first description of the use of DNA probes to detect entertoxigenic *Escherichia coli* in stool samples,[4] it was expected that molecular methods would eventually replace culturing. Indeed, DNA probe technology, which involves the hybridization of specific DNA sequences to complementary sequences of bacterial nucleic-acids, has advantages over more conventional methods. In addition to their absolute specificity they can be used to detect the microorganism directly in a clinical sample, thus eliminating the need for culturing. The US Food and Drug Administration has approved several DNA probes to detect microorganisms, including mycobacteria, *Legionella* species, the pharyngeal pathogen group A *Streptococcus*, and two sexually transmitted disease pathogens: *Chlamydia trachomatis* and *Neisseria gonorrhoeae*.[5] Although innovative, this technology can have low sensitivity, and it often requires the extraction of DNA from a specimen, a procedure that may take several hours. To compensate for the lack of sensitivity, several of the probes commercially available are directed against the 16S rRNA molecule, since this target is present in high copy numbers (10^4–10^5 molecules per bacterial cell). An alternative way of enhancing probe sensitivity is by preceding hybridization with polymerase chain reaction (PCR) amplification of the target region. The disadvantage of PCR is the risk of possible contamination, mainly from DNA carried over from previous amplification reactions, leading to false positives. Detection of amplification products is most commonly done by agarose gel electrophoresis. Although convenient, it is relatively slow, it is not easily automated, and the extra steps involved in the analysis will increase the source of error and contamination. Further concerns about molecular-based methods include the high cost per test, kit shelf-life and the requirement for considerable technical expertise.

The expression of the microbial genome produces greater than 2000 protein molecules within a microbial cell (e.g., the estimated total number of individual proteins expressed from the *Escherichia coli* genome is well over 4000[6]). Hence techniques that can be applied to the characterization of these proteins have enormous potential in diagnostic microbiology. Polyacrylamide gel electrophoresis (PAGE) of whole-organism proteins, originally proposed by Jackman[7], is a powerful, relatively low cost technique.[3] It involves growth of the microorganism in standardised conditions followed by harvesting and cell disruption to release the proteins. After electrophoresis, gels are stained to reveal the characteristic protein-band patterns that can be rapidly analyzed with automatic scanning densitometry and direct computer processing of the data. Results from various taxonomic studies (reviewed by Kersters et al.[8]) reveal that the electrophoretic patterns are discriminatory at species and subspecies levels. Advantages of this approach are the efficiency with which large numbers of strains can be compared and, provided a cell-disruption method can be found, its applicability to different types of microorganisms. Despite these advantages the technique has not been widely adopted because of poor reproducibility due to variations in cell-disruption methods, gel structure, and running conditions: several gels must be run to verify results. Two-dimensional PAGE gives greater resolution, allowing the recognition of several hundred bands compared with about 30 from one-dimensional PAGE, but this version of the technique brings greater problems with reproducibility.

Since the development of the Gram strain over 100 years ago, the nature of the bacterial-cell envelope has been important in microbial characterisation. The walls of many different bacteria have now been analyzed, and various rapid methods, such as high-performance liquid chromatography (HPLC), gas chromatography (GC), and thin-layer chromatography (TLC), have been developed to detect particular components.[3] So far, however, only GC has been commercially developed into a microbial identification system. Originally introduced by Hewlett Packard and now marketed as the MIDI system (Newark, Delaware, US), the basis of the technique is the extraction of whole-cell fatty acids and GC profiling of their methyl esters (FAME profiles). The MIDI system includes a computer for recording data and for matching FAME profiles against a reference library. Among the advantages of the method are that samples are prepared and analyzed in batches, the technique is applicable to a wide variety of microorganisms, and analysis cost per specimen is moderate.[3] However, disadvantages include the possible loss of valuable information as the technique analyzes only a small part of the total bacterium, increased instrument analysis time per specimen, which can take up to 30 minutes, poor long-term reproducibility, which makes the use of a reference library difficult, and no information on whether FAME profiles vary sufficiently to allow typing down to strain level.

It follows that for routine purposes, the ideal method for microbial analysis would include the following criteria:

- Applicability to a wide range of microorganisms using simple, uniform, automated procedures that require minimal sample preparation
- Rapid availability of analytical data and reliable, reproducible results
- High specificity in discrimination or identification, down to species and subspecies level
- Relatively low analytical costs per specimen
- High throughput

With recent developments in analytical instrumentation these criteria are being increasingly fulfilled by physicochemical spectroscopic approaches, often referred to as "whole-organism fingerprinting" methods.[9,10] Such methods involve the concurrent measurement of large numbers of spectral characters that together reflect the overall cell composition. Examples of the most popular methods used in the 20th century include pyrolysis mass spectrometry (PyMS),[11,12] Fourier transform-infrared spectrometry (FT-IR),[13–15] and UV resonance Raman spectroscopy.[16,17] The PyMS technique gives an indirect measure of molecular bond-strength,[18] while FT-IR and Raman spectroscopy predominantly measure the vibrations of molecular bonds.[19,20]

The major advantages that whole-organism fingerprinting methods have over conventional, serological, and DNA-based techniques are that they are rapid with respect to single and multiple samples. For PyMS, typical instrument analysis time is less than 2 minutes per sample and up to 300 samples can be analyzed in one batch. This method also allows automated specimen processing and data capture using simple, non-species-specific methods that have low consumable costs and have been shown to differentiate among culturable microorganisms at genus, species, and subspecies level. In brief, by the mid-1990s PyMS had become a well-established, high-resolution, analytical method that has received a considerable boost from the development of a low-cost, dedicated instrument and advanced computer-based data analytical methods.

15.2 PRINCIPLES OF PYMS

Pyrolysis is a chemical process that involves the thermal degradation of a complex organic material, such as whole organisms, in an inert atmosphere or vacuum. The thermal energy causes molecules to cleave at their weakest points to produce smaller, predominantly volatile fragments, called pyrolysate.[21] Degradation of the sample is reproducible under controlled conditions and the resultant pyrolysate is characteristic of the original material. In Curie-point pyrolysis the sample is dried onto a suitable metal and heated rapidly to the Curie point of the metal. The components of the resultant pyrolysate are then ionized and separated by mass spectrometry on the basis of their mass-to-charge (m/z) ratio (Figure 15.1). A pyrolysis mass spectrum[18] is thus produced and represents a chemical profile or "fingerprint" of the complex material analyzed. The resultant spectral data, when analyzed by suitable

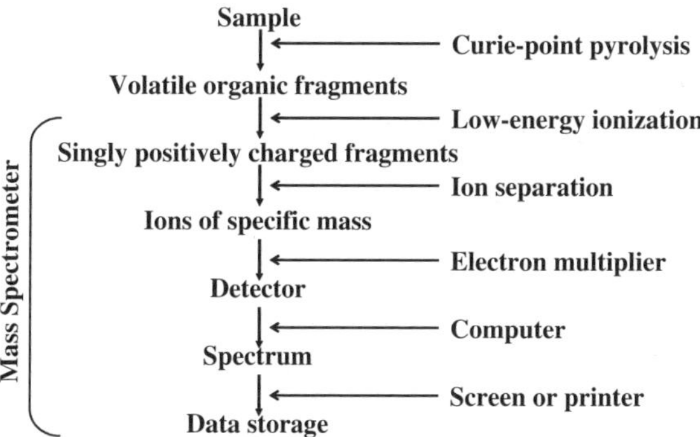

Figure 15.1 Flowchart showing the main stages in pyrolysis mass spectrometry.

multivariate statistical methods, can be used to differentiate, identify, compare and quantify microorganisms.

15.2.1 Curie-Point Pyrolysis MS

Curie-point pyrolysis is a particularly reproducible way to generate volatile pyrolysate. In this method the bacterial sample is coated onto a ferromagnetic foil and heated rapidly by passing a high-frequency, high-amperage oscillating current through the pyrolysis coil for a set time (usually 3 s). This produces an intense magnetic field that penetrates the foil and causes induction heating. When the temperature of the iron-nickel foil reaches the Curie point of the alloy, it ceases to be ferromagnetic; that is, it becomes paramagnetic, and no further current may be induced in it. Consequently the foil cools, inductive heating resumes, and the foil is heated back to the Curie point. The foil temperature is therefore controlled within tight limits by this natural thermostatic effect (Figure 15.2).

The Curie point of the alloy is determined by the iron-to-nickel ratio; an Fe:Ni ratio of 50:50 was used to give a Curie point of 530°C. This pyrolysis temperature was chosen because it has been shown[22] to give a balance between fragmentation from polysaccharides and protein fractions. Foils with Curie points of 300°C to 1000°C are commercially available (Figure 15.2). How-

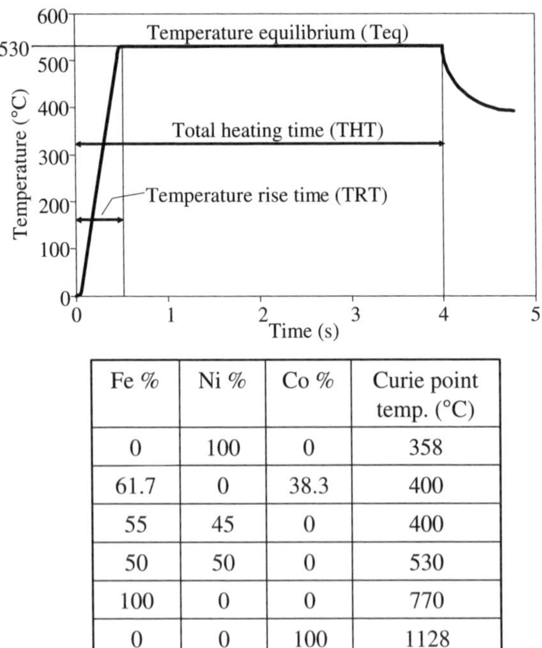

Fe %	Ni %	Co %	Curie point temp. (°C)
0	100	0	358
61.7	0	38.3	400
55	45	0	400
50	50	0	530
100	0	0	770
0	0	100	1128

Figure 15.2 Curie-point pyrolysis: Temperature profile and Curie points of typical ferromagnetic alloys used.

ever, low-temperature pyrolysis yields pyrolysates with a large proportion of high-boiling tarry products. Such products can cause contamination problems by condensing in the sampling tube and expansion chamber. High-temperature pyrolysis gives high pyrolysate yields with a large amount of low molecular weight products, and these are often poorly characteristic and nonreproducible.[22]

The thinly coated sample quickly reaches the temperature of the foil (0.5 s). If oxygen was present, combustion would occur, yielding carbon dioxide and water as major products. However, in the nonoxidizing environment of the vacuum, pyrolysis occurs yielding low molecular weight volatile fragments by cleavage of the covalent bonds. This pyrolysate[21] then diffuses out of the glass tube, into the expansion chamber, and down the molecular beam tube to the ionization chamber of a mass spectrometer. The expansion chamber is designed to allow even mixing of the early low molecular weight products with the late high molecular weight products. The chamber also lengthens the release of products to the mass spectrometer, which has the desired effect of allowing a large number of scans of the pyrolysate. The surfaces of the expansion chamber are usually gold coated to minimize secondary reactions. To prevent condensation of the products, the expansion chamber and molecular beam tube are heated to 150°C, while the sample tube is heated to 100°C.[23]

As the pyrolysate leaves the molecular beam tube and enters the ionization chamber, it encounters a crossing beam of low-energy electrons (30 eV or lower is typical) that are produced by passing a direct current through the "filament" (a tightly coiled, thorium-dioxide impregnated tungsten wire). Collisions with these electrons results in the formation of both molecular and fragment ions, most of which will be single positively charged. To preserve the high-vacuum essential for mass spectrometry, un-ionized products are frozen on a liquid nitrogen cooled trap.

The ions are pushed out of the electron beam, toward the central aperture of the source electrodes by a positively charged repeller plate. They accelerate through the source by a potential gradient applied to the source electrodes, and they are then focused toward the mass analyzer. Typically a quadrupole has been used to separate the ions, although more recently time of flight has been investigated (Letarte et al., 2004).

15.3 EARLY DEVELOPMENTS AND INVESTIGATIONS 1952 TO 1985

Mass spectrometry was first used to analyze pyrolysate by Zemany,[24] who showed that complex biological materials, such as albumin and pepsin, degrade in a reproducible way under carefully standardised pyrolysis conditions. Interest in PyMS declined after this initial study due to the development of the less expensive pyrolysis gas chromatography (PyGC) system. PyGC uses gas chromatography to separate the components of the pyrolysate on the basis of their relative polarity, and hence on the basis of their varying retention times in the chromatography column. The resultant "pyrogram" consists of a series of peaks that, like a pyrolysis mass spectrum, can be seen as a chemical "fingerprint" of the original specimen.

Davison et al.[25] reported the first application of PyGC when they obtained characteristic pyrograms from pyrolyzed chemical polymers. Then in 1962 the US Space Exploration Program provided the driving force for the application of pyrolysis to microbiology when a miniaturized, automated PyGC system was designed to test for the presence of extra terrestrial life.[26] However, this system was never actually used. It was not until the work of Reiner and colleagues[27,28] that the potential of PyGC in microbiology was demonstrated. Subsequent studies (reviewed by Gutheridge and Norris[29]) revealed that PyGC could be used to attain high levels of discrimination between species from a broad range of bacteria. The application of PyGC to microbiology eventually declined, however, due to problems with long-term reproducibility, low-throughput and inadequate data-handling routines.[30] Improvements to the method have since been made, especially in column longevity and performance[31] and in data-handling routines.[9] However, in terms of speed and ease of automation, PyMS is still superior to PyGC.

The first dedicated Curie-point PyMS system was built by Meuzelaar and Kistemaker[11] at the FOM Institute (Institute for Atomic and Molecular Physics) in Amsterdam. This was followed shortly afterward by the construction of the first fully automated instrument, the Autopyms, which used high-speed ion counting and computerized data processing.[18,32] The Autopyms led to the manufacture of two commercial machines: the Extranuclear 5000 (Extranuclear Laboratories, Pittsburgh, PA) and the Pyromass 8–80 (VG Gas Analysis Ltd., Middlewich, Cheshire, UK). However, neither of these machines proved popular, probably because machine cost was in excess of £100000.

At this early stage in PyMS development, various investigations demonstrated the applicability of PyMS to the rapid analysis of mostly clinically significant microorganisms.[29] In particular, the early systems were used to study the slow growing mycobacteria.[32,34,35] Meuzelaar et al.[32] found that most of the 14 species studied could be differentiated. The heterogeneous *Mycobacterium bovis* species, however, showed some overlap with the *M. avium* and *M. xenopi*, giving rise to the possible re-classification of this species. Mycobacteria identification studies by Wieten et al.[34,35] used an approach known as "operational fingerprinting"[18] that involved the PyMS analysis of batches of strains with the inclusion of selected reference strains. The objective was to identify strains as either belonging to the "tuberculosis complex" (*M. tuberculosis*, *M. bovis*, and *M. bovis* BCG) or to other members of *Mycobacterium*. Greater than 90% agreement was obtained with conventional procedures. Other early taxonomic studies included those on *Neisseria*,[36] *Klebsiella*,[18] and *Legionella*.[37] The technique was also used by Windig and colleagues[38,39] in taxonomic studies on the yeast genera *Rhodosporidium* and *Sporidiobolus*. These authors showed a close correlation between the chemical interpretation of the difference spectra and other chemotaxonomic information.

Overall, the early investigations demonstrated the potential of PyMS to reduce greatly the time taken to identify medically important microorganisms. However, several problems still prevented the widespread routine use of this technique. The method was labor intensive due to manual sample-loading, while a prolonged processing time meant that sample throughput per day was low. In addition machine instability, and hence poor reproducibility over time, meant that the comparison of new spectra with those held in spectral libraries was difficult. The application of the technique was further hampered by expensive, awkward hardware and the lack of user-friendly computer software to analyze rapidly the spectral data. As a result of these important limitations the method fell into disrepute.

15.4 THE MID-1980s AND BEYOND

The breakthrough in PyMS development came in 1986 with the introduction of a low-cost, automated pyrolysis mass spectrometer: the Horizon PYMS-200X instrument (Horizon Instruments, Heathfield, East Sussex, UK).[30,40] This machine was superior to earlier models due to the development of a reliable, automated inlet system that allowed high-volume throughput, an improved electron multiplier that led to faster analysis times, enhanced machine reproducibility, and an upgraded statistical package for data analysis. Since the development of the Horizon PYMS-200X machine, and subsequently the RAPyD-400 (launched in 1993) by the same manufacturers, PyMS has been applied to many microbiological studies that are extensively reviewed.[9,12,41–44]

Interpretation of the hyperspectral (or multivariate) data generated by the PyMS system (150 m/z intensities; see Figure 15.3) has conventionally been performed by the "unsupervised" methods of principal component analysis (PCA), discriminant function analysis (DFA, also known as canonical variates analysis [CVA]), and hierarchical cluster analysis (HCA). These methods seek "clusters" in the data,[45] thereby allowing the analyst to group objects together on the basis of their perceived closeness. Although such methods are in some sense quantitative, they are mostly considered as qualitative because their chief purpose is merely to distinguish objects or populations. With PCA, DFA, and HCA there has been considerable focus on the application of PyMS to epidemiological studies of infection outbreaks and, to a lesser extent, classification and identification studies of pathogenic microorganisms. Below are examples of such studies performed since 1986.

15.4.1 Inter-strain Comparison Studies

Studies have shown that although PyMS is not a true epidemiological typing system, since a *permanent* designation is not assigned to the test organism, the methodology is of immense value for the inter-strain comparison of microorganisms involved in outbreaks of infection. Inter-strain comparison is vital to outbreak management as it provides information on whether a cluster of

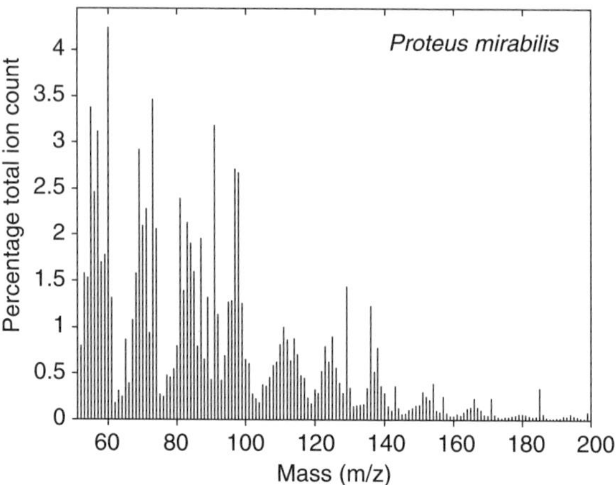

Figure 15.3 Typical PyMS spectrum.

identical strains (an epidemic strain) is involved, or whether the strains are from sporadic, epidemiologically unrelated isolates of the same species. Inter-strain comparison can also highlight the point-source of an outbreak, or alternatively, establish that there was no point source. Examples of pathogenic species and groups to which PyMS has been successfully applied for inter-strain comparison are shown in Table 15.1.

From these studies the high discriminatory power of PyMS has become evident. For example, Goodacre and Berkeley[46] showed that PyMS can be used to distinguish four strains of *E. coli* that differ only in the presence or absence of a single plasmid. The accuracy of discrimination by PyMS has also been demonstrated by many authors[47–52] who have found good congruence between results from differentiation by PyMS and results from routine typing systems. Similarly there appears to be good agreement between results from PyMS analyses and those from analyses by molecular methods.[50,53–55] Additionally the versatility of PyMS for inter-strain comparison studies of *Acinetobacter calcoaceticus*, which is a species not yet amenable to conventional typing systems, has been demonstrated.[56] Overall, from these inter-strain comparison studies it has become evident that the speed and universality of PyMS provide the technique with distinct advantages over conventional typing systems.

15.4.2 Classification and Identification Studies

In general, there has been good agreement between PyMS-derived classification results and results from standard taxonomic methods. This agreement was seen from comparative studies on *Carnobacterium*,[57] *Fusobacterium*,[58] *Peptostreptococcus*,[59] *Photobacterium*,[60] *Prevotella*,[61] *Rothia*,[62] *Streptococcus*,[63] and *Streptomyces*.[64] PyMS was also used, together with molecular and numerical

TABLE 15.1 Inter-strain Comparison Using PyMS

Organism	Reference(s)
Acinetobacter	56
Bacillus	80, 131–135
Bacteroides	136
Candida	137, 138
Carnobacterium	57
Clostridia	53, 54, 131, 139, 140
Enterococcus	141
Escherichia	46, 47
Fusobacterium	58
Haemophilus	50
Klebsiella	49
Legionella	52
Listeria	142
Listonella	51
Mycobacterium	71–73
Peptostreptococcus	59
Photobacterium	60
Prevotella	61
Pseudomonads	55, 143, 144
Rothia	62
Salmonella	145
Staphylococci	146, 147
Streptococci	48, 63, 148
Streptomycetes	64
Yeast	149

methods, to describe the novel species of *Actinomadura latina*,[65] and for the reclassification of genera and species belong to *Bacillus* to other bacterial groups.[66–70]

In other studies PyMS was effective in differentiating closely related species that are difficult to differentiate using conventional taxonomic procedures. For example, the technique was used to distinguish among members of the *M. tuberculosis* complex[71] and to separate *M. xenopi* strains from members of the *M. avium–M. intracellulare* complex.[72] In addition Magee et al.[73] used PyMS to differentiate *M. senegalense* from related mycobacteria. These studies showed that PyMS may allow rapid identification of the slow-growing mycobacteria provided that reference strains with similar growth properties are included for comparison.

15.4.3 Other Microbiological Applications

Using the conventional data analysis methods of PCA, DFA, and HCA, studies have indicated that the role of PyMS in microbiology may extend beyond typing, classification and identification. For example, PyMS studies on changes in bacterial cell composition during exposure to antimicrobial agents demonstrated the applicability of PyMS in rapidly detecting the effects of antimicrobial agents on microorganisms, and in studying the mode of action of antimicrobial agents.[74–76] Furthermore, in industry, PyMS with unsupervised analysis has been used for the selection of novel actinomycetes for pharmaceutical screening program,[77] while in environmental studies, the approach has been used to analyse rhizobial strains isolated from soil in an attempt to follow the spread of genetically engineered microbes.[78,79]

15.5 THE MOVE FROM CLUSTER ANALYSES TO NEURAL NETWORKS

Due to the mainly qualitative nature of PCA, DFA, and HCA, the role of PyMS in microbiology has been somewhat restricted. For example, using these "clustering" methods, the use of PyMS to identify microorganisms can be a *subjective* process because it relies on the interpretation of complex scatter plots and dendrograms. Furthermore the qualitative nature of PCA, DFA, and HCA prevents the application of PyMS to quantitative microbial analysis, while limitations also arise from batch-to-batch variation of PyMS data.[80]

Since 1992 a variety of related but much more powerful data-handling strategies have been applied to the "supervised" analysis of PyMS data. Such methods fall within the framework of chemometrics; the discipline concerned with the application of statistical and mathematical methods to chemical data.[81–85] These methods seek to relate known spectral inputs to known targets, and the resulting model is then used to predict the target of an unknown input.[86]

Artificial neural networks (ANNs) are a novel chemometric approach and have become a well-known means of uncovering complex nonlinear relationships in multivariate data, while still being able to map the linearities.[87–92] Additionally ANNs are considered to be robust to noisy data,[93] such as those that may be generated by PyMS. The application of ANNs to PyMS, in recent years, has led to a number of exciting advances in rapid microbial analysis. These applications within PyMS have been reviewed extensively in Goodacre and Kell,[12] Goodacre, Neal, and Kell,[94] and Keayon, Ferguson, and Ward.[95]

The first application of ANNs to pyrolysis mass spectra from biological samples was by Goodacre, Kell, and Bianchi.[96,97] This study permitted the rapid and exquisitely sensitive assessment of the adulteration of extra-virgin olive oils with various seed oils, a task that previously was labor intensive and difficult. Since this study other laboratories have increasingly sought to apply ANNs to the deconvolution and interpretation of pyrolysis mass spectra, the aim being to expand the application of the PyMS technique from microbial characterisation to the rapid and *quantitative* analysis of the chemical constituents of microbial and other biological samples.

15.5.1 Quantification by PyMS with ANNs

Within our laboratory we have shown that using the combination of PyMS and ANNs it is possible to follow the production of indole in a number of strains of *E. coli* grown on media incorporating various amounts of tryptophan,[98] and to estimate the amount of casamino acids in mixtures with glycogen.[99] It was also shown that it is possible to quantify the (bio)chemical constituents of complex binary mixtures of proteins and nucleic acids in glycogen, and to measure the concentrations of binary and tertiary mixtures of *Bacillus subtilis*, *Escherichia coli*, and *Staphylococcus aureus*.[93,100]

With regard to biotechnology, the combination of PyMS and ANNs has been exploited to quantify the amount of mammalian cytochrome b_5 expressed in *E. coli*[101] and to measure the level of metabolites in fermentor broths.[102] In the latter study samples from fermentations of a single organism in a complex production medium were analyzed quantitatively for a drug of commercial interest; the drug could be quantified in a variety of mutant producing strains cultivated in the same medium, thus effecting a rapid screening for the high-level production of desired substances. In related studies *Penicillium chrysogenum* fermentation broths were analyzed quantitatively for penicillins using PyMS and ANNs,[103] the production of the protein α2-interferon was quantified in recombinant *E. coli* bioprocess samples,[104] and *Gibberella fujikuroi* fermentations were monitored for gibberellic acid production.[105]

Other workers have also demonstrated the potential role of PyMS with ANNs in quantitative biotechnology. Kang et al.[106] used the combination to quantify clavulanic-acid production in a *Streptomyces clavuligerus* fermentation system, while a study by Lee[107] indicated that the quantification of clavulanic-acid production by PyMS-ANNs could be used to quantitatively monitor the morphological differentiation process in a *Streptomyces* fermentation.

15.5.2 Identification and Differentiation of Microbes by PyMS with ANNs

The application of ANNs to the analysis of PyMS spectra has also extended the potential of PyMS in identifying and discriminating microorganisms. ANNs can be encoded simply and the identification results read off in a tabulated format, unlike the complex examination of scatter plots (PCA and DFA) and dendrograms (HCA). Investigations have demonstrated that PyMS-ANNs analyses provide a rapid, accurate and *objective* means of identifying strains of *Eubacterium*,[108] *Mycobacterium*,[109] *Propionibacterium*,[110,111] *Penicillium*,[112] *Streptomyces*,[113] and urinary tract infection isolates.[114] Other studies revealed that PyMS with ANNs could be used for the rapid and accurate detection of the verocytotoxin-production status in clinical isolates of *E. coli*,[115] and for identification of the morphological differentiation state in a *Streptomyces albidoflavus* fermentation.[116] In the study by Kang and colleagues pyrolysis mass spectra were obtained from whole cells at various growth phases. ANNs analysis of the resulting spectra, in which the outputs were encoded as 0 and 1 for mycelium and spore, respectively, allowed the morphological differentiation states to be easily distinguished.

All of the studies above have used back propagation multilayer perceptrons and many other varieties of neural network exist that have been applied to PyMS data. These include minimal neural networks,[117–119] radial basis functions,[114,120] self-organizing feature maps,[110,121] and autoassociative neural networks.[122,123]

Another important advance has been the application of PyMS with ANNs to discriminate between methicillin-resistant and methicillin-sensitive *Staphylococcus aureus*.[24] In this study DFA and HCA showed that the major source of variation between the pyrolysis mass spectra of 15 methicillin-resistant (MRSA) and 22 methicillin-sensitive *Staphylococcus aureus* (MSSA) strains resulted from the phage group of the bacteria, rather than from their resistance or sensitivity to methicillin. By contrast, ANNs could recognize those aspects of the pyrolysis mass spectra that differentiated MRSA and MSSA strains. These results gave the first demonstration that the combination of PyMS with ANNs could provide a rapid and accurate antibiotic-susceptibility testing technique.

15.6 REPRODUCIBILITY OF PyMS

Results of pyrolysis mass spectrometric analyses can be influenced by both phenotypic drift and instrument drift. Phenotypic drift can result from variations in culture growth immediately prior to analysis, and from variations during serial subculturing before analysis.[125, 126] However, this type of drift is not perceived as an obstacle in microbiological work because it can be largely overcome by standardizing culture conditions and by analyzing more than one sample from each culture.

Instrumental drift results from variations in the physical conditions of a pyrolysis mass spectrometer over time.[127] It leads to variation in spectral fingerprints taken from the same material on different occasions. Short-term (<30 days) instrument reproducibility was examined by Manchester et al.[57] who used PyMS to differentiate strains of *Carnobacterium* over a four-week period. Excellent reproducibility was obtained as separation of the five type strains was sustained and spectra did not change significantly over the four weeks.

Long-term reproducibility (>30 days) of PyMS is considered to be a much more serious problem. In a study by Shute et al.[80] reproducibility of PyMS in the differentiation of *Bacillus* species was investigated over a 14-month period. Results showed a change in the overall spectral pattern with time, although the relationships among the samples analyzed at the same time were sustained. A more recent investigation by Chun et al.[128] confirmed these findings but also showed that the nature of the spectral variations between different batches was random, that is to say, there was no consistent time-series drift among the spectra.

Curie-point pyrolysis is very reproducible and pyrolysate transfer to the ion source is tightly controlled. Therefore lack of reproducibility over time is attributed to ion-source aging, which is caused by the accumulation of intractable organic debris around the ion source over long periods of extended use. Aging of the ion source alters ion transmissivity, which in turn causes mass spectral drift.[18,23] Because of this poor long-term reproducibility, within clinical microbiology PyMS has been restricted to the typing of short-term outbreaks where all the microorganisms are analyzed in a single batch.[44]

The first strategy to compensate for mass spectral drift is to tune the instrument. This is typically achieved with the volatile standard, perfluorokerosene, and tuning so that m/z 181 is one-tenth of m/z 69. Unfortunately, this procedure is insufficient to compensate for all the instrumental drift and additional methods are required.

In a attempt to compensate for poor long-term reproducibility in a long-term identification study, Chun et al.[128] applied ANNs to PyMS spectra collected from strains of *Streptomyces* six times over a 20-month period. Direct comparison of the six data sets, by the conventional approach of HCA, was unsuccessful for strain identification, but a neural network trained on spectra from *each* of the first three data sets was able to identify isolates in those three datasets and in the three subsequent datasets.

For PyMS to be used for (1) *routine* identification of microorganisms and (2) in combination with ANNs for quantitative microbiological applications, new spectra must be comparable with those previously collected and held in a data base.[127] Recent work within our laboratory has demonstrated that this problem may be overcome by the use of ANNs to correct for instrumental drift. By calibrating with standards common to both data sets, ANN models created using previously collected data gave accurate estimates of determinand concentrations, or bacterial identities, from newly acquired spectra.[127] In this approach calibration samples were included in each of the two runs, and ANNs were set up in which the inputs were the 150 "new" calibration masses while the outputs were the 150 "old" calibration masses. These associative nets could then by used to transform data acquired on that one day to data acquired at an earlier data. For the first time PyMS was used to acquire spectra that were comparable with those previously collected and held in a database. In a further study this neural network transformation procedure was extended to allow comparison between spectra, previously collected on one machine, with spectra later collected on a different machine;[129] thus calibration transfer by ANNs was affected. Wilkes and colleagues[130] have also used this strategy to compensate for differences in culture conditions to construct robust microbial mass spectral databases.

These advances in correcting for mass spectral drift have far-reaching implications for the potential of PyMS in the microbiology laboratory. However, there are still some problems associated with the technique: (1) it is hardly a nondestructive method, so information on the structure and identity of the molecules producing the pyrolysate will be lost, (2) data acquisition takes 2 minutes per sample, and (3) the technique is labor intensive, tedious, and technically challenging.

It is therefore not surprising that the interest in PyMS as a typing tool diminished at the turn of the twenty-first century and hence why taxonomists have turned to MS-based methods that use soft ionization methods such as electrospray ionization (ESI-MS) and matrix-assisted laser desorption ionization (MALDI MS). These methods generate information-rich spectra of metabolites and proteins, and because the molecular ion is seen, the potential for biomarker discovery is being realized. The analyses of ESI-MS and MALDI-MS data will still need chemometric methods, and it is hoped that researchers in these areas can look back and learn from the many PyMS studies where machine learning was absolutely necessary to turn the complex pyrolysis MS data into knowledge of bacterial identities.

ACKNOWLEDGMENT

The authors are grateful to BBSRC and Wellcome for funding their research into PyMS in the last century.

REFERENCES

1. Casadevall, A. Crisis in infectious disease—Time for a new paradigm, *Clin. Infect. Dis.* 1996, **23**, 790–794.
2. La Scolea, L. Diagnosis of paediatric infections using bacterial antigen detection systems. *Clin. Microbiol. Newsletter* 1998, **10**, 21–23.
3. Gutteridge, C. S.; Priest, F. G. In *Brewing Microbiology*, 2nd ed. Priest, F. G., Campbell, I. (Eds.). London: Chapman and Hall, 1996.
4. Moseley, S. L.; Hug, I.; Alim, A.; Jo, M.; Samapour-Motabeli, M.; Falkow, S. Detection of enterotoxigenic *Escherichia coli* by DNA colony hybridization. *J. Infect. Dis.* 1980, **142**, 892–898.
5. Daly, J. A. Rapid diagnostic tests in microbiology in the 1990s. *Am. J. Clin. Pathol.* 1994, **101**, S22–S26.
6. Blattner, F.; Plunkett, G.; Bloch, C.; Perna, N.; Burland, V.; Riley, M.; Collado-Vides, J.; Glasner, J.; Rode, C.; Mayhew, G.; Gregor, J.; Davis, N.; Kirkpatrick, H.; Goeden, M.; Rose, D.; Mau, B.; Shao, Y. The complete genome sequence of *Escherichia coli* K-12. *Science* 1997, **277**, 1453–1474.
7. Jackman, P. J. H. In *Chemical Methods in Bacterial Systematics*. Goodfellow, M., Minnikin, D. E. (Eds.). London: Academic Press, 1985, pp. 115–129.
8. Kersters, K.; Pot, B.; Dewettinck, D.; Torck, V.; Vancanneyt, M.; Vauterin, L.; Vandamme, P. In *Bacterial Diversity and Systematics*. Priest, F. G., Ramos-Cormenza, A., Tindall, B. (Eds.). New York: Plenum, 1994, pp. 51–66.
9. Magee, J. T. In *Handbook of New Bacterial Systematics*. Goodfellow, M., Donnell, A. G. O. (Eds.). London: Academic Press, 1993, pp. 383–427.

10. Magee, J. T. In *Chemical Methods in Prokaryotic Systematics*. Goodfellow, M., O'Donnell, A. G. (Eds.). New York: Wiley, 1994, pp. 523–553.
11. Meuzelaar, H. L.; Kistemaker, P. G. A technique for fast and reproducible fingerprinting of bacteria by pyrolysis mass spectrometry. *Anal. Chem.* 1973, **45**, 587–590.
12. Goodacre, R.; Kell, D. B. Pyrolysis mass spectrometry and its application in biotechnology. *Curr. Opin. Biotechnol.* 1996, **7**, 20–28.
13. Helm, D.; Labischinski, H.; Schallehn, G.; Naumann, D. Classification and identification of bacteria by Fourier transform infrared spectroscopy. *J. Gen. Microbiol.* 1991, **137**, 69–79.
14. Naumann, D.; Helm, D.; Labischinski, H.; Giesbrecht, P. In *Modern Techniques for Rapid Microbiological Analysis*. Nelson, W. (Ed.). New York: VCH, 1991, pp. 43–96.
15. Naumann, D.; Helm, D.; Labischinski, H. Microbiological characterizations by FT-IR spectroscopy. *Nature* 1991, **351**, 81–82.
16. Nelson, W. H.; Sperry, J. F. In *Modern Techniques for Rapid Microbiological Analysis*. Nelson, W. H. (Ed.). New York: VCH, 1991, pp. 97–143.
17. Nelson, W. H.; Manoharan, R.; Sperry, J. F. UV resonance Raman studies of bacteria. *Appl. Spectrosc. Revs.* 1992, **27**, 67–124.
18. Meuzelaar, H. L. C.; Haverkamp, J.; Hileman, F. D. *Pyrolysis Mass Spectrometry of Recent and Fossil Biomaterials.* Amsterdam: Elsevier, 1982.
19. Griffiths, P. R.; de Haseth, J. *Fourier Transform Infrared Spectrometry.* New York: Wiley, 1986.
20. Colthup, N. B.; Daly, L. H.; Wiberly, S. E. *Introduction to Infrared and Raman Spectroscopy.* New York: Academic Press, 1990.
21. Irwin, W. J. *Analytical Pyrolysis: A Comprehensive Guide.* New York: Dekker, 1982.
22. Windig, W.; Kistemaker, P. G.; Haverkamp, J. Factor analysis of the influence of changes in experimental conditions in pyrolysis-mass spectrometry. *J. Anal. Appl. Pyrolysis* 1980, **2**, 7–18.
23. Windig, W.; Kistemaker, P. G.; Haverkamp, J.; Meuzelaar, H. L. C. The effects of sample preparation, pyrolysis and pyrolysate transfer conditions on pyrolysis mass spectra. *J. Anal. Appl. Pyrolysis* 1979, **1**, 39–52.
24. Zemany, P. D. Identification of complex organic materials by mass spectrometric analysis of their pyrolysis products. *Anal. Chem.* 1952, **24**, 1709–1713.
25. Davison, W. H. T.; Slaney, S.; Wragg, A. L. A novel method of identification of polymers. *Chem. Ind.* 1954, **44**, 135–136.
26. Wilson, M. E.; Oyama, V.; Vango, S. P. United States 1962. New York: Academic Press, pp. 329–338.
27. Reiner, E. Identification of bacteria by pyrolysis gas-liquid chromatography. *Nature* 1965, **206**, 1272–1274.
28. Reiner, E.; Ewing, W. J. Chemotaxonomic studies of some Gram-negative bacteria by means of pyrolysis gas-liquid chromatography. *Nature* 1968, **217**, 191–194.
29. Gutteridge, C. S.; Norris, J. R. The application of pyrolysis techniques to the identification of microorganisms. *J. Appl. Bacteriol.* 1979, **47**, 5–43.

30. Gutteridge, C. S. Characterization of microorganisms by pyrolysis mass spectrometry. *Meth. Microbiol.* 1987, **19**, 227–272.

31. Eudy, L. W.; Walla, M. D.; Hudson, J. R.; Morgan, S. L.; Fox, A. Gas chromatography-mass spectrometry studies on the occurrence of acetamide, propionamide and furyl alchohol in pyrolysates of bacteria, bacteria fractions and model compounds. *J. Anal. Appl. Pyrolysis* 1985, **7**, 231–247.

32. Meuzelaar, H. L.; Kistemaker, P. G.; Eshuis, W.; Boerboom, H. A. Automated pyrolysis-mass spectrometry; application to the differentiation of microorganisms. *Adv. Mass Spectrom.* 1976, **7B**, 1452–1456.

33. Gutteridge, C. S.; Vallis, L.; Macfie, H. J. H. In *Computer-assisted Bacterial Systematics*. M. Goodfellow, D. Jones, Priest, F. G. (Eds.). San Diego: Academic Press, 1985, pp. 369–401.

34. Wieten, G.; Haverkamp, J.; Meuzelaar, H.; Boudewijn Engel, H.; Berwald, L. Pyrolysis mass spectrometry: A new method to differentiate between the mycobacteria of the "tuberculosis complex" and other mycobacteria. *J. Gen. Microbiol.* 1981, **122**, 109–118.

35. Wieten, G.; Haverkamp, J.; Groothuis, D. G.; Berwald, L. G.; David, H. L. Classification and identification of *Mycobacterium africanum* by pyrolysis mass spectrometry. *J. Gen. Microbiol.* 1983, **129**, 3679–3688.

36. Borst, J.; van der Snee-Enkellar, A. C.; Meuzelaar, H. L. C. Typing of *Neisseria gonnorrhoeae* by pyrolysis mass spectrometry. *Antonie van Leewenhoek* 1978, **44**, 253.

37. Kajioka, R.; Tang, P. W. Curie-point pyrolysis mass-spectrometry of *Legionella* species. *J. Anal. Appl. Pyrolysis* 1984, **6**, 59–68.

38. Windig, W.; De Hoog, G. S. Pyrolysis mass spectrometry, II. *Sporidiopolus* and related taxa. *Studies Mycol.* 1982, **22**, 60–65.

39. Windig, W.; Haverkamp, J. Pyrolysis mass spectrometry, I. *Rhodosporidium Studies Mycol.* 1982, **22**, 56–74.

40. Aries, R. E.; Gutteridge, C. S.; Ottley, T. W. Evaluation of a low-cost, automated pyrolysis-mass spectrometer. *J. Anal. Appl. Pyrolysis* 1986, **9**, 81–98.

41. Goodacre, R. Characterisation and quantification of microbial systems using pyrolysis mass spectrometry: Introducing neural networks to analytical pyrolysis. *Microbiol. Eur.* 1994, **2**, 16–22.

42. Goodfellow, M.; Chun, J.; Atalan, E.; Sanglier, J. J. In *Bacterial Diversity and Systematics.* Priest, F. G., Ramos, C. (Eds.). New York: Plenum, 1994, pp. 87–104.

43. Goodfellow, M. Inter-strain comparison of pathogenic microorganisms by pyrolysis mass spectrometry. *Binary Comput. Microbiol.* 1995, **7**, 54–60.

44. Goodfellow, M.; Freeman, R.; Sisson, P. R. Curie-point pyrolysis mass spectrometry as a tool in clinical microbiology. *Zbl. Bakt.* 1997, **285**, 133–156.

45. Everitt, B. S. *Cluster Analysis.* London: Edward Arnold, 1993.

46. Goodacre, R.; Berkeley, R. C. W. Detection of small genotypic changes in *Escherichia coli* by pyrolysis mass spectrometry. *FEMS Microbiol. Lett.* 1990, **71**, 133–137.

47. Freeman, R.; Sisson, P. R.; Jenkins, D. R.; Ward, A. C.; Lightfoot, N. F.; O'Brien, S. J. Sporadic isolates of *Escherichia coli* 0157.H7 investigated by pyrolysis mass spectrometry. *Epidemiol. Infect.* 1995, **114**, 433–440.

48. Freeman, R.; Gould, F. K.; Sisson, P. R.; Lightfoot, N. F. Strain differentiation of capsule type 23 penicillin-resistant *Streptcoccus pneumoniae* from nosocomial infections by pyrolysis mass spectrometry. *Lett. Appl. Microbiol.* 1991, **13**, 28–31.

49. Jackson, R. M.; Heginbotham, M. L.; Magee, J. T. Epidemiological typing of *Klebsiella pneumoniae* by pyrolysis mass spectrometry. *Zbl. Bakt.* 1997, **285**, 252–257.

50. Leaves, N. I.; Sisson, P. R.; Freeman, R.; Jordens, J. Z. Pyrolysis mass spectrometry in epidemiological and population genetic studies of *Haemophilus influenzae*. *J. Med. Microbiol.* 1997, **46**, 204–207.

51. Manfio, G. P.; Goodfellow, M.; Austin, B.; Austin, D. A.; Pedersen, K.; Larsen, J. L.; Verdonck, L.; Swings, J. Typing of the fish pathogen *Listonella (Vibrio) anguillara* by pyrolysis mass spectrometry. *Zbl. Bakt.* 1997, **285**, 245–251.

52. Sisson, P. R.; Freeman, R.; Lightfoot, N. F.; Richardson, I. R. Incrimination of an environmental source of a case of Legionnaires' disease by pyrolysis mass spectrometry. *Epidemiol. Infect.* 1991, **107**, 127–132.

53. AlSaif, N. M.; O Neill, G. L.; Magee, J. T.; Brazier, J. S.; Duerden, B. I. PCR-ribotyping and pyrolysis mass spectrometry fingerprinting of environmental and hospital isolates of *Clostridium difficile*. *J. Med. Microbiol.* 1998, **47**, 117–121.

54. O'Neill, G. L.; Brazier, J. S.; Magee, J. T.; Duerden, B. I. A comparison of PCR ribotyping and pyrolysis mass spectrometry for typing clinical isolates of *Clostridium difficile*. *Anaerobe* 1996, **2**, 211–215.

55. Sisson, P. R.; Freeman, R.; Gould, F. K.; Lightfoot, N. F. Strain differentiation of nosocomial isolates of *Pseudomonas aeruginosa* by pyrolysis mass spectrometry. *J. Hosp. Infect.* 1991, **19**, 137–140.

56. Freeman, R.; Sisson, P. R.; Noble, W. C.; Lightfoot, N. F. An apparent outbreak of infection with *Acinetobacter calcoaceticus* reconsidered after investigation by pyrolysis mass spectrometry. *Zbl. Bakt.* 1997, **285**, 234–244.

57. Manchester, L. N.; Toole, A.; Goodacre, R. Characterization of *Carnobacterium* species by pyrolysis mass spectrometry *J. Appl. Bacteriol.* 1995, **78**, 88–96.

58. Magee, J. T.; Hindmarch, J. M.; Bennet, K. W.; Duerden, B. I.; Aries, R. E. A pyrolysis mass spectrometry study of fusobacteria. *J. Med. Microbiol.* 1989, **28**, 227–236.

59. Murdoch, D.; Magee, J. A numerical taxonomic study of the Gram-positive anaerobic cocci. *J. Med. Microbiol.* 1995, **43**, 148–155.

60. Dalgaard, P.; Manfio, G. P.; Goodfellow, M. Classification of photobacteria associated with spoliage of fish products by numerical taxonomy and pyrolysis mass spectrometry. *Zbl. Bakt.* 1997, **285**, 157–168.

61. Magee, J. T.; Yousefi-Mashouf, R.; Hindmarch, J. M.; Duerden, B. I. A pyrolysis mass spectrometry study of the non-pigmented *Prevotella* species. *J. Med. Microbiol.* 1992, **37**, 273–282.

62. Sutcliffe, I. C.; Manfio, G. P.; Schaal, K. P.; Goodfellow, M. An investigation of the intra-generic structure of *Rothia* by pyrolysis mass spectrometry. *Zbl. Bakt.* 1997, **285**, 204–211.

63. Magee, J. T.; Hindmarch, J. M.; Douglas, C. W. A numerical taxonomic study of *Streptococcus sanguis, S. mitis* and similar organisms using conventional tests and pyrolysis mass spectrometry. *Zbl. Bakt.* 1997, **285**, 195–203.

64. Ferguson, E. V.; Ward, A. C.; Sanglier, J. J.; Goodfellow, M. Evaluation of *Streptomyces* species-groups by pyrolysis mass spectrometry. *Zbl. Bakt.* 1997, **285**, 169–181.

65. Trujillo, M. E.; Goodfellow, M. Polyphasic taxonomic study of clinically significant actinomadurae including the description of *Actinomadura latina* sp. nov. *Zbl. Bakt.* 1997, **285**, 212–233.

66. Heyndrickx, M.; Vandemeulebroecke, K.; Scheldeman, P.; Hoste, B.; Kersters, K.; Devos, P.; Logan, N. A.; Aziz, A. M.; Ali, N.; Berkeley, R. C. W. *Paenibacillus* (Formerly *Bacillus*) *gordonae* (Pichinoty et al. 1986) Ash et al. 1994 is a later subjective synonym of *Paenibacillus* (Formerly *Bacillus*) *validus* (Nakamura 1984) Ash et al. 1994—emended description of *P. validus*. *Int. J. Syst. Bacteriol.* 1995, **45**, 661–669.

67. Heyndrickx, M.; Vandemeulebroecke, K.; Hoste, B.; Janssen, P.; Kersters, K.; Devos, P.; Logan, N. A.; Ali, N.; Berkeley, R. C. W. Reclassification of *Paenibacillus* (Formerly *Bacillus*) *Pulvifaciens* (Nakamura 1984) Ash et al. 1994, a later subjective synonym of *Paenibacillus* (Formerly *Bacillus*) larvae (White 1906) Ash et al. 1994, as a subspecies of *Paenibacillus-larvae*, with emended descriptions of *Paenibacillus-larvae* as *Paenibacillus-larvae* subsp larvae and *Paenibacillus-larvae* subsp pulvifaciens. *Int. J. Syst. Bacteriol.* 1996, **46**, 270–279.

68. Heyndrickx, M.; Vandemeulebroecke, K.; Scheldeman, P.; Kersters, K.; DeVos, P.; Logan, N. A.; Aziz, A. M.; Ali, N.; Berkeley, R. C. W. A polyphasic reassessment of the genus *Paenibacillus*, reclassification of *Bacillus lautus* (Nakamura 1984) as *Paenibacillus lautus* comb nov and of *Bacillus peoriae* (Montefusco et al. 1993) as *Paenibacillus peoriae* comb nov, and emended descriptions of *P. lautus* and of *P. peoriae*. *Int. J. Syst. Bacteriol.* 1996, **46**, 988–1003.

69. Heyndrickx, M.; Lebbe, L.; Vancaneyt, M.; Kersters, K.; DeVos, P.; Logan, N. A.; Forsyth, G.; Nazli, S.; Ali, N.; Berkeley, R. C. W. A polyphasic reassessment of the genus *Aneurinibacillus*, reclassification of *Bacillus thermoaerophilus* (Meier-Stauffer et al. 1996) as *Aneurinibacillus thermoaerophilus* comb nov, and emended descriptions of *A. aneurinilyticus* corrig, *A. migulanus*, and *A. thermoaerophilus*. *Int. J. Syst. Bacteriol.* 1997, **47**, 808–817.

70. Heyndrickx, M.; Lebbe, L.; Kersters, K.; DeVos, P.; Forsythe, C.; Logan, N. A. *Virgibacillus*: A new genus to accommodate *Bacillus pantothenticus* (Proom and knight 1950). Emended description of *Virgibacillus pantothenticus*. *Int. J. Syst. Bacteriol.* 1998, **48**, 99–106.

71. Sisson, P. R.; Freeman, R.; Magee, J. G.; Lightfoot, N. F. Differentiation between Mycobacteria of the *Mycobacterium tuberculosis* complex by pyrolysis mass spectrometry. *Tubercle* 1991, **72**, 206–209.

72. Sisson, P. R.; Freeman, R.; Magee, J. G.; Lightfoot, N. F. Rapid differentiation of *Mycobacterium xenopi* from Mycobacteria of the *Mycobacterium avium-intracellulare* complex by pyrolysis mass spectrometry. *J. Clin. Pathol.* 1992, **45**, 355–357.

73. Magee, J. G.; Goodfellow, M.; Sisson, P. R.; Freeman, R.; Lightfoot, N. F. Differentiation of *Mycobacterium senegalense* from related non-chromogenic mycobacteria using pyrolysis mass spectrometry. *Zbl. Bakt.* 1997, **285**, 278–284.

74. Heginbotham, M. L.; Magee, J. T. Pyrolysis mass spectrometry: A predictor of clinical response to treatment in pulmonary oppurtunist mycobacterial infection: preliminary work with *M. malmoense. Zbl. Bakt.* 1996, **285**, 291–298.
75. Huff, S. M.; Matsen, J. M.; Windig, W.; Meuzelaar, H. L. Pyrolysis mass spectrometry of bacteria from infected human urine. *Biomed. Environ. Mass Spectrom.* 1986, **13**, 277–286.
76. Magee, J. T.; Hindmarch, J. M.; Winstanley, T. G. Applications of pyrolysis mass spectrometry in studies on the mode of action of antimicrobial agents. *Zbl. Bakt.* 1997, **285**, 305–310.
77. Sanglier, J. J.; Whitehead, D.; Saddler, G. S.; Ferguson, E. V.; Goodfellow, M. Pyrolysis mass spectrometry as a method for the classification, identification and selection of actinomycetes. *Gene* 1992, **115**, 235–242.
78. Goodacre, R.; Hartmann, A.; Beringer, J. E.; Berkeley, R. C. W. The use of pyrolysis mass spectrometry in the characterization of *Rhizobium meliloti. Lett. Appl. Microbiol.* 1991, **13**, 157–160.
79. Kay, H. E.; Coutinho, H. L. C.; Fattori, M.; Manfio, G. P.; Goodacre, R.; Nuti, M. P.; Basaglia, M.; Beringer, J. E. The identification of *Bradyrhizobium japonicum* strains isolated from Italian soils. *Microbiology–UK* 1994, **140**, 2333–2339.
80. Shute, L. A.; Gutteridge, C. S.; Norris, J. R.; Berkeley, R. C. W. Reproducibility of pyrolysis mass spectrometry: effect of growth medium and instrument stability on the differentiation of selected *Bacillus* species. *J. Appl. Bacteriol.* 1988, **64**, 79–88.
81. Brereton, R. G. *Multivariate Pattern Recognition in Chemometrics.* Amsterdam: Elsevier, 1992.
82. Brown, S. D.; Blank, T. B.; Sum, S. T.; Weyer, L. G. Chemometrics. *Anal. Chem.* 1994, **66**, R315–R359.
83. Brown, S. D.; Sum, S. T.; Despagne, F.; Lavine, B. K. Chemometrics. *Anal. Chem.* 1996, **68**, R21–R61.
84. Beebe, K. R.; Pell, R. J.; Seasholtz, M. B. *Chemometrics: A Practical Guide.* New York: Wiley, 1998.
85. Lavine, B. K. Chemometrics. *Anal. Chem.* 1998, **70**, R209–R228.
86. Beavis, R. C.; Colby, S. M.; Goodacre, R.; Harrington, P. B.; Reilly, J. P.; Sokolow, S.; Wilkerson, C. W. In *Encyclopedia of Analytical Chemistry*. Meyers, R. A. (Ed.). 2000, Chichester: John Wiley & Sons Ltd., pp. 11558–11597.
87. Wasserman, P. D. *Neural Computing: Theory and Practice*. Van Nostrand Reinhold: New York, 1989.
88. Zupan, J.; Gasteiger, J. *Neural Networks for Chemists: An Introduction*. Weinheim: VCH Verlagsgeesellschaft, 1993.
89. Haykin, S. *Neural Networks*. New York: Macmillan, 1994.
90. Bishop, C. M. *Neural Networks for Pattern Recognition*. Oxford: Clarendon, 1995.
91. Dybowski, R.; Gant, V. Artificial neural networks in pathological and medical laboratories. *Lancet* 1995, **346**, 1203–1207.
92. Ripley, B. D. *Pattern Recognition and Neural Networks.* Cambridge: Cambridge University Press, 1996.
93. Goodacre, R.; Neal, M. J.; Kell, D. B. Rapid and quantitative analysis of the pyrolysis mass spectra of complex binary and tertiary mixtures using multivariate calibration and artificial neural networks. *Anal. Chem.* 1994, **66**, 1070–1085.

94. Goodacre, R.; Neal, M. J.; Kell, D. B. Quantitative analysis of multivariate data using artificial neural networks: A tutorial review and applications to the deconvolution of pyrolysis mass spectra. *Zbl. Bakt.* 1996, **284**, 516–539.
95. Kenyon, R. G. W.; Ferguson, E. V.; Ward, A. C. Application of neural networks to the analysis of pyrolysis mass spectra. *Zbl. Bakt.* 1997, **285**, 267–277.
96. Goodacre, R.; Kell, D. B.; Bianchi, G. Neural networks and olive oil. *Nature* 1992, **359**, 594–594.
97. Goodacre, R.; Kell, D. B.; Bianchi, G. Rapid assessment of the adulteration of virgin olive oils by other seed oils using pyrolysis mass spectrometry and artificial neural networks. *J. Sci. Food Agric.* 1993, **63**, 297–307.
98. Goodacre, R.; Kell, D. B. Rapid and quantitative analysis of bioprocesses using pyrolysis mass spectrometry and neural networks—Application to indole production. *Anal. Chim. Acta* 1993, **279**, 17–26.
99. Goodacre, R.; Edmonds, A. N.; Kell, D. B. Quantitative analysis of the pyrolysis mass spectra of complex mixtures using artificial neural networks—Application to casamino acids in glycogen. *J. Anal. Appl. Pyrolysis* 1993, **26**, 93–114.
100. Timmins, É. M.; Goodacre, R. Rapid quantitative analysis of binary mixtures of *Escherichia coli* strains using pyrolysis mass spectrometry with multivariate calibration and artificial neural networks *J. Appl. Microbiol.* 1997, **83**, 208–218.
101. Goodacre, R.; Karim, A.; Kaderbhai, M. A.; Kell, D. B. Rapid and quantitative analysis of recombinant protein expression using pyrolysis mass spectrometry and artificial neural networks—Application to mammalian cytochrome B5 in *Escherichia coli. J. Biotechnol.* 1994, **34**, 185–193.
102. Goodacre, R.; Trew, S.; Wrigley-Jones, C.; Neal, M. J.; Maddock, J.; Ottley, T. W.; Porter, N.; Kell, D. B. Rapid screening for metabolite overproduction in fermentor broths using pyrolysis mass spectrometry with multivariate calibration and artificial neural networks. *Biotechnol. Bioeng.* 1994, **44**, 1205–1216.
103. Goodacre, R.; Trew, S.; Wrigley-Jones, C.; Saunders, G.; Neal, M. J.; Porter, N.; Kell, D. B. Rapid and quantitative analysis of metabolites in fermentor broths using pyrolysis mass spectrometry with supervised learning: Application to the screening of *Penicillium chrysogenum* fermentations for the overproduction of penicillins. *Anal. Chim. Acta* 1995, **313**, 25–43.
104. McGovern, A. C.; Ernill, R.; Kara, B. V.; Kell, D. B.; Goodacre, R. Rapid analysis of the expression of heterologous proteins in *Escherichia coli* using pyrolysis mass spectrometry and Fourier transform infrared spectroscopy with chemometrics: Application to a2- interferon production. *J. Biotechnol.* 1999, **72**, 157–167.
105. McGovern, A. C.; Broadhurst, D.; Taylor, J.; Kaderbhai, N.; Winson, M. K.; Small, D. A.; Rowland, J. J.; Kell, D. B.; Goodacre, R. Monitoring of complex industrial bioprocesses for metabolite concentrations using modern spectroscopies and machine learning: Application to gibberellic acid production. *Biotechnol. Bioeng.* 2002, **78**, 527–538.
106. Kang, S. G.; Lee, D. H.; Ward, A. C.; Lee, K. J. Rapid and quantitative analysis of clavulanic acid production by the combination of pyrolysis mass spectrometry and actificial neural networks. *J. Microbiol. Biotechnol.* 1998, **8**, 523–530.
107. Lee, K. J. Dynamics of morphological and physiological differentiation in the actinomycetes group and quantitative analysis of the differentiations. *J. Microbiol. Biotechnol.* 1998, **8**, 1–7.

108. Goodacre, R.; Hiom, S. J.; Cheeseman, S. L.; Murdoch, D.; Weightman, A. J.; Wade, W. G. Identification and discrimination of oral asaccharolytic *Eubacterium* spp. using pyrolysis mass spectrometry and artificial neural networks. *Curr. Microbiol.* 1996, **32**, 77–84.

109. Freeman, R.; Goodacre, R.; Sisson, P. R.; Magee, J. G.; Ward, A. C.; Lightfoot, N. F. Rapid identification of species within the *Mycobacterium tuberculosis* complex by artificial neural network analysis of pyrolysis mass spectra. *J. Med. Microbiol.* 1994, **40**, 170–173.

110. Goodacre, R.; Neal, M. J.; Kell, D. B.; Greenham, L. W.; Noble, W. C.; Harvey, R. G. Rapid identification using pyrolysis mass spectrometry and artificial neural networks of *Propionibacterium acnes* isolated from dogs. *J. Appl. Bacteriol.* 1994, **76**, 124–134.

111. Goodacre, R.; Howell, S. A.; Noble, W. C.; Neal, M. J. Sub-species discrimination using pyrolysis mass spectrometry and self-organising neural networks of *Propionibacterium acnes* isolated from normal human skin. *Zbl. Bakt.–Int. J. Med. Microbiol. Virol. Parasitol. Infect. Dis.* 1996, **284**, 501–515.

112. Nilsson, T.; Bassani, M. R.; Larsen, T. O.; Montanarella, L. Classification of species in the genus *Penicillium* by Curie point pyrolysis/mass spectrometry followed by multivariate analysis and artificial neural networks. *J. Mass Spectrom.* 1996, **31**, 1422–1428.

113. Chun, J.; Atalan, E.; Ward, A. C.; Goodfellow, M. Artificial neural network analysis of pyrolysis mass spectrometric data in the identification of *Streptomyces* strains. *FEMS Microbiol. Lett.* 1993, **107**, 321–325.

114. Goodacre, R.; Timmins, É. M.; Burton, R.; Kaderbhai, N.; Woodward, A. M.; Kell, D. B.; Rooney, P. J. Rapid identification of urinary tract infection bacteria using hperspectral whole-organism fingerprinting and artificial neural networks. *Microbiology* 1998, **144**, 1157–1170.

115. Sisson, P. R.; Freeman, R.; Law, D.; Ward, A. C.; Lightfoot, N. F. Rapid detection of verocytotoxin production status in *Escherichia coli* by artificial neural network analysis of pyrolysis-mass spectra. *J. Anal. Appl. Pyrolysis* 1995, **32**, 179–785.

116. Kang, S. G.; Kenyon, R. G. W.; Ward, A. C.; Lee, K. J. Analysis of differentiation state in *Streptomyces albidoflavus* SMF301 by the combination of pyrolysis mass spectrometry and neural networks. *J. Biotechnol.* 1998, **62**, 1–10.

117. Harrington, P. B. Fuzzy Rule-building Expert Systems: Minimal Neural Networks. *J. Chemometrics* 1991, **5**, 467–486.

118. Harrington, P. B. Minimal neural networks: Differentiation of classification entropy. *Chemom. Intell. Lab. Syst.* 1993, **19**, 143–154.

119. Harrington, P. D. Minimal neural networks—Concerted optimization of multiple decision planes. *Chemom. Intell. Lab. Syst.* 1993, **18**, 157–170.

120. Broomhead, D. S.; Lowe, D. Multivariable functional interpolation and adaptive networks. *Complex Syst.* 1988, **2**, 312–355.

121. Kohonen, T. *Self-organization and Associative Memory*. Berlin: Springer-Verlag, 1989.

122. Kramer, M. A. Autoassociative neural networks. *Comput. Chem. Eng.* 1992, **16**, 313–328.

123. Goodacre, R.; Pygall, J.; Kell, D. B. Plant seed classification using pyrolysis mass spectrometry with unsupervised learning; the application of auto-associative and Kohonen artificial neural networks. *Chemom. Intell. Lab. Syst.* 1996, **34**, 69–83.

124. Goodacre, R.; Rooney, P. J.; Kell, D. B. Discrimination between methicillin-resistant and methicillin-susceptible *Staphylococcus aureus* using pyrolysis mass spectrometry and artificial neural networks. *J. Antimicrob. Chemother.* 1998, **41**, 27–34.

125. Freeman, R.; Sisson, P. R.; Ward, A. C.; Hetherington, C. Phenotypic variation as a significant within-strain factor in comparisons of bacteria based on pyrolysis-mass spectrometry. *J. Anal. Appl. Pyrolysis* 1994, **28**, 29–37.

126. Voorhees, K. J.; Durfee, S. L.; Updegraff, D. M, Identification of diverse bacteria grown under diverse conditions using pyrolysis-mass spectrometry. *J. Microbiol. Meth.* 1988, **8**, 315–325.

127. Goodacre, R.; Kell, D. B. Correction of mass spectral drift using artificial neural networks. *Anal. Chem.* 1996, **68**, 271–280.

128. Chun, J.; Ward, A. C.; Kang, S.; Hah, Y. C.; Goodfellow, M. Long-term identification of *Streptomycetes* using pyrolysis mass spectrometry and artificial neural networks. *Zbl. Bakt.* 1997, **185**, 258–266.

129. Goodacre, R.; Timmins, É. M.; Jones, A.; Kell, D. B.; Maddock, J.; Heginbothom, M. L.; Magee, J. T. On mass spectrometer instrument standardization and interlaboratory calibration transfer using neural networks. *Anal. Chim. Acta* 1997, **384**, 511–532.

130. Wilkes, J. G.; Glover, K. L.; Holcomb, M.; Rafii, F.; Cao, X.; Sutherland, J. B.; McCarthy, S. A.; Letarte, S.; Bertrand, M. J. Defining and using microbial spectral databases *J. Am. Soc. Mass Spectrom.* 2002, **13**, 875–887.

131. Sisson, P. R.; Kramer, J. M.; Brett, M. M.; Freeman, R.; Gilbert, R. J.; Lightfoot, N. F. Application of pyrolysis mass spectrometry to the investigation of outbreaks of food poisoning and non-gastrointestinal infection associated with *Bacillus* species and *Clostridium perfringens*. *Int. J. Food Microbiol.* 1992, **17**, 57–66.

132. Synder, A. P.; Thornton, S. D.; Dworzanski, J. P.; Meuzelaar, H. L. C. Detection of the picolinic acid biomarker in *Bacillus* spores using a potentially field portable pyrolysis gas chromatography-ion mobility spectrometry system. *Field Anal. Chem. Technol.* 1996, **1**, 49–58.

133. Helyer, R. J.; Kelley, T.; Berkeley, R. C. W. Pyrolysis mass spectrometry studies on *Bacillus anthracis, Bacillus cereus* and their close relatives. *Zbl. Bakt.* 1997, **285**, 319–328.

134. Basile, F.; Beverly, M. B.; Voorhees, K. J.; Hadfield, T. L. Pathogenic bacteria: Their detection and differentiation by rapid lipid profiling with pyrolysis mass spectrometry. *Trac-Trends Anal. Chem.* 1998, **17**, 95–109.

135. Goodacre, R.; Shann, B.; Gilbert, R. J.; Timmins, É. M.; McGovern, A. C.; Alsberg, B. K.; Kell, D. B.; Logan, N. A. The detection of the dipicolinic acid biomarker in *Bacillus* spores using Curie-point pyrolysis mass spectrometry and Fourier transform infrared spectroscopy. *Anal. Chem.* 2000, **72**, 119–127.

136. Deurden, B. I.; Eely, A.; Goodwin, L.; Magee, J. T.; Hindmarch, J. M.; Bennet, K. W. A comparison of *Bacteroides ureolyticus* isolates form different clinical sources. *J. Med. Microbiol.* 1289, **29**, 63–73.

137. Magee, J. T.; Hindmarch, J. M.; Duerden, B. I.; Mackenzie, D. W. R. Pyrolysis mass spectrometry as a method for inter-strain discrimination of *Candida albicans*. *J. Gen. Microbiol.* 1988, **134**, 2841–2847.

138. Timmins, É. M.; Howell, S. A.; Alsberg, B. K.; Noble, W. C.; Goodacre, R. Rapid differentiation of closely related *Candida* species and strains by pyrolysis mass spectrometry and Fourier transform infrared spectroscopy. *J. Clin. Microbiol.* 1998, **36**, 367–374.

139. Cartmill, T.; Orr, K.; Freeman, R.; Sisson, P. R. Nosocomial infection with *Clostridium difficile* investigated by pyrolysis mass spectrometry. *J. Med. Microbiol.* 1992, **37**, 352–356.

140. Kyne, L.; Merry, C.; O Connell, B.; Harrington, P.; Keane, C.; O Neill, D. Simultaneous outbreaks of two strains of toxigenic *Clostridium difficile* in a general hospital. *J. Hosp. Infect.* 1998, **38**, 101–112.

141. Freeman, R.; Gould, F. K.; Ryan, D. W.; Chamberlain, J.; Sisson, P. R. Nosocomial infection due to enterococci attributed to a fluidized microsphere bed with the help of pyrolysis mass spectrometry. *J. Hosp. Infect.* 1994, **27**, 187–193.

142. Freeman, R.; Sisson, P. R.; Lightfoot, N. F.; NcLaughlin, J. Analysis of epidemic and sporadic strains of *Listeria monocytogenes* by pyrolysis mass spectrometry. *Lett. Appl. Microbiol.* 1991, **12**, 133–136.

143. Corkill, J.; Sisson, P. R.; Smyth, A.; Deveney, J.; Freeman, R.; Shears, P.; Heaf, D.; Hart, C. A. Application of pyrolysis mass spectroscopy and SDS-PAGE in the study of the epidemiology of *Pseudomonas cepacia* in cystic fibrosis. *J. Med. Microbiol.* 1994, **41**, 106–111.

144. Magee, J. T.; Yuang, J.; Ryley, H. C. Typing of clinical and environmental *Pseudomonas cepacia* isolates by pyrolysis mass spectrometry. *Clin. Ecol. Cystic Fibr.* 1993, **10**, 89–93.

145. Freeman, R.; Goodfellow, M.; Gould, F. K.; Hudson, S. J.; Lightfoot, N. F. Pyrolysis-mass spectrometry (Py-MS) for the rapid epidemiological typing of clinically significant bacterial pathogens. *J. Med. Microbiol.* 1990, **32**, 283–286.

146. Freeman, R.; Gould, F. K.; Wilkinson, R.; Ward, A. C.; Lightfoot, N. F.; Sisson, P. R. Rapid inter-strain comparison by pyrolysis mass spectrometry of coagulase-negative staphylococci from persistant CAPD peritonitis. *Epidemiol. Infect.* 1991, **106**, 239–246.

147. Goodacre, R.; Harvey, R.; Howell, S. A.; Greenham, L. W.; Noble, W. C. An epidemiological study of some *Staphylococcus intermedius* strains isolated from dogs, their owners and veterinary surgeons. *J. Anal. Appl. Pyrolysis* 1997, **44**, 49–64.

148. Magee, J. T.; Hindmarch, J. M.; Burnett, I. A.; Pease, A. Epidemiological typing of *Streptococcus pyogenesby* pyrolysis mass spectrometry. *J. Med. Microbiol.* 1989, **30**, 237–278.

149. Timmins, É. M.; Quain, D. E.; Goodacre, R. Differentiation of brewing yeast strains by pyrolysis mass spectrometry and Fourier transform infrared spectroscopy. *Yeast* 1998, **14**, 885–893.

INDEX

Identification of Microorganisms by Mass Spectrometry, Edited by Charles L. Wilkins and Jackson O. Lay, Jr.

CHEMICAL ANALYSIS

A SERIES OF MONOGRAPHS ON ANALYTICAL CHEMISTRY
AND ITS APPLICATIONS

Series Editor
J. D. WINEFORDNER

Vol. 1 **The Analytical Chemistry of Industrial Poisons, Hazards, and Solvents**. *Second Edition.* By the late Morris B. Jacobs
Vol. 2 **Chromatographic Adsorption Analysis**. By Harold H. Strain (*out of print*)
Vol. 3 **Photometric Determination of Traces of Metals**. *Fourth Edition*
Part I: General Aspects. By E. B. Sandell and Hiroshi Onishi
Part IIA: Individual Metals, Aluminum to Lithium. By Hiroshi Onishi
Part IIB: Individual Metals, Magnesium to Zirconium. By Hiroshi Onishi
Vol. 4 **Organic Reagents Used in Gravimetric and Volumetric Analysis**. By John F. Flagg (*out of print*)
Vol. 5 **Aquametry: A Treatise on Methods for the Determination of Water**. *Second Edition* (*in three parts*). By John Mitchell, Jr. and Donald Milton Smith
Vol. 6 **Analysis of Insecticides and Acaricides**. By Francis A. Gunther and Roger C. Blinn (*out of print*)
Vol. 7 **Chemical Analysis of Industrial Solvents**. By the late Morris B. Jacobs and Leopold Schetlan
Vol. 8 **Colorimetric Determination of Nonmetals**. *Second Edition.* Edited by the late David F. Boltz and James A. Howell
Vol. 9 **Analytical Chemistry of Titanium Metals and Compounds**. By Maurice Codell
Vol. 10 **The Chemical Analysis of Air Pollutants**. By the late Morris B. Jacobs
Vol. 11 **X-Ray Spectrochemical Analysis**. *Second Edition.* By L. S. Birks
Vol. 12 **Systematic Analysis of Surface-Active Agents**. *Second Edition.* By Milton J. Rosen and Henry A. Goldsmith
Vol. 13 **Alternating Current Polarography and Tensammetry**. By B. Breyer and H.H.Bauer
Vol. 14 **Flame Photometry**. By R. Herrmann and J. Alkemade
Vol. 15 **The Titration of Organic Compounds** (*in two parts*). By M. R. F. Ashworth
Vol. 16 **Complexation in Analytical Chemistry: A Guide for the Critical Selection of Analytical Methods Based on Complexation Reactions**. By the late Anders Ringbom
Vol. 17 **Electron Probe Microanalysis**. *Second Edition.* By L. S. Birks
Vol. 18 **Organic Complexing Reagents: Structure, Behavior, and Application to Inorganic Analysis**. By D. D. Perrin
Vol. 19 **Thermal Analysis**. *Third Edition.* By Wesley Wm.Wendlandt
Vol. 20 **Amperometric Titrations**. By John T. Stock
Vol. 21 **Reflctance Spectroscopy**. By Wesley Wm.Wendlandt and Harry G. Hecht
Vol. 22 **The Analytical Toxicology of Industrial Inorganic Poisons**. By the late Morris B. Jacobs
Vol. 23 **The Formation and Properties of Precipitates**. By Alan G.Walton
Vol. 24 **Kinetics in Analytical Chemistry**. By Harry B. Mark, Jr. and Garry A. Rechnitz
Vol. 25 **Atomic Absorption Spectroscopy**. *Second Edition.* By Morris Slavin
Vol. 26 **Characterization of Organometallic Compounds** (*in two parts*). Edited by Minoru Tsutsui
Vol. 27 **Rock and Mineral Analysis**. *Second Edition.* By Wesley M. Johnson and John A. Maxwell
Vol. 28 **The Analytical Chemistry of Nitrogen and Its Compounds** (*in two parts*). Edited by C. A. Streuli and Philip R.Averell
Vol. 29 **The Analytical Chemistry of Sulfur and Its Compounds** (*in three parts*). By J. H. Karchmer
Vol. 30 **Ultramicro Elemental Analysis**. By Güther Toölg
Vol. 31 **Photometric Organic Analysis** (*in two parts*). By Eugene Sawicki
Vol. 32 **Determination of Organic Compounds: Methods and Procedures**. By Frederick T. Weiss
Vol. 33 **Masking and Demasking of Chemical Reactions**. By D. D. Perrin

Vol. 34. **Neutron Activation Analysis**. By D. De Soete, R. Gijbels, and J. Hoste
Vol. 35 **Laser Raman Spectroscopy**. By Marvin C. Tobin
Vol. 36 **Emission Spectrochemical Analysis**. By Morris Slavin
Vol. 37 **Analytical Chemistry of Phosphorus Compounds**. Edited by M. Halmann
Vol. 38 **Luminescence Spectrometry in Analytical Chemistry**. By J. D.Winefordner, S. G. Schulman, and T. C. O'Haver
Vol. 39. **Activation Analysis with Neutron Generators**. By Sam S. Nargolwalla and Edwin P. Przybylowicz
Vol. 40 **Determination of Gaseous Elements in Metals**. Edited by Lynn L. Lewis, Laben M. Melnick, and Ben D. Holt
Vol. 41 **Analysis of Silicones**. Edited by A. Lee Smith
Vol. 42 **Foundations of Ultracentrifugal Analysis**. By H. Fujita
Vol. 43 **Chemical Infrared Fourier Transform Spectroscopy**. By Peter R. Griffiths
Vol. 44 **Microscale Manipulations in Chemistry**. By T. S. Ma and V. Horak
Vol. 45 **Thermometric Titrations**. By J. Barthel
Vol. 46 **Trace Analysis: Spectroscopic Methods for Elements**. Edited by J. D.Winefordner
Vol. 47 **Contamination Control in Trace Element Analysis**. By Morris Zief and James W. Mitchell
Vol. 48 **Analytical Applications of NMR**. By D. E. Leyden and R. H. Cox
Vol. 49 **Measurement of Dissolved Oxygen**. By Michael L. Hitchman
Vol. 50 **Analytical Laser Spectroscopy**. Edited by Nicolo Omenetto
Vol. 51 **Trace Element Analysis of Geological Materials**. By Roger D. Reeves and Robert R. Brooks
Vol. 52 **Chemical Analysis by Microwave Rotational Spectroscopy**. By Ravi Varma and Lawrence W. Hrubesh
Vol. 53 **Information Theory as Applied to Chemical Analysis**. By Karl Eckschlager and Vladimir Stepanek
Vol. 54 **Applied Infrared Spectroscopy: Fundamentals, Techniques, and Analytical Problemsolving**. By A. Lee Smith
Vol. 55 **Archaeological Chemistry**. By Zvi Goffer
Vol. 56 **Immobilized Enzymes in Analytical and Clinical Chemistry**. By P. W. Carr and L. D. Bowers
Vol. 57 **Photoacoustics and Photoacoustic Spectroscopy**. By Allan Rosencwaig
Vol. 58 **Analysis of Pesticide Residues**. Edited by H. Anson Moye
Vol. 59 **Affity Chromatography**. By William H. Scouten
Vol. 60 **Quality Control in Analytical Chemistry**. *Second Edition.* By G. Kateman and L. Buydens
Vol. 61 **Direct Characterization of Fineparticles**. By Brian H. Kaye
Vol. 62 **Flow Injection Analysis**. By J. Ruzicka and E. H. Hansen
Vol. 63 **Applied Electron Spectroscopy for Chemical Analysis**. Edited by Hassan Windawi and Floyd Ho
Vol. 64 **Analytical Aspects of Environmental Chemistry**. Edited by David F. S. Natusch and Philip K. Hopke
Vol. 65 **The Interpretation of Analytical Chemical Data by the Use of Cluster Analysis**. By D. Luc Massart and Leonard Kaufman
Vol. 66 **Solid Phase Biochemistry: Analytical and Synthetic Aspects**. Edited by William H. Scouten
Vol. 67 **An Introduction to Photoelectron Spectroscopy**. By Pradip K. Ghosh
Vol. 68 **Room Temperature Phosphorimetry for Chemical Analysis**. By Tuan Vo-Dinh
Vol. 69 **Potentiometry and Potentiometric Titrations**. By E. P. Serjeant
Vol. 70 **Design and Application of Process Analyzer Systems**. By Paul E. Mix
Vol. 71 **Analysis of Organic and Biological Surfaces**. Edited by Patrick Echlin
Vol. 72 **Small Bore Liquid Chromatography Columns: Their Properties and Uses**. Edited by Raymond P.W. Scott
Vol. 73 **Modern Methods of Particle Size Analysis**. Edited by Howard G. Barth
Vol. 74 **Auger Electron Spectroscopy**. By Michael Thompson, M. D. Baker, Alec Christie, and J. F. Tyson
Vol. 75 **Spot Test Analysis: Clinical, Environmental, Forensic and Geochemical Applications**. By Ervin Jungreis
Vol. 76 **Receptor Modeling in Environmental Chemistry**. By Philip K. Hopke

Vol. 77 **Molecular Luminescence Spectroscopy: Methods and Applications** (*in three parts*). Edited by Stephen G. Schulman
Vol. 78 **Inorganic Chromatographic Analysis**. Edited by John C. MacDonald
Vol. 79 **Analytical Solution Calorimetry**. Edited by J. K. Grime
Vol. 80 **Selected Methods of Trace Metal Analysis: Biological and Environmental Samples**. By Jon C.VanLoon
Vol. 81 **The Analysis of Extraterrestrial Materials**. By Isidore Adler
Vol. 82 **Chemometrics**. By Muhammad A. Sharaf, Deborah L. Illman, and Bruce R. Kowalski
Vol. 83 **Fourier Transform Infrared Spectrometry**. By Peter R. Griffiths and James A. de Haseth
Vol. 84 **Trace Analysis: Spectroscopic Methods for Molecules**. Edited by Gary Christian and James B. Callis
Vol. 85 **Ultratrace Analysis of Pharmaceuticals and Other Compounds of Interest**. Edited by S. Ahuja
Vol. 86 **Secondary Ion Mass Spectrometry: Basic Concepts, Instrumental Aspects, Applications and Trends**. By A. Benninghoven, F. G. Rüenauer, and H.W.Werner
Vol. 87 **Analytical Applications of Lasers**. Edited by Edward H. Piepmeier
Vol. 88 **Applied Geochemical Analysis**. By C. O. Ingamells and F. F. Pitard
Vol. 89 **Detectors for Liquid Chromatography**. Edited by Edward S.Yeung
Vol. 90 **Inductively Coupled Plasma Emission Spectroscopy: Part 1: Methodology, Instrumentation, and Performance; Part II: Applications and Fundamentals**. Edited by J. M. Boumans
Vol. 91 **Applications of New Mass Spectrometry Techniques in Pesticide Chemistry**. Edited by Joseph Rosen
Vol. 92 **X-Ray Absorption: Principles,Applications,Techniques of EXAFS, SEXAFS, and XANES**. Edited by D. C. Konnigsberger
Vol. 93 **Quantitative Structure-Chromatographic Retention Relationships**. By Roman Kaliszan
Vol. 94 **Laser Remote Chemical Analysis**. Edited by Raymond M. Measures
Vol. 95 **Inorganic Mass Spectrometry**. Edited by F.Adams,R.Gijbels, and R.Van Grieken
Vol. 96 **Kinetic Aspects of Analytical Chemistry**. By Horacio A. Mottola
Vol. 97 **Two-Dimensional NMR Spectroscopy**. By Jan Schraml and Jon M. Bellama
Vol. 98 **High Performance Liquid Chromatography**. Edited by Phyllis R. Brown and Richard A. Hartwick
Vol. 99 **X-Ray Fluorescence Spectrometry**. By Ron Jenkins
Vol. 100 **Analytical Aspects of Drug Testing**. Edited by Dale G. Deustch
Vol. 101 **Chemical Analysis of Polycyclic Aromatic Compounds**. Edited by Tuan Vo-Dinh
Vol. 102 **Quadrupole Storage Mass Spectrometry**. By Raymond E. March and Richard J. Hughes (*out of print: see Vol. 165)*
Vol. 103 **Determination of Molecular Weight**. Edited by Anthony R. Cooper
Vol. 104 **Selectivity and Detectability Optimization in HPLC**. By Satinder Ahuja
Vol. 105 **Laser Microanalysis**. By Lieselotte Moenke-Blankenburg
Vol. 106 **Clinical Chemistry**. Edited by E. Howard Taylor
Vol. 107 **Multielement Detection Systems for Spectrochemical Analysis**. By Kenneth W. Busch and Marianna A. Busch
Vol. 108 **Planar Chromatography in the Life Sciences**. Edited by Joseph C. Touchstone
Vol. 109 **Fluorometric Analysis in Biomedical Chemistry: Trends and Techniques Including HPLC Applications**. By Norio Ichinose, George Schwedt, Frank Michael Schnepel, and Kyoko Adochi
Vol. 110 **An Introduction to Laboratory Automation**. By Victor Cerdá and Guillermo Ramis
Vol. 111 **Gas Chromatography: Biochemical, Biomedical, and Clinical Applications**. Edited by Ray E. Clement
Vol. 112 **The Analytical Chemistry of Silicones**. Edited by A. Lee Smith
Vol. 113 **Modern Methods of Polymer Characterization**. Edited by Howard G. Barth and Jimmy W. Mays
Vol. 114 **Analytical Raman Spectroscopy**. Edited by Jeanette Graselli and Bernard J. Bulkin
Vol. 115 **Trace and Ultratrace Analysis by HPLC**. By Satinder Ahuja
Vol. 116 **Radiochemistry and Nuclear Methods of Analysis**. By William D. Ehmann and Diane E.Vance

Vol. 117 **Applications of Fluorescence in Immunoassays**. By Ilkka Hemmila
Vol. 118 **Principles and Practice of Spectroscopic Calibration**. By Howard Mark
Vol. 119 **Activation Spectrometry in Chemical Analysis**. By S. J. Parry
Vol. 120 **Remote Sensing by Fourier Transform Spectrometry**. By Reinhard Beer
Vol. 121 **Detectors for Capillary Chromatography**. Edited by Herbert H. Hill and Dennis McMinn
Vol. 122 **Photochemical Vapor Deposition**. By J. G. Eden
Vol. 123 **Statistical Methods in Analytical Chemistry**. By Peter C. Meier and Richard Züd
Vol. 124 **Laser Ionization Mass Analysis**. Edited by Akos Vertes, Renaat Gijbels, and Fred Adams
Vol. 125 **Physics and Chemistry of Solid State Sensor Devices**. By Andreas Mandelis and Constantinos Christofides
Vol. 126 **Electroanalytical Stripping Methods**. By Khjena Z. Brainina and E. Neyman
Vol. 127 **Air Monitoring by Spectroscopic Techniques**. Edited by Markus W. Sigrist
Vol. 128 **Information Theory in Analytical Chemistry**. By Karel Eckschlager and Klaus Danzer
Vol. 129 **Flame Chemiluminescence Analysis by Molecular Emission Cavity Detection**. Edited by David Stiles, Anthony Calokerinos, and Alan Townshend
Vol. 130 **Hydride Generation Atomic Absorption Spectrometry**. Edited by Jiri Dedina and Dimiter L. Tsalev
Vol. 131 **Selective Detectors: Environmental, Industrial, and Biomedical Applications**. Edited by Robert E. Sievers
Vol. 132 **High-Speed Countercurrent Chromatography**. Edited by Yoichiro Ito and Walter D. Conway
Vol. 133 **Particle-Induced X-Ray Emission Spectrometry**. By Sven A. E. Johansson, John L. Campbell, and Klas G. Malmqvist
Vol. 134 **Photothermal Spectroscopy Methods for Chemical Analysis**. By Stephen E. Bialkowski
Vol. 135 **Element Speciation in Bioinorganic Chemistry**. Edited by Sergio Caroli
Vol. 136 **Laser-Enhanced Ionization Spectrometry**. Edited by John C. Travis and Gregory C. Turk
Vol. 137 **Fluorescence Imaging Spectroscopy and Microscopy**. Edited by Xue Feng Wang and Brian Herman
Vol. 138 **Introduction to X-Ray Powder Diffractometry**. By Ron Jenkins and Robert L. Snyder
Vol. 139 **Modern Techniques in Electroanalysis**. Edited by Petr Vaňýek
Vol. 140 **Total-Reflction X-Ray Fluorescence Analysis**. By Reinhold Klockenkamper
Vol. 141 **Spot Test Analysis: Clinical, Environmental, Forensic, and Geochemical Applications**. *Second Edition.* By Ervin Jungreis
Vol. 142 **The Impact of Stereochemistry on Drug Development and Use**. Edited by Hassan Y. Aboul-Enein and Irving W.Wainer
Vol. 143 **Macrocyclic Compounds in Analytical Chemistry**. Edited by Yury A. Zolotov
Vol. 144 **Surface-Launched Acoustic Wave Sensors: Chemical Sensing and Thin-Film Characterization**. By Michael Thompson and David Stone
Vol. 145 **Modern Isotope Ratio Mass Spectrometry**. Edited by T. J. Platzner
Vol. 146 **High Performance Capillary Electrophoresis: Theory, Techniques, and Applications**. Edited by Morteza G. Khaledi
Vol. 147 **Solid Phase Extraction: Principles and Practice**. By E. M. Thurman
Vol. 148 **Commercial Biosensors: Applications to Clinical, Bioprocess and Environmental Samples**. Edited by Graham Ramsay
Vol. 149 **A Practical Guide to Graphite Furnace Atomic Absorption Spectrometry**. By David J. Butcher and Joseph Sneddon
Vol. 150 **Principles of Chemical and Biological Sensors**. Edited by Dermot Diamond
Vol. 151 **Pesticide Residue in Foods: Methods, Technologies, and Regulations**. By W. George Fong, H. Anson Moye, James N. Seiber, and John P. Toth
Vol. 152 **X-Ray Fluorescence Spectrometry**. *Second Edition.* By Ron Jenkins
Vol. 153 **Statistical Methods in Analytical Chemistry**. *Second Edition.* By Peter C. Meier and Richard E. Züd
Vol. 154 **Modern Analytical Methodologies in Fat- and Water-Soluble Vitamins**. Edited by Won O. Song, Gary R. Beecher, and Ronald R. Eitenmiller
Vol. 155 **Modern Analytical Methods in Art and Archaeology**. Edited by Enrico Ciliberto and Guiseppe Spoto

Vol. 156 **Shpol'skii Spectroscopy and Other Site Selection Methods: Applications in Environmental Analysis, Bioanalytical Chemistry and Chemical Physics**. Edited by C. Gooijer, F. Ariese and J.W. Hofstraat

Vol. 157 **Raman Spectroscopy for Chemical Analysis**. By Richard L. McCreery

Vol. 158 **Large (C> = 24) Polycyclic Aromatic Hydrocarbons: Chemistry and Analysis**. By John C. Fetzer

Vol. 159 **Handbook of Petroleum Analysis**. By James G. Speight

Vol. 160 **Handbook of Petroleum Product Analysis**. By James G. Speight

Vol. 161 **Photoacoustic Infrared Spectroscopy**. By Kirk H. Michaelian

Vol. 162 **Sample Preparation Techniques in Analytical Chemistry**. Edited by Somenath Mitra

Vol. 163 **Analysis and Purification Methods in Combination Chemistry**. Edited by Bing Yan

Vol. 164 **Chemometrics: From Basics to Wavelet Transform**. By Foo-tim Chau, Yi-Zeng Liang, Junbin Gao, and Xue-guang Shao

Vol. 165 **Quadrupole Ion Trap Mass Spectrometry**. *Second Edition.* By Raymond E. March and John F. J. Todd

Vol. 166 **Handbook of Coal Analysis**. By James G. Speight

Vol. 167 **Introduction to Soil Chemistry: Analysis and Instrumentation**. By Alfred R. Conklin, Jr.

Vol. 168 **Environmental Analysis and Technology for the Refining Industry**. By James G. Speight

Vol. 169 **Identification of Microorganisms by Mass Spectrometry.** Edited by Charles L. Wilkins and Jackson O. Lay, Jr.